Social Media Archeology and Poetics

Leonardo

Roger F. Malina, Executive Editor

Sean Cubitt, Editor-in-Chief

Video: The Reflexive Medium, Yvonne Spielmann, 2007

Software Studies: A Lexicon, Matthew Fuller, 2008

Tactical Biopolitics: Art, Activism, and Technoscience, edited by Beatriz da Costa and Kavita Philip, 2008

White Heat and Cold Logic: British Computer Art 196–1980, edited by Paul Brown, Charlie Gere, Nicholas Lambert, and Catherine Mason, 2008

Rethinking Curating: Art after New Media, Beryl Graham and Sarah Cook, 2010

Green Light: Toward an Art of Evolution, George Gessert, 2010

Enfoldment and Infinity: An Islamic Genealogy of New Media Art, Laura U. Marks, 2010

Synthetics: Aspects of Art & Technology in Australia, 1956–1975, Stephen Jones, 2011

Hybrid Cultures: Japanese Media Arts in Dialogue with the West, Yvonne Spielmann, 2012

Walking and Mapping: Artists as Cartographers, Karen O'Rourke, 2013

The Fourth Dimension and Non-Euclidean Geometry in Modern Art, revised edition, Linda Dalrymple Henderson, 2013

Illusions in Motion: Media Archaeology of the Moving Panorama and Related Spectacles, Erkki Huhtamo, 2013

Relive: Media Art Histories, edited by Sean Cubitt and Paul Thomas, 2013

Re-collection: Art, New Media, and Social Memory, Richard Rinehart and Jon Ippolito, 2014

Biopolitical Screens: Image, Power, and the Neoliberal Brain, Pasi Väliaho, 2014

The Practice of Light: A Genealogy of Visual Technologies from Prints to Pixels, Sean Cubitt, 2014

The Tone of Our Times: Sound, Sense, Economy, and Ecology, Frances Dyson, 2014

The Experience Machine: Stan VanDerBeek's Movie-Drome and Expanded Cinema, Gloria Sutton, 2014

Hanan al-Cinema: Affections for the Moving Image, Laura U. Marks, 2015

Writing and Unwriting (Media) Art History: Erkki Kurenniemi in 2048, edited by Joasia Krysa and Jussi Parikka, 2015

Control: Digitality as Cultural Logic, Seb Franklin, 2015

New Tendencies: Art at the Threshold of the Information Revolution (1961–1978), Armin Medosch, 2016

Screen Ecologies: Art, Media, and the Environment in the Asia-Pacific Region; Larissa Hjorth, Sarah Pink, Kristen Sharp, and Linda Williams, 2016

Pirate Philosophy: For a Digital Posthumanities, Gary Hall, 2016

Social Media Archeology and Poetics, edited by Judy Malloy, 2016

See <http://mitpress.mit.edu> for a complete list of titles in this series.

Social Media Archeology and Poetics

edited by Judy Malloy

The MIT Press
Cambridge, Massachusetts
London, England

© 2016 Massachusetts Institute of Technology

All rights reserved. No part of this book may be reproduced in any form by any electronic or mechanical means (including photocopying, recording, or information storage and retrieval) without permission in writing from the publisher.

This book was set in Stone Sans and Stone Serif by Toppan Best-set Premedia Limited. Printed and bound in the United States of America.

Library of Congress Cataloging-in-Publication Data

Names: Malloy, Judy, editor.
Title: Social media archeology and poetics / Judy Malloy, ed.
Description: Cambridge, MA : MIT Press, 2016. | Series: Leonardo book series | Includes bibliographical references and index.
Identifiers: LCCN 2015042571 | ISBN 9780262034654 (hardcover : alk. paper)
Subjects: LCSH: Social media—History. | Mass media—History. | Communication—History. | Writing—History.
Classification: LCC HM742 .S6281947 2016 | DDC 302.23/1—dc23 LC record available at http://lccn.loc.gov/2015042571

10 9 8 7 6 5 4 3 2 1

In memory of
Carl Eugene Loeffler (1946–2001),
who brought artists and writers online
to converse, create, publish, and exhibit art and literature
in the early days of Social Media.
"See you online!"

Contents

Series Foreword xi
Acknowledgments xiii

I Introductions 1

1 The Origins of Social Media 3
Judy Malloy

2 The Personal Computer and Social Media 51
Paul E. Ceruzzi

3 Daily Life in Cyberspace: How the Computerized Counterculture Built a New Kind of Place 61
Howard Rheingold

II "Opening the Door to Cyberspace" 87

4 Community Memory: The First Public-Access Social Media System 89
Lee Felsenstein

5 PLATO: The Emergence of Online Community 103
David R. Woolley

6 alt.hypertext: An Early Social Medium 119
James Blustein and Ann-Barbara Graff

7 DictatiOn: A Canadian Perspective on the History of Telematic Art 127
Hank Bull

8 Art and Minitel in France in the 1980s 139
Annick Bureaud

9 Rescension and Precedential Media   147
Steve Dietz

III "See you online!"   161

10 Defining the Image as Place: A Conversation with Kit Galloway, Sherrie Rabinowitz, and Gene Youngblood   163
Steven Durland

11 IN.S.OMNIA, 1983–1993   179
Rob Wittig

12 Art Com Electronic Network: A Conversation with Fred Truck and Anna Couey   191
Judy Malloy

13 System X: Interview with Founding Sysop Scot McPhee   219
Amanda McDonald Crowley

IV Networking the Humanities   225

14 In Search of Identities in the Digital Humanities: The Early History of Humanist   227
Julianne Nyhan

15 Echo   243
Stacy Horn

16 MOOs and Participatory Media   251
Dene Grigar

17 Hacking the *Voice of the Shuttle*: The Growth and Death of a Boundary Object   261
Alan Liu

V Community Networking   273

18 Community Networking: The Native American Telecommunications Continuum   275
Randy Ross (Ponca Tribe of Nebraska and Otoe Missouria)

19 The Art of Tele-Community Development: The Telluride Infozone   277
Richard Lowenberg

Contents

20 Community Networking, an Evolution   291
Madeline Gonzalez Allen

21 Cultures in Cyberspace: Communications System Design as Social Sculpture   297
Anna Couey

VI Social Media Poetics   307

22 Crossing-Over of Art History and Media History in the Times of the Early Internet—with Special Regard to THE THING NYC   309
Susanne Gerber

23 Arts Wire: The Nonprofit Arts Online   333
Judy Malloy

24 Electronic Literature Organization Chats on LinguaMOO   353
Deena Larsen

25 trAce Online Writing Centre, Nottingham Trent University, UK   363
J. R. Carpenter

26 Pseudo Space: Experiments with Avatarism and Telematic Performance in Social Media   377
Antoinette LaFarge

VII Responses   387

A Conversation and Two Epilogues   389
Judy Malloy

27 Expanding on "What Is the Social in Social Media?": A Conversation with Geert Lovink   393
Judy Malloy

28 Epilogue: Slow Machines and Utopian Dreams   399
Judith Donath

29 From Archaeology to Architecture: Building a Place for Noncommercial Culture Online   411
Gary O. Larson

About the Authors   435
Index   445

Series Foreword

Leonardo/International Society for the Arts, Sciences, and Technology (ISAST)

Leonardo, the International Society for the Arts, Sciences, and Technology, and the affiliated French organization Association Leonardo have some very simple goals:

1. To advocate, document, and make known the work of artists, researchers, and scholars developing the new ways that the contemporary arts interact with science, technology, and society.
2. To create a forum and meeting places where artists, scientists, and engineers can meet, exchange ideas, and, when appropriate, collaborate.
3. To contribute, through the interaction of the arts and sciences, to the creation of the new culture that will be needed to transition to a sustainable planetary society.

When the journal *Leonardo* was started some forty-five years ago, these creative disciplines existed in segregated institutional and social networks, a situation dramatized at that time by the "Two Cultures" debates initiated by C. P. Snow. Today we live in a different time of cross-disciplinary ferment, collaboration, and intellectual confrontation enabled by new hybrid organizations, new funding sponsors, and the shared tools of computers and the Internet. Above all, new generations of artist-researchers and researcher-artists are now at work individually and collaboratively bridging the art, science, and technology disciplines. For some of the hard problems in our society, we have no choice but to find new ways to couple the arts and sciences. Perhaps in our lifetime we will see the emergence of "new Leonardos," hybrid creative individuals or teams that will not only develop a meaningful art for our times but also drive new agendas in science and stimulate technological innovation that addresses today's human needs.

For more information on the activities of the Leonardo organizations and networks, please visit our websites at http://www.leonardo.info/ and http://www.olats.org.

Roger F. Malina
Executive Editor, Leonardo Publications

ISAST Governing Board of Directors: Nina Czegledy, Greg Harper, Marc Hebert (Chair), Gordon Knox, Roger Malina, Tami Spector, Darlene Tong

Acknowledgments

In the fall of 2013, I was in residence at Princeton University, as Anschutz Distinguished Fellow, a fellowship appointed by the Princeton Program in American Studies, where past Anschutz fellows have included photographer and labor photography historian Richard Street; Broadway/off Broadway producer David Binder; musician/rock critic Kandia Crazy Horse; poet Kenneth Goldsmith; and Bonnie Marranca, editor of *PAJ: A Journal of Performance and Art*.

Beginning in my own library and in the library of the University of California at Berkeley, in preparing the syllabus for the seminar in *Social Media History and Poetics, and Practice*, which I taught during my fellowship, I reviewed classic texts, including, among many others, Heidi Grundmann's *Art Telecommunication*,[1] Roy Ascott and Carl Eugene Loeffler's *Connectivity: Art and Interactive Telecommunications*,[2] John Quarterman's *The Matrix: Computer Networks and Conferencing Systems Worldwide*,[3] Howard Rheingold's *The Virtual Community: Homesteading on the Electronic Frontier*,[4] and papers and reports ranging from and Casey, Ross, and Warren's *Native Networking: Telecommunications and Information Technology in Indian Country*[5] to Pavel Curtis' Xerox PARC report on LambdaMoo[6] to Steve Durland's "Defining the Image as Place, a Conversation with Kit Galloway Sherrie Rabinowitz & Gene Youngblood."[7]

In archives across the country (including my own in the Rubenstein Rare Book & Manuscript Library at Duke University),[8] there were printouts from the past; however, much of the importance of early social media and its relationship with contemporary social media existed in the memories of the contributors to this book. The need for comprehensive documentation of early social media platforms was apparent.

At Princeton, the students in *Social Media History and Poetics* negotiated the reading list, created extraordinary projects,[9] and gave memorable presentations, particularly on platforms for which in-depth information was available. For instance, actual printouts from Community Memory are available on the website of the Computer History Museum. However, with few resources available on what it was actually like to use pre-Web social media and why these early platforms were so extraordinary, it was difficult to re-create the experience of THE THING, Arts Wire, Art Com Electronic Network

(ACEN), LambdaMoo, and EchoNYC, to name just a few. In contrast, as the class progressed to contemporary platforms, the students had experience with a wide range of contemporary social media platforms and could effectively traverse, study, write about, and present on contemporary social media. It was clear that, although there is also a need for more anthologized documentation of the history of contemporary social media, a book with a focus on pre-Web social media platforms was vitally needed.

In addition to the Princeton students in AMS317, who inspired this book, I would like to thank Professor Dirk Hartog, then Chair of the American Studies Department at Princeton; Program Manager, Judith Ferszt; and Program Assistant, Candice Kessel for their advice and support during the course of my fellowship.

As this book began to take shape, thanks go to distinguished ARPANET pioneers Vint Cerf, Dave Crocker, and Les Earnest for their invaluable help in contributing to and advising on the Introduction.

If the immediate impetus for this book was my term as a Distinguished Fellow at Princeton, it should also be noted that this book has been in the making since April 1986, when I first logged onto ACEN on The WELL. Thus, I would like to thank the spirit of my friend, ACEN founder Carl Loeffler (1946–2001), who in 1986 talked me onto ACEN—saying, "You have got to get online!"—as well as my artist colleagues on ACEN, including artist/sysop Fred Truck; poet Jim Rosenberg; community networker/artist Anna Couey, with whom I later co-hosted the *Interactive Art Conference* on Arts Wire; and Howard Rheingold, who continues to foster, define, and support virtual community.

Working (as Coordinating Editor of *Leonardo*'s fledgling electronic publications beginning in 1988) for *Leonardo* Executive Editor Roger Malina—whose interest in and support of electronic publication continues to be important in documenting art and technology—was an opportunity to edit and write for online publications in the arts at the time of their beginnings. In the early 1990s, Christine Maxwell pioneered publishing information about Internet resources, and the time I spent working for Christine as a consultant on the *Internet Yellow Pages* was valuable in understanding what was available online in that era.

Thanks also to the spirit of Rich Gold (1950–2003) at Xerox PARC for the opportunity to work with social media in the heady atmosphere of the Computer Science Lab (CSL) and for his foresight in creating a program to bring artists and poets to PARC. Thanks to then PARC researcher Pavel Curtis, with whom I worked on LambdaMoo narrative, and to then Xerox PARC hypertext researcher Cathy Marshall, with whom I created the hypertext, *Forward Anywhere*.

And fond thanks to Mark Bernstein, founder and head of Eastgate, publisher of electronic literature, including Michael Joyce's *afternoon*, Carolyn Guyer's *Quibbling*, and my own *its name was Penelope*, as well as titles with social media components, such as the work that Cathy Marshall and I did at PARC. Fond thanks as well as to electronic literature friends and colleagues, including, among many others, Dene Grigar for her scholarly work to bring electronic literature to a wider public; electronic literature pioneer, Stuart Moulthrop, whose work is of continuing interest and who worked with Dene on the *Pathfinders*[10] electronic literature preservation project; Leonardo Flores for his heroic review of 500 works of electronic literature in as many days; Kathi Inman Berens for documenting the work of women digital writers; and Mark Marino for fascinating online conversations on the Critical Code Studies Working Group conference.

Remembering summers spent in Telluride working with MFA students in the documentation of new media art for electronic and social media–based publication, thanks to Dan Collins and Laurie Lundquist, who created and directed the Arizona State University–associated Deep Creek School, where art and technology merged in a high mountain setting; thanks also to Telluride InfoZone Founding Director Richard Lowenberg for the invitation to work on the InfoZone; and to community-based technology transfer guru Madeline Gonzalez, with whom I worked on online arts resources for the Telluride InfoZone.

Thanks to artist/arts advocate Anne Focke, to then New York Foundation for the Arts (NYFA) Executive Director Ted Berger, and to my boss, Michigan-based poet Joe Matuzak, for making Arts Wire (a Program of the New York Foundation for the Arts) possible and for giving me the opportunity to work on Arts Wire for ten very productive years of my life. Thanks to Gary O. Larson, who was an Arts Wire Board member while he was at the National Endowment for the Arts and ever since has continued to share ideas and talk about networking in the arts; and to Arts Wire Board member, Randy Ross (Ponca Tribe of Nebraska and Otoe Missouria), for his important work in bringing Native American culture online, including working to implement the Native Arts Network Association (NANA) on Arts Wire.

For continuing discussion and pub talk about the evolving Internet, I would like to thank my son Sean Malloy, Associate Professor of History/Founding Faculty at the University of California at Merced and a former *HotWired* editor for *Wired* Magazine.

It has been more than thirty years since I began talking information art and new media art theory and practice with interactive artist Sonya Rapoport (1923–2015), who, until her death in June 2015, was an important part of my art world life. Thanks spirit of Sonya for your anchoring presence and for the role model of your dedication to your work. For inspiration, thanks also to my cousin, artist Blyth Hazen, and her partner, artist Jennifer Hall, who created "Do While Studio" in Boston.

Most important, I would like to thank Doug Sery for his interest, guidance, and patient work in bringing this volume to completion and for his continuing scholarly and connected work as New Media, Game Studies, Digital Humanities, and HCI Senior Acquisitions Editor at MIT Press. Thanks also to MIT Press Senior Editor Katherine Almeida for greatly improving this book and to MIT Press Associate Acquisitions Editor Susan Buckley for making the publication process steady and comprehensible.

And many thanks and fond wishes to all of the contributors to this book. It was not an easy job to document systems of the past in the context of contemporary social media. Their enthusiasm and willingness to share their experience in creating, using, and researching early social media are what made this book a pleasure to compile! In order of appearance, they are Paul Ceruzzi, Howard Rheingold, Lee Felsenstein, David R. Woolley, James Blustein, Ann-Barbara Graff, Hank Bull, Annick Bureaud, Steve Dietz, Steven Durland, Kit Galloway, Sherrie Rabinowitz (1950–2013), Gene Youngblood, Rob Wittig, Fred Truck, Anna Couey, Scot Mcphee, Amanda McDonald Crowley, Julianne Nyhan, Stacy Horn, Dene Grigar, Alan Liu, Randy Ross, Richard Lowenberg, Madeline Gonzalez Allen, Susanne Gerber, Wolfgang Staehle, Deena Larsen, J. R. Carpenter, Antoinette LaFarge, Geert Lovink, Judith Donath, and Gary O. Larson.

Notes

1. Heidi Grundmann, *Art Telecommunication* (Vancouver: Western Front; Vienna: BLIX, 1984).

2. Roy Ascott and Carl Eugene Loeffler, Guest eds., "Connectivity: Art and Interactive Telecommunications," *Leonardo* 24, no. 2 (1991).

3. John Quarterman, *The Matrix: Computer Networks and Conferencing Systems Worldwide* (Bedford, MA: Digital Press, 1990).

4. Howard Rheingold, *The Virtual Community: Homesteading on the Electronic Frontier*, rev. ed. (Cambridge, MA: MIT Press, 2000).

5. James Casey (Cherokee), Randy Ross (Ponca Tribe of Nebraska and Otoe Missouria), and Marcia Warren (Santa Clara Pueblo of New Mexico), *Native Networking: Telecommunications and Information Technology in Indian Country* (Washington, DC: Benton Foundation, 1999).

6. Pavel Curtis, *Mudding: Social Phenomena in Text-Based Virtual Reality* (Palo Alto, CA: Xerox PARC, CSL-92-4; April, 1992).

7. Steven Durland, "Defining the Image as Place, a Conversation with Kit Galloway Sherrie Rabinowitz & Gene Youngblood," *High Performance* 37 (1987), 53–59.

8. *Judy Malloy Papers*, Rubenstein Rare Book & Manuscript Library at Duke University. Available at http://library.duke.edu/rubenstein/findingaids/malloyjudy/.

9. "Final Project Showcase, Princeton Program in American Studies: AMS 317, Social Media: History, Poetics, and Practice," 2014. Available at https://www.princeton.edu/ams/program-events/Malloy-student-documentation.pdf.

10. Dene Grigar and Stuart Moulthrop, *Pathfinders: Documenting the Experience of Early Digital Literature* (Scalar, 2015). Available at http://scalar.usc.edu/pathfinders-documenting-the-experience-of-early-digital-literature/.

I Introductions

1 The Origins of Social Media

Judy Malloy

In the formative years of the Internet, researchers in academic and industry Research and Development (R&D) institutions collaboratively addressed the need to connect computing systems with a goal of sharing research and computing resources. The model process with which they created the Internet and its forefather, the ARPANET, was echoed in early social media platforms, where computer scientists, educators, artists, writers, musicians, and community activists explored the promise of computer-based platforms to bring together communities of interest in what would begin to be called "cyberspace."

Internet precursor ARPANET was initially a Defense Advanced Research Projects Agency (DARPA)-funded project[1] to network computers in research labs for the purpose of resource sharing. The early history of ARPANET is well documented.[2,3]

Begin then on the 40th anniversary of the publication of Vint Cerf and Bob Kahn's seminal 1974 paper, "A Protocol for Packet Network Intercommunication."[4] On this anniversary occasion, in an informal yet public conversation, Cerf and Kahn emphasized the collaborative model that was set by the researchers who developed ARPANET and the Internet.

The role of collaboration in the creation of the Internet was highlighted by their presence, sitting informally together at a table in Friend Hall in Princeton, discussing the origins of the Internet.

In 1973, Robert Kahn, who had moved from core DARPA contractor Bolt, Beranek, and Newman (BBN) in Cambridge, Massachusetts, to DARPA headquarters and was working with then DARPA Information Processing Techniques Office (IPTO) (chief Larry Roberts, journeyed to Palo Alto, where he talked with Stanford professor Vint Cerf about the need to develop an open networking concept to link different systems and networks. Working collaboratively from opposite coasts, they developed Transmission Control Protocol (TCP), the protocol that (adding Internet Protocol) metamorphosed to TCP/IP, and—facilitating the movement of information among disparate computer platforms, operating systems, and networks—made possible the Internet.

Throughout the conversation between Cerf and Kahn, the enormously long list of people who contributed to the creation of the Internet, as well as the continuing environment and atmosphere where researchers were welcome to explore and contribute were apparent. Vint Cerf recalled how in the early days of the ARPANET, current ICANN chair and then UCLA graduate student Steve Crocker's approach was that anyone who had a good idea should be able to participate in the project. He set a tone for inviting involvement and participation.[5]

Bob Kahn emphasized the Internet's unprecedented scalability over the course of its existence, observing that this happened because the Internet is about the protocols, not about the underlying networks, so anything can be part of it. In his words at Princeton, "A design at a level that lets lots of things happen underneath, and they have."[6]

The original purpose of the ARPANET, Vint Cerf pointed out in a subsequent telephone interview,[7] was sharing—sharing computing resources. There was a need for remote access to accelerate research. Importantly, because DARPA was funding all of the universities involved in the creation of the Internet, researchers were not competing for research dollars, and the process was collaborative.

It was an environment where discourse, collaboration, innovation, creativity, and the sharing of information were encouraged.

A World Wide Simultaneous Dance

Contingently, when MIT Masters of Science in Visual Studies graduate, Jennifer Hall, founded Do While Studio in Boston in 1985, her creation of a pioneering artists' space for cross-disciplinary collaboration was based partly on what she had learned about collaboration between researchers and artists: "The greatest resource at MIT was the collaboration possible between people," she observed in *Women, Art and Technology*.[8] From this beginning, in its 30 years of existence, as directed by Hall and her partner, Blyth Hazen, over the years, Do While Studio projects have included a wide range of innovative and collaborative projects, such as *World Wide Simultaneous Dance*, produced by Laura Knott, in which live dance and digital connectivity combined to connect people dancing at the same time, all around the world.[9]

"this alternative social world that's always there and never stops and is always validating itself. ..."—Gene Youngblood
Social media platforms, also known as social network services or sites (SNSs), are computer-mediated multi-user communications platforms. "Echo is a social network," Stacy Horn states about the SNS she founded in 1990, "although we didn't call it that at the time, we called them virtual communities (I was never able to come up with a term I liked better)."[10]

The Origins of Social Media

Enabling virtual communication between individuals and groups, social media sites encompass some or all of the following affordances: real-time or deferred computer-mediated conversation; the ability to create online identity; the ability to create virtual friendships and/or virtual community with people with common interests; the ability to exchange information and collaborate with other users of the service; the opportunity to publically disseminate information online, the ability to shape and enhance virtual community with identities, dialogue, images, sound, and video; and the ability to share online resources, such as computer games and archives.

"The point is to have this alternative social world that's always there and never stops and is always validating itself as a possible social world. And then alongside of that are all these other supplementary 'periodic' media," Gene Youngblood theorized presciently in 1987.[11]

Exploring the development and experience of early social media, in *this* book, the words and work of creative computer scientists writers, artists, musicians, historians, and digital humanists—from engineer Lee Felsenstein in Northern California to critic Annick Bureau in Paris to digital humanists Alan Liu in Santa Barbara and Julianne Nyhan in London; from artists Hank Bull in Vancouver, B.C., and Wolfgang Staehle in New York City to curator Steve Dietz in Minnesota and researcher/artist Judith Donath in Massachusetts—bring different histories and perspectives, asking and answering persistent questions through the lens of their development and/or their creative uses of early social media systems. Their chapters not only demonstrate how social media evolved from both individual and collaborative efforts but also document the lineage of a wide range of social media affordances. In narratives, such as Arts Wire's odyssey from a text-based conferencing system to a graphical user interface (GUI) system, this book also documents how some social media platforms adapted to the World Wide Web, as well as how text-based social media—such as the Electronic Literature Organization chats, hosted by Deena Larsen on LinguaMOO—persisted successfully in parallel to the Web.

"There is a communitarian dimension to all media used to make art," observes social media–linked performance pioneer, Antoinette LaFarge:

the crowd-feeling of a theater audience, the ebb and flow of a painter's opening, the collaborative spirit of a filmmaking team; and this is apart from the social dimensions of economics, class, gender, race, and so on. But by calling only certain forms (and specifically only certain forms of *software*) "social media," we do not just underline the degree to which sociality and the social are at the heart of the work produced through such media. We also mark how the affordances of such media for real-time, distributed, pseudonymous, improvisational agency have been redefining the social itself for three decades, since the first experiments in electronic community arose in the 1970s.[12]

What Have We Carried Forth from Historic Social Media Platforms? What Have We Lost in the Process?

Has the energy that arose around The Electronic Café, Art Com Electronic Network (ACEN), and System X vanished in the selfie culture of contemporary social media? Or does this energy continue in heroic individual uses of social media, such as English professor Jeff Nunokawa's literate essays, posted early every morning on Facebook since 2007?[13]

Is there a difference between the depth of imagination and immersion that text-based command line interfaces inspired before the advent of GUIs? If so, can we take Sherry Turkle's *Life on the Screen* exploration of this issue as regards personal computer platforms[14] a step further into the cognitive differences between text-based early social media and graphic intensive contemporary social media?

Or, is the gift of pervasive online images so puissant that the resultant diminishment of word-based discussion is immaterial?

Contingently, have exploratory and formative discussions—for instance, the discussions about "Software as Art" that took place on ACEN in the 1980s—disappeared in 140-character bursts of linked information? Or have serious conversations migrated to gated platforms, such as the Critical Code Studies Working Group's online conference that every other year hosts intense discussions on exploring code with humanities methodologies?[15]

Have the artist and arts community–centered missions of early social media platforms, such as Arts Wire, vanished in self-representation on Facebook castles?

Arts Wire's mission began as follows: "The mission of Arts Wire is to provide the arts community a communications network that has, at its core, a strong composite voice of artists and community-based cultural groups"; it concluded by emphasizing how Arts Wire worked

to create a place for artists and the arts community in the national communications infrastructure as part of a larger effort to develop for artists a more integral place in society as a whole.[16]

Does the inestimable value of the representation of artists and writers and arts organizations on Facebook and/or their own homepages compensate for the loss of such ideals and community?

What can we learn from early social media that will inspire us to connect the not necessarily different missions of community and representation—and in the process envision a greater cultural presence on social media?

"Get Lamp"

In this age of ubiquitous contemporary social media, we may never again experience the first-time delight of virtually picking up a lamp to explore uncharted territory in BBN computer scientist Will Crowther's 1970s Interactive Fiction, *Adventure*. We may never again experience the magic of participating in Bill Bartlett's 1979 *Interplay*, in which—as artists in Canberra, Edmonton, Houston, New York, Toronto, Sydney, Vancouver, and Vienna discoursed one after another online—printouts of their continuing dialogue on computer culture emerged from terminals in every city that participated.

Nevertheless, early social media is set forth in this book with the premise that the documentation of pre-Web social networking history is of interest to understanding and participating in contemporary social media—present and future.

To introduce the chapters in this book, it is important to go back a little further in time and set the stage for connectivity in the arts by exploring the work of the researchers who not only created the Internet but also pioneered the shape of early social media. Thus, in this introduction, we follow ARPANET researchers from online conferencing around the development of protocols, early mailing lists, networked email, and *Adventure* to the summer of 1976, when the Internet was demonstrated in a "bread truck" at a historic tavern in the hills of California.

Along the way, from individual researchers and artists, there was Community Memory; there was PLATO; there were the *Satellite Arts Project* and *Send/Receive*; there was ARTEX; there was Minitel-based art and literature; there were bulletin boards, the Electronic Café, The WELL, ACEN, Humanist, THE THING, Arts Wire, and many others. As pioneer Canadian telematic artist, Hank Bull, observes in this book:

Small, independent groups often worked outside official structures, using consumer-level equipment and home-made electronics. Certainly, for the events I was involved in, most of the effort was spent just getting the connection to work. To hear a voice, read a message, or see a face on the screen, beamed in from afar, seemed like some kind of miracle. To have both sound and picture at once was exhilarating. And an event involving several points on a global network, all in touch with each other at the same time, with simultaneous sound, image, and text, was a complex, relatively rare phenomenon that took a great deal of planning and preparation.[17]

Creative Collaboration in the Formative Years of the Internet: The Request for Comment Sets the Stage

The people who received RFC 1 felt they were being included in a fun process rather being dictated to by a bunch of protocol czars. It was a network they were talking about, so it made sense to try to loop everyone in. —Walter Issacson[18]

As it progressed, the collaborative process of the creation of ARPANET was not only enhanced but also documented by the Request for Comments (RFC) that accompanied every advance and decision. As Vint Cerf observed in an RFC created for the 30th Anniversary of RFCs:

RFCs document the odyssey of the ARPANET and, later, the Internet, as its creators and netizens explore, discover, build, re-build, argue and resolve questions of design, concepts and applications of computer networking.[19]

For example, the role of RFCs in the ARPANET R&D process is illustrated by the documentation of the File Transfer Protocol (FTP), beginning in the early 1970s. The working group that developed FTP was chaired by MIT graduate student Abhay Bushan. Members of the team and respondents included Bob Braden (UCLA); Alex McKenzie and Nancy Neigus (BBN);[20] Mark Krilanovich (UCSB); Jim White (SRI-ARC); Greg Hicks (Utah); and the originator of the RFC process, Steve Crocker (UCLA), described by Braden at one meeting as "in on his flying carpet"[21]

As reported in *Where Wizards Stay Up Late: The Origins Of The Internet*, of particular importance is how ARPANET Network Working Group meetings—which were an integral part of the process that created FTP—set a precedent for online conferencing:

In the six months it spent working on the file-transfer protocol, the team usually met face to face in regular sessions of the Network Working Group. But it also frequently used real-time computer teleconferencing. Members of the team would all log on at once, from Palo Alto and Cambridge and L.A. and Salt Lake, and for an hour or two at a stretch trade comments back and forth. Conversing through keyboards and terminals was less spontaneous than speaking, but Bhushan believed it forced clarity into their thinking. It also had the advantage of creating a record of their work.[22]

Games and Mailing Lists in the Early Days of Connectivity

In the 1950s, in the formative years of computer science as a discipline, Bloomsbury-bred British computer scientist Christopher Strachey, who later developed an early concept for time-sharing, was working for the National Physical Laboratory (NPL), when, as a visitor to the lab where Alan Turing worked, he began writing drafts-playing programs for the University of Manchester Computer (MUC). It wasn't long before—utilizing Turing's hardwired, noise-based random number generator (which improved pseudo-random results)—Strachey created a love-letter generator and produced a series of ardent letters signed, "Yours — (adv.) M. U. C."

Harbingers of the culture of contemporary social media, both the intertwined roles of collaboration and creativity in early computer science and the playful spirit of the letters themselves metaphorically initiate the social media aspects of the ARPANET era:

HONEY DEAR

MY SYMPATHETIC AFFECTION BEAUTIFULLY ATTRACTS YOUR AFFECTIONATE ENTHUSIASM. YOU ARE MY LOVING ADORATION: MY BREATHLESS ADORATION. MY FELLOW FEELING BREATHLESSLY HOPES FOR YOUR DEAR EAGERNESS. MY LOVESICK ADORATION CHERISHES YOUR AVID ARDOUR.

YOURS WISTFULLY

M.U.C.[23]

"You are standing at the end of a road before a small brick building. Around you is a forest. ..." **—Will Crowther**

In the 1970s, examples of how the environment of ARPANET research set the stage for social media–based communication are the "finger" utility—created by Stanford Artificial Intelligence Laboratory (SAIL) Executive officer Les Earnest—that developed into proto-blogging, as researchers posted commentary in a program meant to keep track of their whereabouts,[24] the early mailing lists, and the online camaraderie around games, such as Will Crowther's *Adventure*.

MsgGroup, Yumyum, and SF-lovers

The *MsgGroup* mailing list, initiated by Steve Walker, then at ARPA, and based on the early TENEX email program SNDMS, explored the use of the ARPANET for email-based conferencing. Walker was interested not only in discussion around the MSG email utility but also in exploring how dialogue could be created using email. Although some of the process was done by hand, *MsgGroup* allowed messages that were sent to the list to be distributed to all members. It was possible, even at this time (1975), to not only read messages but also reply to them with a message that would in turn be sent to the entire list.

Official communication was not the only purpose of 1970s mailing lists. In conversation, Vint Cerf identified several historic cultural lists, including the *Yumyum* list and *SF-lovers*.[25]

Beginning in 1972, *Yumyum*, hosted on SAIL's PDP-10 and accessible from any ARPANET-connected node, utilized a contemporary online-to-print model to create a restaurant guide, initially for the San Francisco Bay Area. With early input from then Stanford computer science professor, Tom Binford, and SAIL computer scientist, Jack Holloway, when Les Earnest was Local Arrangements Chair for the Stanford-hosted 1973 *International Joint Conference on Artificial Intelligence*, he solicited reviews from SAIL personnel and expanded the coverage. Kasee Menke, the secretary for SAIL, came up with the name *Yumyum* and compiled the first print publication, which was sold at the Stanford Book Store.[26,27]

Earnest continued to host *Yumyum*, eventually changing the name to *California Yumyum*, to reflect its geographical coverage. By email he notes, "Still later, when there came to be more people online, I posted it in a public area on SAIL so that anyone could read it for free. Because our files have been archived on saildart.org we still have access to all the editions that were produced, though since restaurants generally change over time those are not very useful listings."[28]

By legend, an unofficial mailing list started by science fiction fans began at MIT in circa 1975 on the MIT-ITS system. At that time, it was distributed by using "TO": to mail to multiple recipients. In 1979, Richard Brodie, who would later create Microsoft Word,[29,30] telneted to MIT-ITS systems from his Alto during his first week working at Xerox PARC and created *SF-lovers* from the MIT addresses. With a series of hosts, operators, distributors, and archives—including Usenet, via the work of Mary Ann Horton at Berkeley, including sflovers.rutgers.edu—*SF-lovers* is one of the first, if not the first, online film, television, and literature-focused social media platform.

Other early lists included *HUMAN-NETS*, a list to explore human factors in networking and computer science, and the wine lovers list started by John Hollerbach.[31]

Adventure!
ARPANET researchers also played games. One of the most influential and popular games was Will Crowther's *Adventure*,[32] a game that inspired the Interactive Fiction, *Zork*, and the Interaction Fiction publisher InfoCom. Crowther, it should be noted, was a member of the BBN team that made the Interface Message Processor (IMP). In *Where Wizards Stay Up Late: The Origins of the Internet*, Katie Hafner and Matthew Lyon describe his ARPANET work:

> Writing small intricate bits of code was one of Crowther's greatest pleasures in life. His code was among the leanest anyone who worked with him had ever seen. Since the IMP's were to be hooked up to phone lines, with data constantly flowing in and out, the idea was to have a data packet arrive at an IMP, be processed, and immediately either be sent forward or placed in a queue to await its turn—all in less time than it took to snap your fingers.[33]

Based on his caving adventures at Mammoth Cave in Kentucky and the maps he had made of caves, as well as the Dungeons and Dragons games played with researcher friends and other early games,[34] in his spare time, Will Crowther wrote *Adventure* to entertain his children after a divorce. He initially released *Advent* (as it was then called due to character length restrictions) in circa 1975. Coded in FORTRAN, probably for BBN-TENEX running on the PDP-10, the words with which *Adventure* begins are legendary: "You are standing at the end of a road before a small brick building. Around you is a forest …"

There are things you need to explore the cave: "get lamp," you type; "taken," the system replies. "get keys," you type; "taken," the system replies. There are "twisty little passages"; there is treasure; there are dwarves, dragons, and a giant clam (or is it an oyster?). There are magic words and random elements.

As Crowther showed his game to colleagues and friends, everyone wanted to play the game. The use of the network to spread this game among the ARPANET community marked a point in the use of computer networks, where ARPANET was becoming a more informal set of protocols with which people could share anything and through which people could gain access to anything. — Stanford Computer Science[35]

Vint Cerf remembers playing *Adventure* on the TENEX operating system and recalls that a user could not only log on to the machine hosting *Advent*, but also run a command and identify researchers with whom—by typing comment-flagged text back and forth to each other—to discuss progress in the game.[36]

After Crowther moved to Xerox PARC in 1976, *Adventure* was retooled (with added non-player characters, locations, and puzzles) by SAIL computer programmer, Don Woods, who discovered the original work on ARPANET and with Crowther's permission developed and distributed the Crowther and Woods version (1977) that is generally in contemporary circulation.[37]

Thus, out of ARPANET came not only the router that Will Crowther programmed for BBN but also a game that spawned textual adventures and eventually crossed boundaries into electronic literature territory.

"Please go on"
In this era of parallel explorations of communications, play, and social networking, interactive applications were a component of a seminal *International Conference on Computer Communications* (ICCC), held on October 24–26, 1972, in Washington, DC, instigated by ARPA, directed by Bob Kahn, and reportedly featuring more than 40 terminals, most of which worked. As Kahn wrote in RFC 371:

I am organizing a computer communication network demonstration to run in parallel with the sessions. This demonstration will provide attendees with the opportunity to gain first hand experience in the use of a computer network. The theme of the demonstration will be on the value of computer communication networks, emphasizing topics such as data base retrieval, combined use of several machines, real-time data access, interactive cooperation, simulation systems, simplified hard copy techniques, and so forth. I am hoping to present a broad sampling of computer based resources that will provide attendees with some perspective on the utility of computer communication networks.[38]

Creative applications demonstrated at ICCC included two programs written in response to Joseph Weizenbaum's ELIZA.[39] PARRY, written by Stanford psychiatrist Kenneth Colby, simulated a paranoid schizophrenic, and DOCTOR was BBN Programmer Bernie Cosell's LISP conversion of ELIZA. In September 1972, Vint Cerf documented a recording of a conversation between PARRY and the DOCTOR. Cerf's RFC 439 documentation of this conversation reads in part:

*You should pay more attention.
Suppose you should pay more attention.
*You're entitled to your own opinion.
What makes you think I am entitled to my own opinion?
*It bothers me just to be around people in general.
Please go on.[40]

The First Use of Network Email Announces Its Own Existence

In research institutions, exchanging messages with other users on the same machine had been in practice since the 1960s. At MIT, for instance, by the early 1970s, more than a thousand users of MIT's *Compatible Time-Sharing System* (CTSS) used MAIL to coordinate their research and exchange information.[41] MIT researcher Tom Van Vleck recalls his creation of MAIL with colleague Noel Morris in this way:

Noel and I wrote a version of MAIL for CTSS in the summer of 1965. Noel saw how to use the features of the new CTSS file system to send the messages, and I wrote the actual code that interfaced with the user. (We made a few changes from the proposal during the course of implementation: e.g. to read one's mail, users just used the PRINT command instead of a special argument to MAIL.)

The MAIL command created or inserted messages into a file called MAIL BOX in the recipient's home directory. Privileged users could send URGENT MAIL instead, and could send mail even if the user's disk quota was exhausted. The LOGIN command was modified to print

YOU HAVE MAIL BOX

or

YOU HAVE URGENT MAIL[42]

Historic email programs were host-specific. But in the early 1970s at BBN, programmer Ray Tomlinson—who had already worked on CPYNET (a TENEX operating system file-transfer program that transferred files through the ARPANET) and was working on improving the time sharing system mail program, SNDMSG—merged code from CPYNET with SNDMSG and its associated mail reading program, READMAIL, and in the process, he enabled messages to be sent to people on *other* computers. Fortuitously, he selected the now ubiquitous @ sign to indicate the recipient's host. In his words:

The first message was sent between two machines that were literally side by side. The only physical connection they had (aside from the floor they sat on) was through the ARPANET. I sent a number of test messages to myself from one machine to the other. The test messages were entirely forgettable and I have, therefore, forgotten them. Most likely the first message was

The Origins of Social Media

QWERTYUIOP or something similar. When I was satisfied that the program seemed to work, I sent a message to the rest of my group explaining how to send messages over the network. The first use of network email announced its own existence.[43]

In the years that followed, a series of collaborative, often RFC-documented advances contributed to the incremental development of email. Dave Crocker, who was instrumental in the development of email standards, points out via email that, "Individual software was developed by one or a few folk, but all of the 'standards' work was through highly collaborative groups, as well as on-going discussion groups, such as MSGGRP and Header-People."[44]

Thus, the affordances of email were created in a continuing series of advances that, building on each other, created the flexible system that we now take for granted, and along the way there were many struggles and storied developments. "The evolution of electronic mail development is a story of collaboration filled with personalities, software systems and acronyms," Crocker observes in a *Washington Post* article on email history.[45]

For instance, physicist and ARPA director from 1971 to 1975, Stephen Lukasik, not only promoted the advantages of using email communication but also in the process instigated improvements. In his words:

And I said to Larry Roberts, who by that time was head of the office, "Larry, what do I do with all these messages? They just accumulate." And so he literally went home one night and wrote a little system called RD, which was read your mail, throw it away, put it in this category, put it in this category, put it in that—I mean the first beginnings of a mail system, as opposed to just sending telegrams—up until that time, electronic mail was sending telegrams to people, and they threw them away.[46]

Building on Larry Roberts' RD, Marty Yonke created a network mail utility, eventually called BANANARD. Yonke's utility was followed by John Vittal's MSG, a stand-alone package that, in addition to other improvements, included an "answer" (now reply) command:

The wonder of Answer is that it suddenly made replying to email easy. Rather than manually copying the addresses, the user could just type Answer and Reply. Users at the time remember the creation of Answer as transforming—converting email from a system of receiving memos into a system for conversation. —Craig Partridge.[47]

To move mail through the system, mail transport protocols were also developed with input from many researchers. For instance, at USC, Suzanne Sluizer and Jonathan B. Postel developed Mail Transport Protocol (MTP),[48,49] which provided a foundation for the Simple Mail Transfer Protocol (SMTP) standard for email transmission. At U.C. Berkeley, Eric Allman developed sendmail—"intended to help bridge the gap between the totally ad hoc world of networks that know nothing of each other and the clean, tightly-coupled world of unique network numbers."[50] And at the University of Delaware, working as part of the ARPANET community, Dave Crocker and others developed the necessary standards to make email a reliable, international system—writing the RFC for the Internet standards for electronic mail in 1982.[51]

Facilitating interactive virtual discourse with individuals and/or communities of interest, as well as allowing the creation of mailing lists and personal networks, email was and continues to be an essential social media application. As Janet Abbate observes in *Inventing the Internet:*

> Email laid the groundwork for creating virtual communities through the network. Increasingly, people within and outside the ARPA community would come to see the ARPANET not as a computing system but rather as a communications system.[52]

"We chose to work in email, a medium so familiar and by now so ordinary that it was almost invisible." —Cathy Marshall
About 20 years later, as then Xerox PARC researcher Cathy Marshall noted in the introduction to our collaboratively created hypertext, *Forward Anywhere*:

> We could have met in a MOO, or used digital video to tell each other our stories; these technologies would have had some cachet, especially in the technologically charged environment at Xerox PARC. But instead, we chose to work in email, a medium so familiar and by now so ordinary that it was almost invisible—a part of our everyday lives.[53]

Community Memory and PLATO Open the Door to Public Cyberspace

> I say that Community Memory "opened the door to cyberspace and found that it was hospitable territory," due to the creativity it facilitated on the part of the users. —Lee Felsenstein[54]

Not all the researchers who created the foundations of social media in the 1970s were working on connecting laboratories. Some were creating systems and inviting the public to participate.

On August 8, 1973, computer-mediated public social media began in Leopold's Records, one block from the campus of the University of California at Berkeley, where Community Memory was placed on the table next to the bulletin board of tacked paper notices. As Lee Felsenstein—the driving force behind Community Memory and one of the founders of the legendary Homebrew Club, which spawned the Apple II among other early personal computers—explains in his paper in this book:

> A handmade poster, with psychedelic lettering, read "Community Memory." Inside the box was a teleprinter—a Teletype Model 33 ASR which had gone through three years' service as a commercial time-sharing computer terminal. Urethane foam glued inside of the cardboard muffled the whirr of the teleprinter's motor and the "chunk-chunk-chunk" of its print head.

Standing beside the terminal was a young person dressed similarly to most of the students and other people entering the store. As they came toward the terminal, this person would say, "Would you like to use our electronic bulletin board? We're using a computer."[55]

While musicians, poets, and the Berkeley community peopled a working Community Memory, in the summer of the same year, 1973, at the University of Illinois Computer-based Education Research Laboratory (CERL), in the words of programmer, David Woolley:

two decades before the World Wide Web came on the scene, the PLATO system pioneered online forums and message boards, email, chat rooms, instant messaging, remote screen sharing, and multiplayer games, leading to the emergence of the world's first online community.[56]

Beginning in 1960, Programmed Logic for Automatic Teaching Operations (PLATO) was an innovative computer-based educational system that in the 1970s not only utilized plasma display for bitmapped graphics but also supported touch and hosted early multiplayer online games.

In 1973, Paul Tenczar asked 17-year-old university student, David Woolley, to write a program that would let PLATO users to report system bugs online. The program that Woolley wrote, Plato Notes, metamorphosed into Personal Notes, a system of user communication. It was followed by Talkomatic, designed in the fall of 1973 by Doug Brown, which allowed several users to chat together as a group.

In the process, communication between users became an important aspect of the project. As Woolley observes:

The sense of an online community began to emerge on PLATO in 1973–74, as Notes, Talkomatic, "term-talk," and Personal Notes were introduced in quick succession. People met and got acquainted in Talkomatic, and carried on romances via "term-talk" and Personal Notes.[57]

How Mail Art Paralleled Widening Circles of Communication

Although obviously they are not precisely similar, the comparison of social media and mail art, emphasized in Annmarie Chandler and Norie Neumark's *At a Distance*,[58] is of interest. From the points of view of widening circles of communication and communication with people whom sometimes one never meets, there are cogent similarities.

"It was possible to communicate with someone on one level without knowing them on any other, and this added to the mystery of it all," Edward Plunkett observed in *Art Journal* in 1977.[59]

Called the "Eternal Network" in the 1960s by French artist Robert Filliou and brought further into an artist's activity by Ray Johnson, whose mail art circles Ed Plunkett named "The New York Correspondence School," mail art began to widen in international circles in the 1970s—serendipitously paralleling the development of ARPANET.

As mail art networks expanded, like Twitter streams in the digital humanities, no one artist had the same circle. Furthermore, as Hank Bull points out in this book, "The virtuality of mediated, distance communication favored the construction of fictional identities such as Genesis P. Orridge, Anna Banana, Mr. Peanut, and Doctor Brute."[60]

Thus, while work on ARPANET progressed, artists created a parallel social medium where they shared their work. If somehow mail art made the transition to the Internet seamless for some (as it did for me), I am not sure I know why, but perhaps it was the magical communication with people one did not know. Receiving email seldom approximates the packages and envelopes that showed up in my mail, particularly the beautiful artists' stamps from Ed Higgins, Steve Durland, and Anna Banana. But as Higgins once noted:

Some people think it ironic artists' stamps can't always get work through the mail. But that's not their point. Again, the point is art, the communication of ideas.[61]

This, if ARPANET had communication between researchers at its heart, mail art had communication between artists at its heart. Only instead of file transfers, mail art was an exchange of artworks mailed to other artists. It might be small collages, works with Fluxus or dada roots, self-published small books or zines, artist-made sheets of stamps, hand-lettered concrete poetry, even unwrapped objects with addresses stenciled or pasted on their surface.

Like the "six degrees of separation" that informed early social media, mail art circles slowly widened, with (in the case of mail art) exhibitions greatly increasing participation. In New York City, mail art circles broadened with the Whitney Museum of American Art exhibition, curated by Ray Johnson and Marcia Tucker, from September 2 to October 6, 1970. In California in 1978, mail art circle expansion was keyed by the *Artwords & Bookworks* exhibition, curated by Judith Hoffberg and Joan Hugo at the Los Angeles Institute of Contemporary Art (LACMA) from February 28 to March 30, 1978. *Artwords & Bookworks* included works by approximately 600 artists and a catalog with their addresses. As a result, the mail artists included in the exhibition used the catalog to contact artists who interested them, and the network widened.[62]

The Internet Is Gestated in a "Bread Truck" at a Historic Tavern

On August 17, 1976, a "bread truck"—slang for a mobile satellite communications truck—was parked at a historic once stage stop/once speakeasy/now student and researcher watering hole in the hills above Stanford. Under the auspices of ARPA, Vint Cerf, and Bob Kahn, in the hills of California, networked global communication was gestated in a series of successful experiments that utilized TCP/IP to send packets across different networks. BBN computer programmer, Ginny Strazisar, wrote the gateway software. In the words of Don Nielson, Stanford Research Institute principal investigator for ARPA:

Since the days when it was a stagecoach stop between San Francisco and Monterey, Rossotti's was a well-known San Francisco mid-peninsula "watering hole" nestled in the second bank of foothills west of San Francisco Bay. In the 1970s, it had a casual atmosphere and some outdoor seating—a good location for the small ceremony about to take place. No one would mind if we parked SRI's "bread truck" van alongside the courtyard and ran a few wires to one of the tables. It was far enough from SRI (Stanford Research Institute) to qualify as "remote," but close enough to have good radio contact with them through a repeater station atop a hill above Stanford.[63]

Nielson continues his Computer History Museum account by explaining that the terminal was placed on one of the wooden tables in the outdoor courtyard of the tavern and then connected to the van. The message sent on this historic occasion was a weekly report—a long email—on the work of ARPA contractors.

A year later in November 22, 1977, the "bread truck" sent a message from a Northern California coastal road and connected ARPANET to Europe by way of another ARPA network, the Atlantic Packet Satellite Network.

It hopped from California to Boston, on to Norway and Great Britain, and then back to California by way of a small town in West Virginia. —*Wired*[64]

The "bread truck" was unused for 10 years until Don Alves and Nielson got it running, restored it and its equipment, and got it to the 1997 Supercomputer Conference in San Jose for the celebration of the 20th anniversary of the first three-network test of the Internet.

While not beautiful, it did seem to carry some symbolism for many who saw it. So, rather than returning it to certain deterioration and scrap, SRI offered it to the [Computer History] Museum, where it lives today.[65]

"Within the terrific 'electronic space'" —Carl Loeffler

Combining telecommunications and *informatique* (information), the word telematic, from the French word *telematique*, was first used in 1978 by Simon Nora and Alain Minc in *L'informatisation de la* Societe.[66] In its relationship to ARPANET, the telematic art that paralleled the development of the Internet was perhaps the table where adults placed the children during large family Thanksgiving dinners. In my family, this table was situated in the sun porch, along with musical instruments and a view to the garden.

Beginning in the 1970s, running parallel to Internet research was a strain of experimental art and theory that was deeply imbued with a parallel idea of interactive communication.

Gene Youngblood in Heidi Grundmann's seminal book, *Art Telecommunication*, said:

> I want continuous and pervasive access to creative conversations that describe the kind of world I desire, and I want you to have the same access to your reality, the reality that you desire at the same time that I am having access to mine. ... I want to see the rise of what I call AUTONOMOUS REALITY COMMUNITIES. This would be communities of people of politically significant magnitude, in terms of numbers, defined not by geography because they are realized through telecommunications networks and not through neighborhood relationships, defined not by geography but by consciousness, ideology and desire, which is of course the only human community after all.[67]

Hank Bull, in "A Conversation with Hank Bull" on the Interactive Art Conference on Arts Wire, observed:

> the kind of interactivity that interests me is between people. Projects like Shanghai Fax or Hypernation are intended essentially to become conversations, open networks, as full of interruptions, lost threads and ambiguities as real communication. It is important for me to make these conversations global and planetary.[68]

In addition to a focus on communications theory, another concept that artists brought to this table was an emphasis on looking at technology as something to be used creatively. Individual artists expressed this differently, but it was a core precept. Roy Ascott, in *Art Telecommunication*, noted that:

> for art, telematics, while being the product of considerable technological innovation, equally can be seen to carry forward an aesthetic of participation and interconnectedness implicit in the developing strands of art practice during this century. Indeed, it constitutes in many ways a return to values expressed in the culture of the very distant past.[69]

Satellite Arts Project and Send/Receive

In 1977, a year or so after ARPANET tested the Internet from a "bread truck" in Northern California, two teams of artists—working with NASA support to enable live video connection via satellite—demonstrated how global communications would enable the creation of telematic art in a virtual environment. The two projects were the *Satellite Arts Project* (July 26; November 21, 22, and 23, 1977) and *Send/Receive* (September 10 and 11, 1977).

Each of these projects was based on the idea of artist-initiated communication unfettered by geography. *Satellite Arts Project* directors, Kit Galloway and Sherrie Rabinowitz, sought, they told Steve Durland, "the ability to support real time conversation regardless of geography." They set forth a challenge for social media in the arts:

> if the arts are to take a role in shaping and humanizing emerging technological environments, individuals and arts constituencies must begin to imagine at a much larger scale of creativity.[70]

The Origins of Social Media

Contingently, in the words of Carl Loeffler, who was instrumental in the West Coast component of the *Send/Receive* Project:

But while the world clearly seems to grow smaller because of our advanced rapid-communication capability, this same technology also serves to highlight the rich diversity of languages and cultures in the world. Thus, the more efficaciously people can communicate on a global level, the greater will be their sense of understanding and their appreciation of uniqueness and specificity. In the future, we will acknowledge our commonality and delight in our differences.[71]

For *Send/Receive*, artists on the East Coast collaborated with artists on the West Coast to create a two-way 15-hour interactive transmission/satellite-enabled performance event between New York City and San Francisco. In New York, the artists who initiated the project included Willoughby Sharp, Liza Bear, and Keith Sonnier, and the event was produced by Bear and Sonnier. In San Francisco, the producers included artist Sharon Grace and Carl Loeffler, director of La Mamelle/Art Com. In an interview, Loeffler noted that the equipment included:

a Communication Technology Satellite, a Mobile Satellite tracking/ receiving station that they parked along the Hudson River, the teleconferencing station at Ames Research Center, the huge satellite dish behind Stanford University, and many other things.[72]

He described the project in this way:

Within the terrific "electronic space," we conducted dance pieces using split screen, where dancers in NYC and SF would interact. Music experiments with performers playing to activity in SF and the reverse. Experiments in looping the signal around the system a few times and taking delight in the noise and delay. A quasi-moon walk segment, using a moon suit we got from NASA. All sorts of other investigations.[73]

For *Satellite Arts Project*, with the support of NASA, an interactive composite-image satellite dance performance was produced by Kit Galloway and Sherrie Rabinowitz. Dancers 3,000 miles apart at the Goddard Space Flight Center in Maryland and the Educational TV Center in Menlo Park, California, were

electronically composited into a single image that was displayed on monitors at each location, creating a "space with no-geographical boundaries" or virtual space in which the live performance took place.[74]

From these beginnings and others,[75] a vibrant group of art-centered early social media platforms arose in the 1980s, including Galloway and Rabinowitz's Electronic Café, which began at the 1984 Los Angeles Olympics, and ACEN on the WELL, initiated by Carl Loeffler and Fred Truck in 1986.

As Dieter Daniels notes in *Net Pioneers 1.0*,

artists built, designed, and operated their own permanent structures for simultaneously social, discursive, and technical networks. Even more important than the technical innovation involved was the integration of these networks into the participants' everyday lives and the communities that emerged within the projects, as well as an international exchange among the projects.[76]

Early BBS Systems Connect Diverse Communities

January 16, 1978, was a very snowy day. Couldn't get dug out, so called Randy. I had the CACHE message recorder phone line in my house, and Hayes had "invented" the hobbyist modem. That's about it. XMODEM was born of the necessity of transferring files mostly between Randy and myself, at some means faster than mailing cassettes (if we'd lived less than the 30 miles apart we did, XMODEM might not have been born). CBBS was born of the conditions "all the pieces are there, it is snowing like @#$%, lets hack."[77] —Ward Christensen

In 1978, during an epic snowstorm in Chicago, Ward Christensen (software) and Randy Seuss (hardware) created the historic Chicago Computerized Bulletin Board System(CBBS). They published their results in a "how to" in *Byte* in November of the same year.[78]

Soon, in workrooms and garages throughout the country, there were thousands and then tens of thousands of often one or two person-hosted Bulletin Board Systems (BBSs)—running on early PCs connected by telephone lines. Initially, many did not have screens but displayed the output on printouts. Communications speeds were slow, with 300-baud modems most prevalent. Users could exchange messages, read news, download software, and on some systems even play games online. BBSs with multiple phone lines sometimes had chat capability.

If the development of the contemporary Internet can be thought of as a series of small streams and rivers that flowed into a larger navigable river—where cultures of all kinds would eventually participate, sometimes flowing "underground"[79]—in this process, BBSs played an important role. As Paul Ceruzzi observes in his chapter in this book:

BBSs offered a vision to computer users of computing beyond games, spreadsheets, and word processing. They provided a way for owners of modest personal computers to enter a world, which hitherto had been accessible only to those at the top military bases or university labs. They helped novice computer users learn how to manage the intricate details of digital communication. And they created a genuine sense of community among the users.[80]

Because it was possible to host a BBS on only one computer and because they were often free to use (except for the telephone bills, which if the system was not local could be very high), BSSs were core in the development of community networking. Although initially many BBS Sysops were computer hobbyists, who learned about BBSs from computer clubs, soon BBSs emerged for diverse constituencies, such as Native American, African American, rural, and LGBT communities.

The Origins of Social Media

In an interview about the BBS distribution of her LGBT game, *Caper in the Castro*, created in the late 1980s and distributed via BBSs, CM Ralph observes:

> Back then we had BBS—Bulletin Board Systems. Computers connected directly to each other over the phone. There was no internet. But these BBSs were the forerunner to what forums are today. They were used primarily to distribute programs and information. The LGBT BBS network was very small and very underground—just a few in countries around the world—but we all knew each other.[81]

Artist-operated BBSs included System X, described in detail in Amanda McDonald Crowley's interview with Scot McPhee[82]; ARTBASE, hosted in Minnesota by Bob Gale; ARTLINK, hosted by Alan Sandman in Atlanta, GA; and Matrix Artists Network, hosted by Jeff Mann in Toronto, who, while writing for the *Connectivity* issue of *Leonardo* that Roy Ascott and Carol Loeffler edited, expressed a common goal:

> The synthesis of computer and telecommunications technologies has created a new medium for human communication, one that opposes the one-way flow of information pervasive in our mass-media culture. This medium is decentralized and accessible and encourages collaboration, collective thought and the formation of human relationships. It is an electronic geography where community is defined by common interest.[83]

How BBSs Fostered Diversity and Rural Connectivity

In the early 1990s, Randy Ross (Ponca Tribe of Nebraska and Otoe Missouria), at that time Vice-President of American Indian Telecommunications, worked tirelessly to bring the Indian Nations online, and there was substantial Native presence in early BBS-based community networks.

Among the many Native American BBSs, which former Director of the Indian-owned nonprofit community computer network INDIANnet, George Baldwin (Osage and Kansa tribes of Oklahoma) lists in "American Indian and Alaskan Native Network Information Systems"[84] are the Center for American Indian Economic Development BBS (Flagstaff, AZ); the Zuni BBS; the Cherokee BBS; the Michigan-based Native American BBS; the Russell County BBS, run by Cynthia Denton in Hobson, Montana; Spirit Knife (New Mexico); and out of Rapid City, South Dakota, the Dakota BBS (George Baldwin Project Director; Sysop Anne Fallis). Dakota BBS hosted email, information about tribal activities, and information for tribal planners. As did Russell BBS, Dakota BBS also utilized the North American Presentation Level Protocol Syntax (NAPLPS) to host graphic images of Native American art.

However, Randy Ross emphasizes that despite strong Native American participation in early BBSs, the Internet did not bring increased opportunity or connection for Native Americans:

> For Indian Nations, their economies continue to be the worst in the United States. Out of the top ten poorest counties in the entire United States; six of them are in South Dakota. On Shannon

County or what is most of the Pine Ridge Indian reservation, the annual income of family of 4 is less than $6,000 per annum. The unemployment has remained steady at 80% to 90% since 1984 and running steady through to today, 2014. Access to broadband and information services has remained unattainable by many households on these reservation areas.[85]

African American BBSs were connected by AfroNet, originally conceived by Ken Onwere, implemented by Idette Vaughan (The BlackNet) and others, and currently documented by African American cyberspace activist Art McGee, who explains that:

For those unaware, Afronet is an echomail backbone supported by African and African-American BBS Sysops across North America. The goal is to distribute conferences with African and African-American themes throughout North America.[86]

In Montana, Big Sky Telegraph (set up by Dave Hughes and Frank Odasz and run by Odasz) linked rural schools, including Native American Schools, as well as rural libraries, women's centers, and disability organizations. Elsewhere, BBSs for people with disabilities included the Western New York Disabilities Forum hosted by Paul R. Sadowski, which was available on both Buffalo Free-Net in New York and the Project Enable BBS in West Virginia.

It was a time when dedicated community networking "circuit riders" talked and implemented the excitement of computer-mediated communications as if they were riding the trail 100 years ago:

The key was that the Fido BBS could be called nightly at 2AM by each of the 29 Fido BBSs in the remote schools in turn. In less than 2 minutes each call, all the messages generated on the school Fidos would pour into the memory of that one Fido BBS. And IF the Fido had on its disk messages addressed to any one of the Fido's at the schools, they would be "sent" at the same time. In less than an hour in the middle of the night, all the traffic was received at Big Sky or sent from it.

Then the Fido BBS at Big Sky would shut down automatically at 3AM and come up as an Ufgate Server, connected by serial cable across the room to a port on the SCO Unix system. So all the traffic from the Fido's in Montana went into the Unix system using UUCP protocol that was natural to the Unix machine, but matched under UFGATE on the Fido system.

THEN all the traffic from the Montana school BBSs destined to any Fido BBS in the world (I recall the kids communicated with Germans, Norwegians, and Japanese who had access to a local Fido—each with its unique Fido Address identity, while message from a schoolboy or girl had their name as the ultimate recipient or author) went from that SCO Unix system by telephone modem, to my Old Colorado City Communications Unix server using UUCP protocols. And my system would pass all the traffic received from Montana by UUCP hop hop through a Hewlett Packard local office's Unix system, to a Fido system in Silicon Valley, which "knew" how to forward and route replies to other global Fidos.

So the primary heart of Big Sky Telegraph was low cost Fido BBSs linked to UUCP Unix boxes, over which the EDUCATIONAL content—teaching and learning—flowed.

That was the greatest revolution in Rural Communications since the crank party line telephones of the 1930s. —Dave Hughes[87]

The core role of the men and women who traveled and worked tirelessly to help set up networks in communities and rural areas is documented in this book by Madeline Gonzalez Allen, a Bell Labs expatriate, who used her knowledge to travel around the West, helping communities—Telluride, Boulder, the Southern Ute Tribe, among many others—connect to the Internet. In her words from her chapter "Community Networking, a Personal Journey":

I followed a vision for community networking, of communities coming together and deciding for themselves how they wanted to use the then-nascent public Internet for the benefit of their own communities, rather than this technology being shaped and driven solely by those privileged few who at the time were "in the know" (and would eventually lead to the whole ".com" phenomenon). At that critical point in time, as the Internet was becoming a public medium, I felt a calling to do all I could so that *everyone*—regardless of their educational background, income level, employment status, ethnicity, gender or any other "classification"—could have the same opportunity to learn about and shape and benefit from this emerging technology.[88]

Not only did early BBSs create a vital underground network culture in the 1980s, but also they set the stage for conferencing systems, such as The WELL, EchoNYC, THE THING, and Arts Wire; for commercial systems, such as CompuServe and AOL; and ultimately for a Web2 resurgence of community networking, for example, AsianAvenue (1997), BrownPride.com (1997), and BlackPlanet.com, which was begun in 1999 by Omar Wasow.

ARTEX Sends Dialogue and Narrative around the World via I.P. Sharp

In parallel with the development of BBSs, in the late 1970s and early 1980s, at the heart of international artist-centered computer-mediated communication was the Canadian timesharing company I.P. Sharp (IPSA), where Bob Bernecky, IPSA's chief APL programmer, had the interest and foresight to envision the importance of artist-generated computer-mediated communications projects. Using IPSA's systems, Canadian artists, including Robert Adrian X, Bill Bartlett, and Norman White, produced collaborative dialogue and narrative that materialized at nodes all around the world.

Via IPSA, Bill Bartlett created the telematics project *Interplay* for the Computer Culture Exposition at the 1979 *Toronto Super8 Film Festival*. During *Interplay*, as artists from all around the world discoursed online, printouts of their dialogue emerged from terminals in every city involved in the project. The subject was computer culture.[89]

IPSA also began facilitating the work of Robert Adrian X, who lived in Vienna and produced *The World in 24 Hours* in 1982. In 1985, Norman White produced *hearsay*, a tribute to Hungarian-born poet Robert Zend, who had died that summer. In *hearsay*, following the sun, a text written by Zend was sent around the world in 24 hours via IPSA. Alluding perhaps to the feats of messengers and the initial loss of content in the difficulties of telematics, at each of the eight participating nodes, the message was translated into a different language. The message (in the original capital letters that the system probably required) was:

THE MESSENGER ARRIVED OUT OF BREATH. THE DANCERS STOPPED THEIR PIROUETTES, THE TORCHES LIGHTING UP THE PALACE WALLS FLICKERED FOR A MOMENT, THE HUBBUB AT THE BANQUET TABLE DIED DOWN, A ROASTED PIG'S NUCKLE FROZE IN MID-AIR IN A NOBLEMAN'S FINGERS, A GENERAL BEHIND THE PILLAR STOPPED FINGERING THE BOSOM OF THE MAID OF HONOUR.

"WELL, WHAT IS IT, MAN?" ASKED THE KING, RISING REGALLY FROM HIS CHAIR. "WHERE DID YOU COME FROM? WHO SENT YOU? WHAT IS THE NEWS?" THEN AFTER A MOMENT, "ARE YOU WAITING FOR A REPLY? SPEAK UP MAN!"

STILL SHORT OF BREATH, THE MESSENGER PULLED HIMSELF TOGETHER. HE LOOKED THE KING IN THE EYE AND GASPED: "YOUR MAJESTY, I AM NOT WAITING FOR A REPLY BECAUSE THERE IS NO MESSAGE BECAUSE NO ONE SENT ME. I JUST LIKE RUNNING."[90]

Because IPSA's email function was expensive for artists, working with Robert Adrian X, Gottfried Bach, in the Vienna IPSA office, created ARTBOX, a less expensive, more user-friendly version of the IPSA "Mailbox." After several builds, in 1983, ARTBOX officially became ARTEX: the Artists' Electronic Exchange program, and in July 1983—with participant characters in 11 cities including Pittsburg, Vancouver, Vienna, San Francisco, and Toronto—the collaborative fairy tale, *La Plissure du Text*, was produced on the ARTEX network by British artist and telematics theorist Roy Ascott. At that time, Ascott, "The Magician," was based in Paris. As *La Plissure du Texte* traveled from artist to artist, an improvised, collaboratively authored text was augmented and printed out at each node in its journey. Ascott explains,

The title of the project alludes, of course, to Roland Barthes' book "Le Plaisir du Texte" but pleating (plissure) is not intended to replace pleasure (plaisir) only to amplify and enhance it.[91]

CSNET Connects Researchers; Usenet and Listservs Connect Many

Networked communication seemingly slid so easily into many lives, almost as if it had always been there. But behind the scenes, along the trail there were struggles, innovations, and receptive policymakers—as circles of networked communication expanded.

For instance, in 1979, at the University of Wisconsin, Lawrence Landweber, the chair of the Computer Science Department, put forth a proposal to the National Science Foundation (NSF) to fund a CSNET network that would allow all computer science departments to link to ARPANET. It took a few years, but with the help of Vint Cerf and Dave Farber (who was then at the University of Delaware), and with funding from the NSF, CSNET began operation in June 1981.[92]

Usenet

Meanwhile, in 1979, Unix-to-Unix Copy (UUCP), which connected platforms using the Unix operating system and was written by Mike Lesk at Bell Labs in New Jersey in 1978, was released as part of Version 7 Unix. With V7 installed at Duke University in Durham, North Carolina, graduate students, Tom Truscott and Jim Ellis, along with Steve Bellovin, a graduate student at the University of North Carolina at Chapel Hill, were looking to improve information exchange between UNIX computers on the two campuses. The "Newsnet" that they engineered grew from online connection between two universities into Usenet, a global system of more than 100,000 newsgroups.

Combining email functions with organized discussion and information exchange, Usenet was a widely distributed social media platform—furthered at U.C. Berkeley when Mary Ann Horton, Eric Allman, and Eric Schmidt set up a gateway between Usenet and ARPANET and made it work.[93]

Usenet exchanges were sometimes informative and collegial, sometimes disjointed and repetitive, sometimes contentious. But once in a while something wonderful occurred. For instance, as James Blustein and Ann-Barbara Graff observe in their chapter "alt.hypertext: An Early Social Medium," on August 6, 1991, in response to a question about development efforts on distributed hypertext links, Tim Berners-Lee posted a response on alt.hypertext, and in so doing publicly introduced his work on the World Wide Web.[94]

As Usenet usage expanded—with access by users of commercial services, such as America Online (AOL)—in the early 1990s, the Usenet experience was colored by the influx of strident, repetitive voices and the daunting filtering necessary to discover useful information. On some Usenet lists, community building was like attempting to build community in Grand Central Station. Nevertheless, a substantial number of Usenet lists for the arts emerged.

Art, literature, and cultural community-centered discussions on Usenet included, among many others, rec.arts.fine, rec.music.opera, rec.arts.theater, soc.culture.native, soc.culture.african.american, soc.culture.asian.american, and the LGBT site soc.motss, which, along with BBSs, was instrumental in creating LGBT culture online.[95,96] Many Usenet arts groups—such as alt.art.modern, alt.art.post-moder, alt.art.video, alt.art-com, and alt.finearts.rec—were clustered in the less rigorously controlled "alt" hierarchy groups, where at times an underground energy fueled outposts of experimental art and alternative culture, while at other times there was no escaping long ranting posts and continually looping discussions.

Usenet groups have been archived on Google Groups, and many are still accessible for posting. However, "trolls" on some of these lists have completely obliterated their initial vitality and purpose.

Listservs

Listservs, now managed by L-Soft, the company founded by Listserv inventor, Éric Thomas, have withstood the test of time in the arts and humanities. These automated mailing lists—that simplified the arduous process of maintaining computer-mediated mailing lists—often continue to be centrally created, hosted, and frequented by cohesive communities, professionals and/or professional organizations.

Listservs can be moderated or unmoderated, with moderated Listservs requiring approval of postings before they are distributed. Examples of early Listservs, which created viable communities in the arts and humanities, are MCN-L, maintained by the Museum Computer Network for discussion of issues of information technology in cultural heritage; *NINCH-Announce*, produced by the National Initiative for a Networked Cultural Heritage (NINCH)[97] and edited by NINCH executive director David Green; and *FineArt Forum* (FAF). Begun by Ray Lauzzana in 1987—with the goal of disseminating information about the use of computers in the Fine Arts—FAF and its spinoffs, *Leonardo Electronic News* and *Leonardo Electronic Almanac*, were/are hosted by Leonardo/The International Society for the Arts, Sciences and Technology (Leonardo/ISAST) and were also distributed via host-maintained email lists.

Other historic cultural Listservs are *Postmodern Culture* (PMC), founded in 1990 and published by the Johns Hopkins University Press; the many discipline-specific lists under H-Net (for humanities scholars); Museum_L (for museum professionals); *Edupage* in the field of information technology in higher education; and Humanist, created by Willard McCarty in 1987 and documented in this book by Julianne Nyhan's chapter, "In Search of Identities in the Digital Humanities: The Early History of Humanist," in which she observes:

> Within the context of DH, Humanist can arguably be categorized as a proto-social media platform due to the ways it has enabled information, knowledge and social connections to be made (and perhaps unmade) and transferred. More to the point, however, is that newer and slicker social media have come (and, in some cases, gone), but Humanist has endured. Indeed, it arguably remains digital humanities' most vital locus of long-form questioning, imagining and reflecting on and about itself and its many interdisciplinary intersections.[98]

Creative Social Interaction on Early Cultural Conferencing Systems

Logging onto the WELL in 1986 (or watching my Twitter stream flow by in 2015), we—my friends and colleagues online—could be a diverse group of people who might never otherwise meet—creating a framework of stories as we travel together to Canterbury Cathedral. We could be lords and ladies in a court in medieval England, where a master of ceremonies has asked us to throw dice in order to assume the persona of one of 56 stories in from *Chaunce of the Dyse*.[99] Or, we could be in a castle in the magical woods of Italo Calvino's *The Castle of Crossed Destinies*, each of us telling stories that in real

The Origins of Social Media

life we might not always speak (as today on Pinterest, with images, we tell stories to strangers). Or perhaps we are in Zimmermann's café in the 18th century, challenging our parents with the drinking of coffee and listening to the Collegium Musicum. Or perhaps we ourselves are students in Leipzig, playing music instead of pleasing our parents by studying law.

If we are in France, in the 20th century, perhaps we are at an Oulipo party, in the garden of François Le Lionnais—talking the future of literature/not literature with Raymond Queneau, George Perec, and Italo Calvino. Or perhaps we are a gallery, where an array of minitels is spread out to create a distributed narrative.[100]

If we are in Los Angeles, we could be at the 1984 Olympics where The Electronic Café linked diverse communities in cafés across the city.[101]

We/I could be recollecting what it was like when—with a backdrop of the Continental Divide—we strung wires across the trees to set up modem connections at Deep Creek in Telluride so that art students could describe environmental artworks on *Leonardo Electronic News*. Or I could be in a basement in Berkeley in November 1994, when Tim Collin's and Reiko Goto's Carnegie Mellon University "Art Systems" class "entered" the collaborative narrative I had created at Xerox PARC on LambdaMOO.[102] Or I could be watching the beautiful snow fall outside my window in Princeton, NJ, moving seamlessly between laptop and online resources as I write the introduction to this book.

Art Com Electronic Network on The WELL

At the moment, however, in my imagination, I/we are on The WELL—an acronym for *Whole Earth 'Lectronic Link*, founded in 1985 by Whole Earth Catalog/Review editor, Stewart Brand, and public health expert and advocate, Larry Brilliant.[103]

Initially, The WELL's server was a VAX, located beside Sausalito harbor, at the foot of a dock, where colorful houseboats were inhabited by artists and poets. The conference software—the Unix-based PicoSpan computer conferencing system, written by Marcus D. Watts in 1983 and provided by Brilliant's conferencing software company, Network Technologies International (NETI)—both enabled and organized virtual conversation. Although some users found PicoSpan's learning curve difficult, once mastered, its flexibility and clarity fostered an environment of online community.

Art Com Electronic Network (ACEN), founded by Carl Loeffler, the Director of the San Francisco alternative Artspace La Mamelle/Art Com, was implemented on The WELL by digital and information artist Fred Truck and began operation in April 1986.[104] In the decade before the World Wide Web, Loeffler's vision was to create an online environment for contemporary art that included the electronic publication of art journals, a conferencing system for art-centered discussion, and the interactive online publication of computer-mediated artworks and electronic literature. As Fred Truck explains, "He wanted an electronic version of the Art Com/La Mamelle space, and all the activities it engaged in."[105]

ACEN brought artists online; published text art and electronic literature online—including John Cage's *The First Meeting of the Satie Society*, my interactive *Uncle Roger*, and Jim Rosenberg's *Diagram Poems*—and attracted, as was Carl's vision, a wider audience than was usual in alternative art spaces.

If you wrote some words into a topic on ACEN, anyone could respond, and they did. In 1986, as the text flashed slowly across the screen, the idea that people from all over the world were typing these words into The WELL's Conferencing system was so magical that the "lag" and the expense were fleeting concerns. On The WELL, the conversation—in addition to ACEN, there were more than twenty conferences devoted to "Arts and Letters," including ARTS, BOOKS, DESIGN, PHOTOGRAPHY, and POETRY—was both intellectual and jovial, and on ACEN, it was likely to concern the virtual environment that we were not only exploring but also creating. For instance, online "talk" in the "Software as Art" topic on ACEN centered on the sharing of ideas about programming new media art.

The excitement of working in this new field and sharing ideas with others, who were approaching the work in different ways and had different backgrounds, is memorable. On ACEN, after producing the collaborative work, *Bad Information*, I began writing the individual hyperfiction, *Uncle Roger*, in 1986 and was amazed at how, in a parallel topic created by Howard Rheingold, people wrote discussion and criticism of *Uncle Roger*, and I responded to their questions[106] (twenty years before Goodreads was founded).

Surprisingly, the citizens online on The WELL were little concerned by the artificial barriers in real life between South of Market art spaces and people from other walks of life. And the audience gradually widened.

As Fred Truck observes:

ACEN was inhabited by people from all different fields, not just art. One thing we all had in common was an interest in digital technology. Digging a little deeper, another thing we had in common was ACEN, a place we could do what we wanted in any way we wanted. It is a little idealistic to compare ACEN to Camelot, but I can tell you, it was a realized dream that will never happen again in my life. We took that moment, went with it and never looked back.[107]

In his concluding words (circa 1988) in "Telecomputing and the Arts: A Case History," Carl Loeffler wrote:

We hope that our excitement about telecomputing is communicated to you by this text. Sitting here in San Francisco we are highly motivated by the technological advances being made in Silicon Valley, and desire to put these advances in service of contemporary art. For anyone reading this, we offer an open channel for your interest and participation. See you online![108]

Interlude: Multi-user Dungeons (MUDs) and Muds Object Oriented (MOOS)

A user who opens the door to the closet (the standard entry way to this environment) enters a well defined room that is based on Pavel Curtis' real-life living room. "It is very bright, open, and airy here, with large plate-glass windows looking southward over the pool to the gardens beyond. On the north wall, there is a rough stonework fireplace. The east and west walls are almost completely covered with large, well-stocked bookcases."[109]

Originating as a multi-user "adventuring" program written in 1979 by Roy Trubshaw and Richard Bartle, who were students at the University of Essex in England,[110] MUDs are cohesive text-based social media environments where conversation and virtual adventures occur daily, usually observed only by the participants—just as they do in real life. LambdaMOO, a MUD that uses an object-oriented programming language (a cross between C++ and LISP), was developed at Xerox PARC by Pavel Curtis,[111-113] whose theatrical and computer science backgrounds merged fortuitously in the creation of this extraordinary virtual environment:

In October of 1990, I began running an Internet-accessible MUD server on my personal workstation here at PARC. Since then, it has been running continuously. With interruptions of only a few hours at most. In January of 1991, the existence of the MUD (called LambdaMOO) was announced publicly via the Usenet group rec.games.mud. As of this writing, well over 3,500 different players have connected to the server from over a dozen countries around the world.
—Pavel Curtis[114]

Beginning in the early 1990s and continuing through the early years of the World Wide Web, LambdaMOO and other MOOS that utilized LambdaMOO code were core parts of cultural cyberspace—inspiring performative narratives, such as the continuing body of work created by Antoinette LaFarge's Plaintext Players[115]; hosting collaboratively created narrative environments, such as Carolyn Guyer's *HI-Pitched Voices*; hosting Dene Grigar's MOO-situated defense of her thesis[116]; and providing virtual meeting places, such as the Electronic Literature Organization (ELO) Chats, hosted by Deena Larsen,[117] where writers and artists could converse via text.

The fact that MOOs persisted at least through the 1990s, while many text-based early social media platforms did not, points to the importance of immersive real-time response to fellow participants; to the value of creating or collaboratively participating in the creating of one's own social media environment; and to the potentially rich experience of an informal virtual setting, where discussions take place wherever a host with descriptive imagination desires—at an opening at an alternative art space with virtual pizza and beer; at a faculty room with virtual wine and cheese; at an art school with a view of the Continental Divide; or in Pavel Curtis' bookshelf-lined living room.

And you really feel you are there.

THE THING and Arts Wire

A lot of things happened next, as it turned out, but Arts Wire and THE THING, like so many other early online arts experiments chronicled here and elsewhere, have left palpable voids in the online landscape.[118] —Gary O. Larson

THE THING

In its heyday in the 1990s, THE THING was an extraordinary artists project that—with nodes in Cologne, Berlin, Vienna, London, and Stockholm, among others—was a virtual space for the creation, exhibition, and documentation of contemporary art and net art projects:

Here, it is inspirational to look back at the early era of the Internet. THE THING NYC exhibited a playful approach and, far ahead of its time, exemplified what is capable in networked communication. —Susanne Gerber[119]

Begun by German-born, New York City–based artist, Wolfgang Staehle, THE THING began operating out of a Tribeca basement in New York City in November 1991. Working with art spaces and galleries, print publications, conferences, video, and the Internet, THE THING straddled the 1990s art world—incorporating, for instance, in its activities publication of the *Yellow Reader*, a print publication with excerpts from online discussions; electronic dissemination of art magazines and journals, such as The *Journal of Contemporary Art*; Quicktime Movies, by John Baldessari (produced for THE THING, by David Platzker.as a part of the exhibition of Baldessari's *Books and Ephemera* at Printed Matter and Dia Art Foundation); and the postcolonial BindiGirl web project by Prema Murthy, in which Murthy's avatar juxtaposed her own images with images from ancient Indian texts. As Wolfgang Staehle relates in an Interview with Susanne Gerber:

There was this contagious enthusiasm for what was suddenly possible, combined with a sense of play and a curiosity for what would happen next. There was this inspiring community feeling, that we were able to reframe the discourse, that we had the means to realize our ideas independently of the traditional art system. It was a never-ending current of discussions, talks and debates around the globe. People from all over would come to our office in Manhattan to work with us for a while or realize some specific project. And then, of course, there were the parties. When I see the whole project in its totality today, I see a huge sculpture with concentrations in some parts and countless lines of flow of people and information. The whole thing was a social and technological work of art - a creation and a creature. Did we dream enough?

"Can you ever dream enough?" Gerber responds.[120]

Arts Wire

From 1992 to 2001, providing an online platform for artists, arts administrators, and arts funders from all around the country, Arts Wire, a program of the New York Foundation for the Arts (NYFA), offered a forum to share ideas, work, programs, and core advocacy concerns.

The vitality, diversity, and cultural significance of its individual artist and nonprofit organization members were at the core of its Arts Wire's collective vision. Among Arts Wire members were writers, visual artists, musicians, dancers and theatre artists. Arts Wire members also included critics, arts administrators, arts funders, such as the National Endowment for the Arts and the Andy Warhol Foundation, and arts organizations—from Out North in Anchorage, Alaska to DiverseWorks in Houston, Texas and the Frank Silvera Writers' Workshop Foundation in New York City; from American Indian Telecommunications in Rapid City, South Dakota to Opera America in New York City. Creating a diverse cultural presence on the Internet, Arts Wire members also included the National Association of Latino Arts and Cultures, the Asian Cultural Council, the Kitchen, and PS122, as well as the Association of Independent Video and Filmmakers, the American Music Council, The Joyce Theater, and Americans for the Arts, among many others.[121]

Arts Wire began in 1988 when a group of artists and arts workers got together at Orcas Island to talk about new ways to support artists. Many ideas emerged from this NYFA-organized conference; the participants wanted to keep talking. Seattle-based artist and arts administrator, Anne Focke, came up with the name and idea, and Arts Wire began to take shape after a meeting with NYFA's Executive Director, Ted Berger, and others.

Initially under the leadership of Focke and later under the leadership of Michigan-based poet director, Joe Matuzak, beginning in 1992, Arts Wire hosted online discussion among artists and arts administrators, worked with artists and arts organizations to provide Internet presence, produced *Arts Wire Current* (later *NYFA Current*), an influential weekly electronic newsletter about issues in the arts, and worked to expand diversity and collaboration in an art-centered Internet environment.

On Arts Wire, the idea—initially fostered by BBSs—of online systems created for diverse communities and cultures was expanded on a platform that worked to create community for the nonprofit arts as a whole; provide information, such as AIDSwire, hosted by Michael Tidmus; and host online conferencing, such as the Native Arts Network Association, where a group of native arts organizations, from New York City to California, worked together online, with core involvement from Atlatl in Phoenix.

Coordinated by musician/composers Pauline Oliveros, Douglas Cohen, and David Mahler, Arts Wire's NewMusNet was a virtual place for information, discussion, and publication of new music and issues about/for composers, performers, and presenters of experimental music. In parallel, hosted by Anna Couey and Judy Malloy, Arts Wire's Interactive Art Conference was an online laboratory for focused discussion, the production of interactive art, and interviews with artists and writers working in new media.

Arts Wire also created online components of conferences, such as the *Fourth National Black Writers Conference*, held at Medgar Evers College of CUNY with a theme of "Black Literature in the 90's: A Renaissance to End All Renaissances?" Keynote speakers were Paule Marshall and Amiri Baraka; texts of many of the speeches and panels were available on Arts Wire.

If Arts Wire was ahead of its time, nevertheless, participation on Arts Wire provided confidence and experience in working online that greatly contributed to the rich and diverse presence of the arts in the contemporary Internet.[122]

"A design at a level that lets lots of things happen underneath, and they have" —Bob Kahn

It took 20 years, from when ARPANET researchers gathered in Washington in 1972 and put on the Mother of Researcher Parties, but in 1993, from July 23 to 25, community networkers, artists, techno-creators, hackers, and educators gathered in Telluride to celebrate the newly implemented Telluride community InfoZone, community networking as a whole, and the cultural Internet.

Not only did *Tele-Community 93* introduce the Telluride InfoZone and promote community networking, but it also brought together people working at the forefront of nonprofit telematics and—with some excitement—demonstrated the capabilities of community and cultural networking to the public.

Headed and inspired by InfoZone Program Director, Richard Lowenberg, and the Telluride Institute, the gathering included Madeline Gonzales, Lee Felsenstein, Howard Rheingold, Anne and Lewis Branscomb, John Naisbitt, Dave Hughes, Gene Youngblood, Judy Malloy, Lewis Rossetto, Steve Cisler, Dan Collins, Randy Ross, and many others.[123]

Friday afternoon. The basement of the Sheridan Opera House is humming with monitors, Centri, modems, phone lines (too many/not enough) and a lone Minitel terminal. Mission Control will house ongoing demos of MUDs and MUSEs, Internet Relay Chat, The WELL, Community Memory, and a temporary franchise of the Electronic Cafe. The cable-stringers (Gene Cooper, Brett Dutton) greet the altitude-queasy newcomers, and the non-electronic hum turns to all of the TBAs still on the agenda. I walk two blocks to the Elementary School and meet with Bell Labs expatriate Madeline Gonzalez and text artist Judy Malloy to set up and maintain the portfolio of network channels that will allow outsiders to participate.[124] —Eric Theise

Leading up to *Tele-Community '93*, Richard Lowenberg invited composers/artists Mark Coniglio and Dawn Stoppiello, Paul DiMarinis, David Dunn, John Lifton, Richard Povall, David Rosenbaum, Morton Subotnick, and Steina and Woody Vasulka to give a series of talks, performances, and telematic events *in Composer to Composer* (C2C). During *Tele-Community '93*, on main street, groups of people gathered around Kit Galloway and Sherrie Rabinowitz to witness and participate in the Electronic Café's multimedia implementation of *Tele/Comm/Unity*, a project to connect artists, universities, and cultural centers around the world.

There were parties in lofts, a picnic on the West Meadows, and lunches and dinners with people only formerly known by their logins. At the same time, the goals were serious and forward-looking:

As government and corporate interests form alliances and position themselves to create a new National Information Infrastructure, there is a growing movement among regional and local communities and dedicated individuals to shape a more humane, socially serving direction for our tele-media-ted future. Participate in this vital conversation and help promote an ecology of the information environment."[125] —Richard Lowenberg

From Opening the Door to Cyberspace to Social Media Poetics

How the energy, collaboration, and innovation of those who created ARPANET and the early Internet was carried into early social media platforms and how early social media platforms, particularly—as set forth in this book—those created by artists, humanists, and community networkers can and should inform us as we converse, create, and collaborate on contemporary social media—one of the most democratic of public media that ever existed.

Ideally, in the arts, the documentation of early art-centered social media enhances the contemporary cultural Internet, in the same way that post-1950s experimental art (e.g., FLUXUS) continues to inform and infiltrate our culture. Exploring the lineage of contemporary social media, we recollect, discover, and set forth historic social media platforms in the expectation that this media archeology will enrich our vibrant contemporary social media in the same way that the archeology of classic sculpture informed the inquisitive spirit of the Renaissance.

What precisely can we learn by studying the lineage of social media? The answers to this question appear in every chapter in this book.

How This Book Proceeds: Introductions

The focus of this book is primarily on platforms that began before the era of the World Web, although some either migrated to the Web or ran parallel to the Web on text-based systems. This introduction on "The Origins of Social Media," Smithsonian Curator of Aerospace Electronics and Computing, Paul Ceruzzi's "The Personal Computer and Social Media," and legendary cyberspace sage Howard Rheingold's classic paper on

"Daily Life in Cyberspace" set the stage for a semi-chronological treasure chest of both personal and scholarly accounts of the early history of social media.

"Opening the Door to Cyberspace"

The next section begins with electronic engineer and inventor Lee Felsenstein's defining phrase "Opening the Door to Cyberspace." In addition to Felsenstein's own paper, "Community Memory—The First Public Access Social Media System," this section includes PLATO programmer David Woolley's paper on "PLATO: The Emergence of Online Community"; and Canadian computer scientist and scholar, James Bluestein, and Nova Scotia College of Art and Design Vice-President, Ann-Barbara Graff's paper on the Usenet group, alt.hypertext, "alt.hypertext: an Early Social Medium." It moves from pioneer telematic artist Hank Bull's chapter on "DictatiOn: A Canadian Perspective on the History of Telematic Art" to Parisian critic Annick Bureaud's "Art and Minitel in France in the 1980s," and it concludes with curator Steve Dietz's account of how five innovative exhibitions he created at the Walker Art Center documented telematic art and net art and incorporated social media affordances in the process.

"See you online!"

For anyone reading this, we offer an open channel for your interest and participation. See you online! —Carl Loeffler (1946–2001)[126]

"See you online!" looks at social media platforms created by artists, writers, and musicians in the 1980s, beginning with artist/writer, co-director, Art in the Public Interest, and former editor-in-chief of *High Performance*, Steve Durland's classic paper on the Electronic Café: "Defining the Image as Place, a Conversation with Kit Galloway, Sherrie Rabinowitz & Gene Youngblood." Electronic Café co-founder, Sherrie Rabinowitz, died in 2013. Given the core role the Electronic Café played in developing and generating interest and excitement in what they call "Aesthetic Research in Telecommunications,"[127] it is particularly important that, in dialogue with media theorist Gene Youngblood, this interview—originally conducted in March 1987 for *High Performance*—presents the voices of Galloway and Rabinowitz together.

"See you online!" also includes "netprov" artist and professor Rob Wittig's chapter on the Seattle-based "IN.S.OMNIA, 1983–1993"; my chapter, "Art Com Electronic Network on theThe WELL: A Conversation with Fred Truck and Anna Couey"; and Australian curator and former executive director of Eyebeam in New York City, Amanda McDonald Crowley's "Interview with Scot McPhee, Founding Sysop of the Australian-based System X."

System X enabled visual artists, sound artists, and musicians to share work and collaborate on the system. When asked about the legacy of System X, McPhee observed that:

I guess we gave initial impetus to many early Internet art pioneers in Australia. I think that we also brought to people the idea that it could be a viable platform for the construction of artistic meaning. In the late 1980s and early 1990s, when I used to talk to people about this stuff, the most common response was "why would I want to talk to a bunch of computer nerds?" But within 10 years, nearly everyone in the developed world under the age of 40 had an email address and a website, and artists' web projects, art resource sites, mailing lists, and so on proliferated everywhere. I guess we forged the start of a path that others could follow.[128]

Networking the Humanities

"Networking the Humanities" presents four different views on how humanities scholars and cultural innovators created a foothold for the humanities online, beginning with Willaed McCarty's 1987 creation of Humanist—explored in depth by London-based digital humanities professor and scholar Julianne Nyhan, who, in her chapter, "In Search of Identities in the Digital Humanities: The Early History of Humanist," traces Humanist's pivotal role in the development of the Digital Humanities discipline, with a focus on the year of its founding.

In "MOOS and Participatory Media," Director of the Digital Technology and Culture Program at Washington State University Vancouver, Professor Dene Grigar, situates her 1995 defense of her thesis in LinguaMOO in the context of contemporary social media. In "Hacking the Voice of the Shuttle: The Growth and Death of a Boundary Object," University of California, Santa Barbara, English Professor, Alan Liu, sets forth the history of his seminal *Voice of the Shuttle (VoS)*, which created a unifying platform for humanities scholarship online.

From an urban cultural community point of view, New York City–based Echo founder and sysop Stacy Horn's lively chapter, "EchoNYC," explores how in her words:

I wanted Echo to have a more diverse voice. Having more women would simply make Echo better. I also wanted fewer tech-oriented discussions and more talk about the things I was interested in. To make this happen, we opened conferences like Art, Movies & Television, Books, and Culture, and invited organizations like PEN and newspapers and magazines like *Ms.*, the *Village Voice*, and *High Time*s to host conferences as well.[129]

Community Networking

Most telecommunications service providers currently refer to the home, neighborhoods, and communities as the "last mile." They indicate that providing "last mile"–enhanced connectivity, especially in rural areas, is not economically viable. They've got their economic models backward. The greatest source of value in most people's lives is local, derived from self, family, and community. In a globally networked and communicative society, local environments will have the opportunity to provide and benefit from new means of livelihood, knowledge resources distribution, and smarter economic and social valuation. The local area should be considered as the "first mile"[130] —*The Applied Rural Telecommunications Investment Guide*

Focusing on networking local community, the Community Networking movement included many platforms that centrally invited communities to explore local culture. For instance, Blacksburg Electronic Village, a project of Virginia Tech, included a strong section on museums and libraries; Dakota BBS hosted Native American art; Boulder Community Network hosted and continues to host an extensive arts section. In this book, the "Community Networking" section focuses on networks with core artist involvement, as well as on the role of dedicated individuals in fostering community networking.

Randy Ross (Ponca Tribe of Nebraska and Otoe Missouria), former vice-president of American Indian Telecommunications, opens the "Community Networking" section with a thoughtful essay on the impact of telecommunications on Native communities:

From 1984 to 2014, this three-decade span has impacted Indian Country for good and the bad. It is time to explore that pathway and to attempt to reconcile early thought patterns about shifts in tribal life-way and new federalism under a new economy driven by emergent broadband technologies and informatics.[131]

This section also includes artist and community networking innovator Richard Lowenberg's documentation of "The Telluride InfoZone"; former founding Boulder Community Network Coordinator/former founding Association for Community Networks Director, Madeline Gonzalez Allen's "Community Networking, a Personal Journey"; and artist and social networker Anna Couey's "Cultures in Cyberspace: Communications System Design as Social Sculpture," the documentation of her 1992 project that explored issues of networking and diversity across systems and cultures.

Social Media Poetics
German artist, critic, and scholar Susanne Gerber's innovative paper, "Crossing-Over of Art-History and Media-History in the Times of the Early Internet with Special Regards to THE THING NYC"—which includes theory, an interview with THE THING Founder, Wolfgang Staehle, and a timeline appendix—introduces the "Social Media Poetics" section. In these chapters, artists, scholars, and critics explore the 1990s platforms that began in the pre-Web environment and moved to the Web (THE THING, Arts Wire) or carried pre-Web ideals forward in the new Web environment (TrAce) or continued to explore the capabilities of text-based interfaces in parallel with the Web (the Electronic Literature Organization Chats). At the heart of this section are three chapters by creators of electronic literature: my chapter on "Arts Wire: The Nonprofit Arts Online"; Deena Larsen's "Electronic Literature Organization Chats on LinguaMOO"; and artist, writer, researcher, and performer J. R. Carpenter's scholarly history of "TrAce Online Writing Centre Nottingham Trent University."

"Social Media Poetics" concludes with artist and University of California–Irvine Professor Antoinette LaFarge's "Pseudo Space Experiments with Avatarism and Telematic Performance in Social Media." In her words:

I belong to a transitional generation that did not grow up with computers; I adopted them as my main tool and medium as soon as the desktop computer appeared in the late 1980s. But it wasn't until I discovered social media in the early 1990s that I became completely fascinated, and this was because it was only then that I saw real potential for changing the way I could be in the world as an artist and a person.[132]

"accumulating layers of information at electronic speed"—DAX

We are now living and working in a shared electronic space, and we have compressed time and space such that we must deal with our resulting altered consciousness. We have broken through the boundary of "thingness." The environment in which we now probe feels more like water because every thought is like an immersion. We are traversing a complex system, accumulating layers of information at electronic speed—discovering internal landfalls to aid us in our search for the critical path. —Digital Art Exchange Group[133]

"Social Media Archeology and Poetics" does not cover with individual chapters all the arts systems and forums that arose in the days before the World Wide Web, although many are mentioned in various places throughout this book. A few that should be highlighted are the Australian-based Artsnet Electronic Network, directed by Sue Harris and Phillip Bannigan; the Australian Network for Art and Technology (ANAT); the Digital Art Exchange Group (DAX), which under the leadership of Bruce Breland connected performing artists in Pittsburgh and Dakar, Senegal, in *DAX Dakar d'Accord* (among other projects); NETJAM, founded by Craig Latta in 1991, a computer network that fostered International collaboration on musical composition, and the Australian-based collective, VNS Matrix (Josephine Starrs, Julianne Pierce, Francesca da Rimini, and Virginia Barratt), that infiltrated Internet-based forums to distribute "A Cyberfeminist Manifesto for the 21st Century."

Additionally, individual artists—such as Brazilian poet Eduardo Kac, who created a series of works addressing telepresence—produced microsocial media platforms to facilitate collaboration and cultural presence. Karen O'Rourke's 1989 *City Projects* connected artists, students, and teachers in 11 universities on three continents.[134] Artur Matuck brought together artists' projects in his Reflux, a network to "allow for the active use of telecommunications for language and art research, providing a vehicle for intercultural expression."[135] Tim Perkis, John Bischoff, and other musicians created The Hub as a platform for the composing and playing of music. "The result is a really new kind of collective composition, a new social way of making music that didn't exist before. We have a good time," Perkis noted on *Making Art Online*.[136]

In the 1980s, Joe Davis, Dana Moser, and Charles Kelley produced *Nicaraguan Interactions*, a series of slow-scan video transmissions between Boston and Nicaragua.

In the last transmission in the series, a group of children in Boston engaged in a slow scan conversation with their grandmother in Nicaragua. In a sublime moment of irony, innocence and realism, they waved at the image of their grandmother on the video monitor, as she, crying, admonished them to "write."[137]

Into the Present; Into the Future

Internationally, leading into the future, <nettime>,[138] a moderated mailing list for net criticism and Internet cultural politics initiated in 1995 by Geert Lovink and Pit Schultz, served and continues to serve the core roles of bringing global voices to the forefront, situating art discourse in an open yet informed community of artists and writers and theorists, demonstrating the continued influence of interactive text-based mailing lists and spawning other lists, including fiberculture and -empyre-.

Continuing with energy into the present, *Rhizome*,[139] founded by Mark Tribe in 1996 and now physically located within the New Museum in New York City, is an important bridge application that started in the early Web and has continued to sustain community with changing content and a lively focus on the new media arts.

In the UK, *Furtherfield*, founded by artists Ruth Catlow and Marc Garrett in 1997, has created online and physical spaces, fostering both online discussion and physical exhibition of contemporary arts and digital technologies. In July 2015, for instance, at their gallery in Finsbury Park, they hosted a "MoCC Free Market," where visitors were invited to "Leave your money at home and use your personal data to buy, sell, or barter for a delicious range of commodity experiences."[140]

Indeed, the energy of artist and/or activist and/or scholar-initiated, immersive interactive narrative continues in contemporary social media platforms, although it is not always easy to find. For instance, Second Life was an integral part of Joseph DeLappe's "The Salt Satyagraha Online: Gandhi's March to Dandi in Second Life."[141] Princeton Professor of English, Jeff Nunokawa, uses Facebook to host his literate essays on 19th-century literature and contemporary life.[142] Tumblr is giving voice to diverse, postcolonial voices, and Twitter has become a gathering place for digital humanities scholars.

Twitter also serves as a creative platform for works such as Dene Grigar's *The 24-Hr. Micro-Elit Project*, in which—evoking historic global telematic narratives—she edited her stories about living in Dallas, Texas, into twitter-sized literary texts and posted one every hour for 24 hours, inviting others to contribute.[143]

In-depth conferencing continues in projects such as the Critical Code Studies Working Group.[144] Contemporary literature is explored on Goodreads and in Google Hangouts; new platforms are opening for academic discussion. The Modern Language

Association's MLA Commons, for instance, launched at the MLA Convention in Boston in 2013,[145] creates an online community for MLA members—providing opportunities for scholarly discussion and fostering collaboration and the sharing of ideas and resources.

Social media–based artists' explorations of communication continue. As An Xiao notes about her *Morse Code Tweets @ The Brooklyn Museum*, for which she created Morse code tweets in conjunction with a narrative of communicating with her grandmother in Manila:

"So in my life there is this little personal microcosm of the larger story of instant communications."[146]

Responses

Presenting the different ideas of three individuals whose history and expertise are important in beginning to address issues raised in *this* book, the chapters in the final section of "Social Media Archeology and Poetics" are responses not conclusions. They open with Dutch media theorist and Professor of New Media, Geert Lovink's responses to questions that expand on his 2012 essay in *e-flux*: "What Is the Social in Social Media?"[147]

"But how can we teach students to create in a difficult medium that so beautifully (and relentlessly) combines text, image, design, interactivity and collaboration?" I ask.

"Taken my historical baggage into consideration, there is no doubt: the Web is a social sculpture," Lovink responds. "With this in mind, art materializes itself in the social, not in the material remnants of the action. The Internet is no different from participatory arts that Claire Bishop describes in her epic *Artificial Hells* (Verso Books, 2012). As Beuysian artwork it exists in the living connections between the users and their data. The role of teachers then is to bring together the development of a signature of the art student, through skills, and combine this (or, should we say: collide) with the world of discourse and debate."[148]

In "Epilogue: Slow Machines and Utopian Dreams," Faculty Fellow, Berkman Center for Internet & Society, Harvard University, former director of the Sociable Media Group at MIT Media Lab, and author of *The Social Machine: Designs for Living Online*, Judith Donath creatively explores the collective meaning of social media archeology, observing that:

The world of dial-up modems and floppy disks and ASCII bulletin board systems seems very long ago. But the ideals of that time, in spite of their naiveté, indeed because of it, are very valuable. Untainted by cynicism or corrupted by practicalities, they remind us of what the social net ought to be; they remind of the direction to head in, even if we will not quite get there.[149]

"Social Media Archeology and Poetics" concludes with writer Gary O. Larson's scholarly review of cultural Internet policy: "From Archaeology to Architecture: Building a Place for Noncommercial Culture Online." As Larson, whose books include *The Reluctant Patron: The U.S. Government and the Arts, 1943–1965* and *American Canvas: An Arts Legacy for Our Communities*, observes:

Dial-up doldrums and command-line confusion notwithstanding, there *was* something genuinely exciting about forging connections with art and artists in the "prehistoric" era before the advent of the World Wide Web, a sense of discovery and wonder that today's broadband environment cannot match. Which is not to say that we shouldn't try, however, both to recapture that spirit and to create spaces online in which such experimentation can continue to be celebrated and preserved.[150]

Passing the Baton

I pass the baton to all the extraordinary writers in this book!

Notes

1. Advanced Research Projects Agency, "ARPA Changes Names." Available at http://www.darpa.mil/about-us/timeline/arpa-name-change.

2. Katie Hafner and Matthew Lyon, *Where Wizards Stay Up Late: The Origins of the Internet* (New York: Simon & Schuster, 1998).

3. Janet Abbate, *Inventing the Internet* (Cambridge, MA: MIT Press, 1999).

4. Vinton G. Cerf and Robert E. Kahn, "A Protocol for Packet Network Intercommunication," *IEEE Transactions on Communications*, 22:5 (1974): 637–648.

5. "Perspectives on the Internet and Its Evolution. Duet Speakers: Vinton Cerf, Robert Kahn," Princeton University, March 12, 2014.

6. Ibid.

7. Telephone interview with Vint Cerf, May 13, 2014.

8. Jennifer Hall and Blyth Hazen, "Do While Studio," ed. Judy Malloy, *Women, Art & Technology* (Cambridge, MA, 2003), 292.

9. Ibid., 300.

10. Stacy Horn, "EchoNYC" [*Social Media Archeology and Poetics*].

11. Steven Durland, "Defining the Image as Place, a conversation with Kit Galloway, Sherrie Rabinowitz, and Gene Youngblood" [*Social Media Archeology and Poetics*]. First published in *High Performance* 37 (1987): 52–59.

12. Antoinette LaFarge, "Pseudo Space: Experiments with Avatarism and Telematic Performance in Social Media" [*Social Media Archeology and Poetics*].

13. Sherry Turkle, *Life on the Screen: Identity in the Age of the Internet* (New York: Simon & Schuster, 1995).

14. Jeff Nunokawa, *Note Book* (Princeton, NJ: Princeton University Press, 2015).

15. Mark C. Marino, "Field Report for Critical Code Studies," *Computational Culture* (November 9, 2014). Available at http://computationalculture.net/article/field-report-for-critical-code-studies-2014%E2%80%A8/.

16. Judy Malloy, "Arts Wire: The Non-Profit Arts Online" [*Social Media Archeology and Poetics*].

17. Hank Bull, "DictatiOn, a Canadian Perspective on the History of Telematic Art" [*Social Media Archeology and Poetics*].

18. Walter Isaacson, *The Innovators* (New York: Simon & Schuster, 2014), 255.

19. Vint Cerf, "RFCs—The Great Conversation," Network Working Group, *30 Years of RFCs*, Request for Comments: 2555, pt. 4 (April 7, 1999). Available at https://tools.ietf.org/html/rfc2555/.

20. Women who worked on ARPANET included Suzanne Sluizer (USC), an author of the Mail Transport Protocol; Nancy Neigus (BBN), who wrote or co-wrote many RFCs for FTP; Ginny Strazisar (BBN), who wrote the 1st gateway software for TCP/IP; Kasee Menke, an editor of Stanford AI's 1973 compilation of the *Yumyum* mailing list; and Elizabeth Feinler, director of the Network Information Center (NIC).

21. Bob Braden, "[ih] Early FTP development in the ARPAnet" (September 7, 2002). Available at http://mailman.postel.org/pipermail/internet-history/2002-September/000117.html.

22. Hafner and Lyon, 1998, 174.

23. Christopher Strachey, "The 'Thinking Machine'," *Encounter* 3 (1954): 25–31, 26. The original was probably all in capital letters, somewhat as printed in this paper.

25. Cerf, 2014.

26. Stanford Artificial Intelligence Laboratory, *YUMYUM,* October 1973. Available at http://www.saildart.org/YUMYUM.DOC%5BP,DOC%5D22/.

27. A 1996 issue of *Yumyum*, edited by Les Earnest and Patte Wood, is linked from the *Kaleberg Symbiont Archives* and available at http://www.kaleberg.com/yumyum/index.html.

28. Email from Les Earnest, April 20, 2015.

29. Len Shustek, "Microsoft Word for Windows Version 1.1a Source Code," The Computer History Museum Historical Source Code Series [2014]. Available at http://www.computerhistory.org/atchm/microsoft-word-for-windows-1-1a-source-code/.

30. Martin Veitch, "Meet the Man Who Wrote Microsoft Word," *IDG Connect* (September 19, 2014). Available at http://www.idgconnect.com/abstract/8829/meet-man-who-wrote-microsoft-word/.

31. Patrick Henry Winston, "Here's to you, World Wide Web," *Slice of MIT* (February 19, 2012). Available at http://slice.mit.edu/2012/02/19/heres-to-you-world-wide-web/.

32. *ADVENTURE, The Interactive Original by Will Crowther (1976) and Don Woods*. (Reconstructed by Donald Ekman, David M. Baggett, and Graham Nelson.) Available at http://www.web-adventures.org/cgi-bin/webfrotz?s=Adventure&n=1052/.

33. Hafner and Lyon, 1998, 99.

34. For instance, Gregory Yob's *Hunt the Wumpus* was created in BASIC in circa 1974. See Gregory Yob, "Hunt the Wumpus," *The Best of Creative Computing* 1 (1976): 247.

35. Stanford Computer Science, "The Origins of E-mail," September 17, 1999. Available at http://cs.stanford.edu/people/eroberts/courses/soco/projects/1999-00/internet/email.html.

36. Cerf, 2014.

37. Dennis J. Jerz, "Somewhere Nearby is Colossal Cave: Examining Will Crowther's Original 'Adventure' in Code and in Kentucky," *Digital Humanities Quarterly* 1:2 (2007). Available at http://www.digitalhumanities.org/dhq/vol/001/2/000009/000009.html.

38. Robert Kahn, "Demonstration at International Computer Communications Conference," RFC 371, July 12, 1972. Available at http://tools.ietf.org/html/rfc371/.

39. Joseph Weizenbaum, "ELIZA—A computer program for the study of natural language communication between man and machine," *Communications of the ACM* 9:1 (1966): 36–45.

40. Vint Cerf, "PARRY Encounters the DOCTOR," RFC 439, January 21, 1973. Available at http://www.faqs.org/rfcs/rfc439.html.

41. Tom Van Vleck, "The History of Electronic Mail," last update August 8, 2013. Available at http://www.multicians.org/thvv/mail-history.html.

42. Ibid.

43. Ray Tomlinson, "The First Network Email." Available at http://openmap.bbn.com/~tomlinso/ray/firstemailmain.html.

44. Dave Crocker, email correspondence, April 24, 2015.

45. Dave Crocker, "A History of E-mail: Collaboration, Innovation and the Birth of a System," *Washington Post* (March 20, 2012). Available at http://www.washingtonpost.com/national/on-innovations/a-history-of-e-mail-collaboration-innovation-and-the-birth-of-a-system/2012/03/19/gIQAOeFEPS_story.html.

46. "An Interview with STEPHEN LUKASIK Conducted by Judy O'Neill on October 17, 1991. Redondo Beach, CA," Charles Babbage Institute Center for the History of Information Processing,

University of Minnesota, Minneapolis. Available at http://conservancy.umn.edu/bitstream/handle/11299/107446/oh232sl.pdf?sequence=1&isAllowed=y/.

47. Craig Partridge, "The Technical Development of Internet," *IEEE Annals of the History of Computing* 30:2 (April–June 2008): 3–29. Available at http://www.ir.bbn.com/~craig/email.pdf.

48. Suzanne Sluizer and Jonathan B. Postel, "Mail Transfer Protocol," RFC 780, Information Sciences Institute University of Southern California, May 1981. Available at https://tools.ietf.org/html/rfc780/.

49. S. Sluizer and J. Postel, "Mail Transfer Protocol ISI TOPS20 File Definition," RFC 785, July 1981. Available at https://tools.ietf.org/html/rfc785/.

50. Eric Allman, "SENDMAIL—An Internetwork Mail Router," University of California, Berkeley, Mammoth Project. Available at http://docs.freebsd.org/44doc/smm/09.sendmail/paper.pdf.

51. Dave Crocker, "Standard for the Format of ARPA Internet Text Messages," RFC 822, August 13, 1982. Available at https://www.ietf.org/rfc/rfc0822.txt Crocker also collaborated with Vittal, Pogran, and Henderson on an initial document, RFC 733, available at https://tools.ietf.org/html/rfc733/.

52. Abbate, 1999, 111.

53. Judy Malloy and Cathy Marshall, *Forward Anywhere* (Cambridge, MA: Eastgate, 1996).

54. Lee Felsenstein, "Community Memory—The First Public Access Social Media System." [*Social Media Archeology and Poetics*].

55. Ibid.

56. David R. Woolley, "PLATO: The Emergence of Online Community." [*Social Media Archeology and Poetics*].

57. Ibid.

58. Annmarie Chandler and Norie Neumark, eds. *At a Distance: Precursors to Art and Activism on the Internet* (Cambridge, MA: MIT Press, 2006).

59. Edward M. Plunkett, The New York Correspondence School, *Art Journal*, Spring 1977.

60. E. F. Higgins III, "Artists' Stamps," *Print Collectors Newsletter*, Nov.–Dec. 1979.

61. Hank Bull, "DictatiOn, a Canadian Perspective on the History of Telematic Art" [*Social Media Archeology and Poetics*].

62. In 1977, at Judith Hoffberg's invitation, I exhibited a large unfolding hand-drawn map in *Artwords & Bookworks*. To my surprise, mail art began arriving. The first work I received was from Lon Spiegelman. I wasn't sure what to send in return, but I wanted to receive more interesting things in the mail, so I began to create a small series of stapled cartoon narratives. This was the beginning of the "Lucy" series, which, although Carl Loeffler talked me into creating

performance works around these narratives, was primarily written for and in the spirit of The Eternal Network.

63. Don Nielson, "The SRI Van and Computer Internetworking," *Core* 3.1, February 2002. Available at http://www.computerhistory.org/core/media/pdf/core-2002-02.pdf.

64. Cade Metz, "Bob Kahn, the Bread Truck, and the Internet's First Communion," *Wired* (August 13, 2012). Available at http://www.wired.com/2012/08/bob-kahn-internet-hall-of-fame/.

65. Nielson, 2002.

66. Simon Nora and Alain Minc, *L'informatisation de la Societe*, Paris, La Documentation française, 1978; English version. Simon Nora and Alain Minc, The Computerization of Society: Report to the President of France (Cambridge, MA: MIT Press, 1980).

67. The words were spoken by Gene Youngblood in a radio conversation with Paul Brennan. They are quoted in Heidi Grundmann, *Art Telecommunication* (Vancouver: Western Front; Vienna: BLIX, 1984), 23.

68. Anna Couey and Judy Malloy, "A Conversation with Hank Bull on the Interactive Art Conference on Arts Wire," November 1996. Available at http://www.well.com/~couey/interactive/bull.html.

69. Roy Ascott, "Art and Telematics," ed. Heidi Grundmann, *Art Telecommunication* (Vancouver: Western Front; Vienna: BLIX, 1984), 29.

70. Steven Durland, 1987.

71. Carl Loeffler, "Modem Dialing Out," ed. Roy Ascott and Carl Eugene Loeffler *Connectivity: Art and Interactive Telecommunications. Leonardo* 24:2 (1991): 113–114.

72. Judy Malloy, "Keeping the Art Faith, Interview with Carl Loeffler," Topic 90 on Art Com Electronic Network, (ACEN), on The WELL, January 11, 1988. Probably one of the first social media platform–situated interviews. Available at http://www.well.com/user/jmalloy/artcom.html.

73. Ibid.

74. Durland, 1987, 2015.

75. In his paper in this book, Hank Bull notes that Douglas Davis, Nam June Paik, and Joseph Beuys produced an International live satellite telecast as part of the opening of *Dokumenta* 6.

76. Dieter Daniels, "Reverse Engineering Modernism with the Last Avant Garde," eds. Dieter Daniels and Gunther Reisinger, *Net Pioneers 1.0: Contextualizing Early Net-Based Art* (Berlin: Sternberg Press, 2010), 22–23.

77. Ward and Randy, "The Birth of the BBS," Chinet, 1989.

78. Ward Christensen and Randy Seuss, "Hobbyist Computerized Bulletin Board," *Byte* 3:11 (November 1978): 150–157.

79. In circa 1989, U.C. Berkeley Forest pathologist Fields Cobb—http://ourenvironment.berkeley.edu/people_profiles/fields-w-cobb-jr/—who had many words of wisdom, called my attention to this concept of the role of researchers in the greater picture.

80. Paul Ceruzzi, "The Personal Computer and Social Media" [*Social Media Archeology and Poetics*].

81. Luke Winkle, "A Q&A With The Designer of the First LGBT Computer Game: Paste" (June 9, 2014). Available at http://www.pastemagazine.com/articles/2014/06/caper-in-the-castro-interview.html.

82. Amanda McDonald Crowley, "System X: Interview with Founding Sysop Scot McPhee" [*Social Media Archeology and Poetics*].

83. Jeff Mann, "The Matrix Artists' Network: An Electronic Community," *Leonardo*, 24:2 (1991): 230–231.

84. George Baldwin (Osage and Kansa tribes of Oklahoma), "American Indian and Alaskan Native Network Information Systems," The Race and Ethnicity Collection. Available at http://race.eserver.org/native-indian-and-alaskan.html.

85. Randy Ross, "Issues in Native American Telecommunications" [*Social Media Archeology and Poetics*].

86. Arthur McGee, AFRONET BBS List. Available at http://www.africa.upenn.edu/BBS_Internet/afro_bbs.html.

87. Dave Hughes, "Big Sky Telegraph (1)." Available at http://davehugheslegacy.net/index.php?option=com_content&view=article&id=424:big-sky-telegraph-1&catid=102&Itemid=210/. Note that with Dave's permission, I made a few spelling changes in his text.

88. Madeline Gonzalez Allen, "Community Networking, a Personal Journey" [*Social Media Archeology and Poetics*].

89. Documentation of ARTEX Projects, including those by Robert Adrian X and Norman White, is available in *Telematic Connections: The Virtual Embrace*, Walker Art Center, 2001. Curated by Steve Dietz. Available at http://telematic.walkerart.org/overview/index.html.

90. Ibid. The text is from Robert Zend, *From Zero To One* (Victoria, BC: The Sono Nis Press, 1973). It is available in Norman White's account of *hearsay* at http://telematic.walkerart.org/timeline/timeline_white.html.

91. Roy Ascott, 1984, 35.

92. Abbate, 1999, 183.

93. Ronda Hauben, "The Early Days of Usenet: the Roots of the Cooperative Online Culture." Available at http://www.columbia.edu/~rh120/ch106.x10]/.

94. Jamie Blustein and Ann Barbara Graff, "alt.hypertext: an early social medium" [*Social Media Archeology and Poetics*].

95. David Auerbach, "The First Gay Space on the Internet," *Slate* (August 20, 2014). Available at http://www.slate.com/articles/technology/bitwise/2014/08/online_gay_culture_and_soc_motss_how_a_usenet_group_anticipated_how_we_use.html.

96. David Auerbach, "When AOL Was GayOL: How LGBTQ Nerds Helped Create Online Life as We Know It," *Slate* (August 21, 2014). Available at http://www.slate.com/articles/technology/bitwise/2014/08/lgbtq_nerds_and_the_evolution_of_life_online.html.

97. More information about the National Initiative for a Networked Cultural Heritage (NINCH) is available at http://www.ninch.org/index.html.

98. Julianne Nyhan, "In Search of Identities in the Digital Humanities: The Early History of Humanist" [*Social Media Archeology and Poetics*].

99. *Chaunce of the Dyse,* MS Fairfax 16 and MS Bodley 638, University of Oxford, Bodleian Library.

100. Annick Bureaud, "Art and Minitel in France in the 1980s" [*Social Media Archeology and Poetics*].

101. Durland, 1987, 2015.

102. Judy Malloy, "Narrative Structures in LambdaMOO," ed. Craig Harris, *In Search of Innovation—The Xerox PARC PAIR Experiment* (Cambridge, MA: MIT Press, 2000), 102–117.

103. Recently purchased by The Well Group, Inc., a private investment group composed of long-time WELL members, The WELL (http://www.well.com) has for years featured art, literature, theater, and music conferences. I founded the Arts Conference in 1993, and it is still a place for art talk on The WELL.

104. Judy Malloy, "Art Com Electronic Network: An Interview with Fred Truck and Anna Couey" [*Social Media Archeology and Poetics*].

105. Ibid.

106. Judy Malloy, "Uncle Roger, an Online Narrabase," in *Connectivity: Art and Interactive Telecommunications*, eds. Roy Ascott and Carl Loeffler, *Leonardo* 24:2 (1991): 195–202.

107. Malloy, "Art Com Electronic Network: An Interview with Fred Truck and Anna Couey."

108. Carl Loeffler, "Telecomputing and the Arts: A Case History," *Art Com Media Distribution Catalog* 1 (n.d.): 66–71.

109. Malloy, 2000, 104.

110. Richard Bartle, "Interactive Multi-User Computer Games," MUSE Ltd Research Report, December 1990.

111. Pavel Curtis, "Mudding: Social Phenomena in Text-Based Virtual Realities," Xerox PARC CSL-92–4, April 1992.

112. Pavel Curtis and David A. Nichols, "MUDs Grow Up: Social Virtual Reality in the Real World," Xerox PARC, May 5, 1993.

113. Pavel Curtis, *LambdaMOO Programmer's Manual* for LambdaMOO Version 1.8.1, May 2004.

114. Curtis, 1992, 4.

115. Antoinette LaFarge, "Pseudo Space Experiments with Avatarism and Telematic Performance in Social Media" [*Social Media Archeology and Poetics*].

116. Dene Grigar, "MOOS and Participatory Media" [*Social Media Archeology and Poetics*].

117. Deena Larsen, "Electronic Literature Organization Chats on LinguaMOO" [*Social Media Archeology and Poetics*].

118. Gary O. Larson, "From Archaeology to Architecture: Building a Place for Noncommercial Culture Online" [*Social Media Archeology and Poetics*].

119. Susanne Gerber, "Crossing-Over of Art-History and Media-History in the Times of the Early Internet with Special Regards to THE THING NYC" [*Social Media Archeology and Poetics*].

120. Ibid.

121. Judy Malloy, "Arts Wire: The Non-Profit Arts Online" [*Social Media Archeology and Poetics*].

122. Ibid.

123. Richard Lowenberg, "The Telluride InfoZone" [*Social Media Archeology and Poetics*].

124. E. S. Theise, "Megabytes @ Kilofeet: Telluride Ideas Festival & InfoZone," *FringeWare Review* 3 (January 1994): 36–39.

125. Richard Lowenberg, "Telluride IDEAS FESTIVAL 1992," email announcement. Available at http://www.bio.net/bionet/mm/bioforum/1993-July/005290.html.

126. Loeffler, n.d.

127. "Telecollaborative Art Projects of Electronic Café International Founders Kit Galloway & Sherrie Rabinowitz." Available at http://www.ecafe.com/museum/history/ksoverview2.html.

128. Crowley [*Social Media Archeology and Poetics*].

129. Horn [*Social Media Archeology and Poetics*].

130. *The Applied Rural Telecommunications Investment Guide* (1995). Sponsored by the Colorado Advanced Technology Institute and funded by the Economic Development Administration of the U.S. Department of Commerce. Developed under contract by the Telluride Institute InfoZone Program; Project Director: Richard Lowenberg; with web publishing assistance provided by Lee Taylor, Telluride Woodcraft. It is no longer available online.

131. Randy Ross [*Social Media Archeology and Poetics*].

132. Antoinette LaFarge [*Social Media Archeology and Poetics*].

133. The Digital Art Exchange, "Philosophy." Available at http://www.digitalartexchange.net/e/philosophy.html.

134. Karen O'Rourke, "*City Projects*: An Experience in the Interactive Transmission of Imagination," in Roy Ascott and Carl Eugene Loeffler, Guest eds., *Connectivity: Art and Interactive Telecommunications. Leonardo* 24:2 (1991): 215–219.

135. Artur Matuck, "The Reflux Project—Event of a Geotronic Revolution." Available at http://www.colabor.art.br/arturmatuck/sitev1/portfolio/telecom_arts/reflux_article.pdf.

136. Tim Perkis, in Judy Malloy, ed., *Making Art Online*, Telematics Timeline, Walker Art Center, 2001. Originally exhibited by Reflux in 1991 and by ANIMA in 1994. Available at http://telematic.walkerart.org/timeline/timeline_malloy.html.

137. Dana Moser, "Notes on Telecommunications Art: Shifting Paradigms," Ascott and Loeffler, eds., 1991, 213–214.

138. Information about the "nettime mailing lists" is available at http://www.nettime.org/

139. *Rhizome* is available at http://rhizome.org/.

140. Furtherfield is available at http://furtherfield.org/.

141. Joseph DeLappe, "The Salt Satyagraha Online: Gandhi's March to Dandi in Second Life." Available at http://www.delappe.net/game-art/mgandhis-march-to-dandi-in-second-life/.

142. Nunokawa, 2015.

143. Dene Grigar, "On the Art of Producing a Phenomenally Short Fiction Collection over the Net using Twitter: The 24-Hr. Micro-Elit Project," 2009. Available at http://www.nouspace.net/dene/24hr/24-Hr._Micro-Elit_Project/Home.html.

144. Marino, 2014.

145. MLA Commons is available at https://commons.mla.org/.

146. An Xiao, *Morse Code Tweets @ The Brooklyn Museum*. Available at https://vimeo.com/13056836/.

147. Geert Lovink, "What Is the Social in Social Media?" *e-flux*, 2012. Available at http://www.e-flux.com/journal/what-is-the-social-in-social-media/.

148. Judy Malloy, "Expanding on 'What Is the Social in Social Media?'—A Conversation with Geert Lovink" [*Social Media Archeology and Poetics*].

149. Judith Donath, "Slow Machines and Utopian Dreams" [*Social Media Archeology and Poetics*].

150. Gary O. Larson, "From Archaeology to Architecture: Building a Place for Noncommercial Culture Online" [*Social Media Archeology and Poetics*].

Bibliography

Abbate, Janet. 1999. *Inventing the Internet*. Cambridge, MA: MIT Press.

Ascott, Roy. 2007. *Telematic Embrace: Visionary Theories of Art, Technology, and Consciousness*, edited and with an essay by Edward A. Shanken. Berkeley, CA: University of California Press.

Ascott, Roy, and Carl Eugene Loeffler eds. 1991. Connectivity: Art and Interactive Telecommunications. *Leonardo* 24:2.

Casey, James, and Randy Ross. 1999. *(Ponca Tribe of Nebraska and Otoe Missouria) and Warren, Marcia (Santa Clara Pueblo of New Mexico), Native Networking: Telecommunications and. Information Technology in Indian Country*. Washington, DC: Benton Foundation.

Cherny, Lynn, and Elizabeth Reba Weise, eds. 1996. *Wired Women: Gender and New Realities in Cyberspace*. Seattle, WA: Seal Press.

Coleman, Beth. 2011. *Hello Avatar, Rise of the Networked Generation*. Cambridge, MA: MIT Press.

Curtis, Pavel, "Mudding: Social Phenomena in Text-Based Virtual Realities," Xerox PARC CSL-92–4 (April 1992).

Daniels, Dieter, and Gunther Reisinger, eds. 2010. *Net Pioneers 1.0: Contextualizing Early Net-Based Art*. Berlin: Sternberg Press.

Grundmann, Heidi. 1984. *Art Telecommunication. Vancouver: Western Front*. Vienna: BLIX.

Hafner, Katie, and Matthew Lyon. 1998. *Where Wizards Stay Up Late: The Origins of the Internet*. New York: Simon & Schuster.

Haynes, Cynthia, and Jan Rune Holmevik, eds. 2001. *High Wired: On the Design, Use, and Theory of Educational MOOs*. Ann Arbor, MI: University of Michigan Press.

Isaacson, Walter. 2014. *The Innovators*. New York: Simon & Schuster.

Malloy, Judy. 2003. *Women, Art and Technology*. Cambridge, MA: MIT Press.

McDonald, Christopher Felix. *Building the Information Society: A History of Computing as a Mass Medium*, Ph.D. Dissertation, Department of History, Princeton University, 2011.

Moser, Mary Anne, and Douglas MacLeod. 1996. *Immersed in Technology: Art and Virtual Environments*. Cambridge, MA: MIT Press.

Nakamura, Lisa. 2002. *Cybertypes: Race, Ethnicity, and Identity on the Internet*. New York, London: Routledge.

Quarterman, John. 1990. *The Matrix: Computer Networks and Conferencing Systems Worldwide*. Bedford, MA: Digital Press.

Rettberg, Scott. 2011. All Together Now: Hypertext, Collective Narrative, and Online Collective Knowledge Communities. In *New Narratives: Stories and Storytelling in the Digital Age*, ed. Ruth Page and Bronwen Thomas. Lincoln, NE: University of Nebraska Press.

Rheingold, Howard. 2000. *The Virtual Community: Homesteading on the Electronic Frontier*. rev. ed. Cambridge, MA: MIT Press.

Turkle, Sherry. 1995. *Life on the Screen: Identity in the Age of the Internet*. New York: Simon & Schuster.

2 The Personal Computer and Social Media

Paul E. Ceruzzi

Gordon Moore's famous paper, in which he predicted a regular doubling of electronic components on an integrated circuit, was accompanied by a cartoon. It showed a department store clerk selling "Handy Home Computers" next to counters selling "Notions" and "Cosmetics."[1] Moore's paper was published in 1965, and as we all know, the prediction proved remarkably accurate, although there is evidence that "Moore's Law" is ending. But what of the cartoon? Moore predicted that by 1973 or 1974, a few integrated circuits would contain as many components as the UNIVAC, one of the first commercial computers, introduced in the early 1950s. The "Handy Home Computer" did in fact appear beginning in the late 1970s. Its appearance was anything but rational or orderly, as several histories have pointed out. Even less orderly was the way in which the personal computer facilitated social interaction—today one of the primary uses of personal computing devices.

A desire for computer-mediated social interaction preceded the invention of the personal computer. Human beings are social animals and will exploit whatever means they can to facilitate social activities. Tom Standage even argued that many of the social interactions we associate with the Internet were found among 19th-century users of the Morse telegraph.[2] Other essays in this volume, such as Lee Felsenstein's paper on the Community Memory and David R. Wooley's paper on PLATO, illustrate how visionaries sought to facilitate social interaction using centralized computers that were "time-shared": a large, centrally located computer that allocated its processor time among users. If the computer's processing time and remote connection were fast enough compared to the human user's reaction time, the user had the illusion that he or she was having a "conversation" with the machine: to query a database or to iterate a possible solution to a mathematical problem. Although not initially envisioned, before long users extended this metaphor to include a *literal* conversation to others logged on at the time.

The excitement that accompanied the announcements of the first personal computers was driven by a perception that these personal devices would allow users to get away from the time-sharing model of interactive computing. Often, despite other uses, the time-sharing systems described above were intended mainly for commercial users, for which the services charged a fee based on three factors: connection time, bulk storage use, and time used by the computer's central processor and memory. The latter fee, for seconds of processor time, could be very high. The early adopters of personal computers were amazed by their ability to run a program for hours without incurring such fees. Social uses on commercial time-sharing systems were not practical given that fee structure. If there were such users, they were found only at a few university or military laboratories, where time on the computer was set aside for research purposes. That often meant that the high fees were incurred, but the Defense Department paid the bill. On many university campuses, not all students or faculty had such access. Those fortunate few who did have access quickly found ways to communicate to others logged onto the same computer. That may have been the first use of computers for social interaction and has been cited as a forerunner of email, but it obviously had limits.

One exception to this structure was set up in the mid-1960s at Dartmouth College. Teletype terminals were installed around the campus, and they were connected to a large General Electric mainframe computer. Liberal arts, science, or engineering majors could access the system using the BASIC programming language, without the per-minute charges. But the Dartmouth model of free, unfettered access did not prevail, although BASIC, developed specifically for that system, was later adopted by personal computer enthusiasts. The Dartmouth system was heavily used, but not for social interaction.

The first personal computers were legendary for their primitive features, lack of systems software, and unreliable hardware. So strong was the desire to have ownership that people tolerated these shortcomings—and worked tirelessly to overcome them. Historians are not in agreement as to which was the "first" personal computer, but one can say that the era began with the announcement, on the cover of *Popular Electronics* in January 1975, of the "Altair" by the Albuquerque New Mexico firm MITS. I say "announcement" because it was quite a while before Altair kits or assembled computers were delivered to customers in any quantity. Nonetheless, the Altair has its rightful place in history for no other reason than its role in persuading Bill Gates and Paul Allen to leave Cambridge, Massachusetts, move to Albuquerque, and devote themselves to overcoming the machine's shortcomings (the first product of their company, initially called "Micro-Soft," was an adaptation of Dartmouth's BASIC for the Altair). Other similar machines followed the Altair, all of them taking advantage of the recently invented microprocessor: the computer-on-a-chip that was foreseen by Gordon Moore a decade earlier. The next several years, from 1975 to about 1981, were among the most effervescent and exciting the computer industry ever saw (or ever will see), with scores of machines coming on the market, accompanied by software of ever-increasing sophistication. Key events in this era were the 1977 introduction of the Apple II computer, along with personal computers from Radio Shack and Commodore, and microprocessor-controlled video game consoles from Atari. Finally, a personal computer from IBM in 1981 rounded out this era.

These personal computers took advantage of the microprocessor, but they were similar in design to the small "minicomputers" that were already in common use. Minicomputers were being sold by a number of companies, primarily Data General and Digital Equipment Corporation, both headquartered in Massachusetts. Among the design features carried over from minicomputers were a series of input/output ports, by which the computers communicated with the outside world. That required specialized chips or circuits that translated information, as it was used inside the machine, into a form that could be reliably sent out of the machine.[3] Another device used by minicomputers that was carried over to the personal computer was the Teletype: an electromechanical terminal developed for telecommunications long before the minicomputer was invented. Its adoption, which saved the computer manufacturers the trouble of developing their own terminals, was used at Dartmouth and by the ARPANET, the predecessor to the Internet. (Among the few nonalphabetic symbols on a Teletype keyboard was the @ sign, adopted later for email and now the icon of the digital age.) Later on the Teletype was replaced by video monitors and keyboards, dot-matrix printers, and other devices.[4]

Absent from this discussion is the use of one of those ports for communication with other computers. The port that sent data "serially," or in a linear stream, could be connected to a device called a "modem" (modulator-demodulator), which converted computer data into audio tones that could be transmitted over ordinary telephone lines. That use of the PC was slow in coming for a number of reasons. One was that much of the excitement around the early personal computers centered on their ability to do other things. They could play games interactively—something that no one near an IBM mainframe was supposed to do. They could perform spreadsheet calculation—again something previously restricted to the "priesthood" of elite mainframe users. They could also perform word processing—another revolutionary application. One could set up a personal computer to access commercial time-sharing services, but a connection with large commercial services was precisely what early adopters of the PC were trying to avoid. The desire to own and control one's computer, even if only to use it as a doorstop, cannot be underestimated. There were technical hurdles: the modems were not cheap, and they required communications software that had to be specially configured. Before long, free or inexpensive software for PCs began to appear, such as "XMODEM" (later "MODEM7"), which made setup a lot easier. U.S. telephone service at the time was a regulated monopoly. Telephone equipment was not bought but rather rented from AT&T. Modifications were forbidden, although court cases from the 1950s and 1960s did allow the limited connection of third-party equipment, as long as it did not damage the network. Around the time of the personal computer revolution, AT&T made two decisions that would have a far-reaching impact. The first was a decision that

AT&T would not discriminate between voice and data sent over its lines. Both were in the audio range, and data were in principle no different, even if it sounded funny to a person listening in on the line. The second was to introduce a jack, the "RJ11," which allowed a customer to plug and unplug telephones without the need for a company technician to visit one's home. That allowed one to connect a modem directly to the phone network, rather than having to dial a number, wait for a high-pitched tone indicating a connection, and place the receiver into a cradle that acoustically coupled the modem to the phone line. Not long after the personal computers appeared came another key innovation from the Georgia electronics company Hayes Microcomputer Products. This innovation was a "smart" modem that one could leave connected to the line and would automatically connect and disconnect the computer without the need for a person to intervene.[5]

Additional factors affected the use of PC for social interaction, and these were more economic and political than technical. The U.S. phone network was arguably the finest in the world. As a regulated monopoly, it had characteristics that were the product of political as well as technical decisions. Specifically, in the United States, a local phone call was not billed by the minute but was part of a flat rate charged to the customer. Long-distance calls were billed by the minute. They were priced higher at peak business times, lower at night and on weekends. Personal computer users who wished to communicate with their colleagues could stay connected as long as they wished, as long as the connection was a local call (and as long as one did not wish to use the phone simultaneously for voice phone calls). Long-distance charges, even late at night, could mount rapidly during a session.

Out of this complex milieu of technical, social, and political factors emerged the "bulletin board system" (BBS). The metaphor was one of a cork board on a college campus, where students posted notes of all sorts of topics. BBSs were typically grassroots activities, often set up by small businesses or electronics hobbyists. The hardware requirements were minimal, typically an extra phone line and a modest personal computer, only with the addition of a hard disk drive in addition to the usual floppy drives. Bulletin board software was available for free or at a modest cost. A standard joke was that a person would buy an IBM XT computer that had a 10 megabyte hard drive, buy a bracket that allowed one to stand it on its side rather than rest it horizontally on a desk, and, voilà, one had a "server." And one was no longer a "user" but a "sysop" (system operator). The content of these BBSs varied according to the whims of the sysop. A shopping mall might have a list of shops, phone numbers, and show times for movies. A small business would have information about its services, hours, location, and so on. Computer or electronic repair shops set them up with multiple phone lines, and these evolved into network hubs for a community. One could send and receive simple text messages. Initially, in many cases, there was barely enough interaction to qualify these as social media, but their impact was large. BBSs offered a vision to computer users of computing beyond games, spreadsheets, and word processing. They provided a way for owners of modest personal computers to enter a world, which hitherto had been accessible only to those at the top military bases or university labs. They helped novice computer users learn how to manage the intricate details of digital communication. They created a genuine sense of community among the users. Legends grew up (unverified) that MIT computer science professors or Apple co-founder Steve Wozniak could be found on local bulletin boards late into the night. Aerospace historian Joe Corn relates how people swooned and "went mad" the first time they saw a human being take to the air in a heavier-than-air machine[6]; similar if less dramatic emotional reactions came with one's first encounter with the potential of computer-mediated communication from one's home.

BBS sysops developed a way to circumvent the long-distance telephone charges, by gathering messages and sending them to various distant destinations late at night, when phone rates were low. Because the messages were prestored, the call could be completed quickly. The developer of the most popular store-and-forward software called it "FIDONET," suggesting a dog that dutifully fetched one's newspaper from the front lawn. By the mid-1980s, the local BBS, plus "FIDONET," was the poor man's Internet, at a time when the Internet was in its infancy. FIDONET was especially influential in Third World countries that lacked the sophisticated telecommunications infrastructure of the United States.

As the personal computer matured, it settled on standard hardware that was based on the IBM architecture and Microsoft operating system software, and (with exceptions, such as the arts community) Apple products a distant second. The initial drive to acquire a PC as a rebellion against the mainframe mentality receded. The machines became more reliable, the software more capable, and users simply wanted them to do useful or creative work. At the other end of the spectrum, large-scale commercial time-sharing businesses began to offer time on their systems to personal computers, with a rate structure that was more favorable than what they charged commercial customers. One of the most influential was CompuServe, founded in 1969 as a commercial service. It began offering time to individuals in the 1980s, initially during off-hours to utilize otherwise underutilized capacity. At its headquarters in Columbus, Ohio, banks of Digital Equipment Corporation PDP-10 computers offered a variety of services: mail, news, weather, stock market prices, sports scores, winning lotto ticket numbers, and so on. The company established local phone numbers in major locations across the country so that users did not incur long-distance charges. The local phone lines were connected to Columbus by a packet-switched network, similar to the ARPANET.

CompuServe also offered an unusual feature called "CB Simulator," which allowed users to chat with one another, in the manner of users of the then-popular Citizens Band radio. This was one of the beginnings of what has since become a defining social feature of modern networks: chat rooms and instant messaging. CB radio had exploded in popularity, in part as a result of the imposition of a nationwide 55-mile-an-hour speed limit in 1973. Truckers embraced the medium as a way of informing each other of the presence of "Smokeys" (state police, who wear Smokey-the-Bear hats) enforcing what the truckers felt was an absurd law. The CompuServe product followed closely the metaphor of the radio, with "channels" and other similar controls. Anyone who remembers the CB craze, or who has seen the 1977 Hollywood film *Smokey and the Bandit*, starring Burt Reynolds and Sally Field, knows that the level of discourse on the CB channels could be rather low.[7] So it was with chat on computer networks, and, in many circles, the phenomenon shows no sign of abating. It is not possible to keep up with the latest outrage that parents have over the online antics of their teenage children, as new forms of instant communication appear on an almost weekly basis. CompuServe was founded to serve businesses; as a result, the personal activities on it tended to be more serious. That was in contrast to other services that came along, which will be discussed next. The Citizens Band metaphor did not apply to CompuServe's service as it evolved. One attribute of the CB radio phenomenon did carry over and is a characteristic of social media to the present day. In the early days of CB radio, users gave their real names or call signs, as amateur radio operators did. Before long, however, the operators adopted pseudonyms, called "handles," which disguised their identities.[8] That allowed for a more freewheeling interchange on the radio, but it also led to a coarsening of the social interaction. This coarsening is a characteristic of many, but not all, current social interactions on the Internet. That is balanced by an ability to express one's opinions without fear of retaliation or ridicule, especially valuable for people living under repressive political regimes. There is, however, always the risk that their identity can be unmasked. Some services, especially Facebook, make strenuous efforts to ensure that the person logged in is a real person logged on under his or her real name. Facebook's success at this is in dispute.

The chat rooms and discussion forums for other similar early services exhibited these traits much more. Other papers in this book, such as Rob Wittig's "IN.S.OMNIA, 1983–1993" and Judy Malloy's "Art Com Electronic Network on The WELL: Interview with Fred Truck and Anna Couey," address visionary 1980s nonprofit-oriented and/or artist-initiated online forums. But for commercial forums, the most popular was America Online, later known as AOL. AOL had a complex history, emerging out of a failed service called "The Source," founded in the late 1970s by a visionary named William von Meister. His vision of what would later be called "cyberspace" was remarkably clear, if his execution of that vision fell short, and the service was short-lived. Unlike CompuServe, The Source was intended from the start for home use using personal computers. According to a well-known story, Steve Case, a marketing executive for Pizza Hut, found himself logging on to The Source using a Kaypro personal computer late into the night and saw the potential for the service.[9] After many false starts, in 1985,

Case cofounded Quantum Computer Services, later renamed America Online. Initially, it offered services for owners of the popular Commodore 64 personal computer. The Source was gone by this time, having been bought by CompuServe as a hedge against competition. But according to one observer of the scene, "The Source was more than just another online system. It was the system that started it all. ... It constituted a *vision* of what an online utility could and should be."[10] One of AOL's defining features was its chat rooms, gently moderated by volunteers who contributed their labor in exchange for playing a seminal role in a momentous social phenomenon. AOL eventually had millions of paying subscribers who dialed into the service from personal computers in the home.

All of this growth took place while the Internet was in its infancy. As the Internet was emerging out of the military-funded ARPANET and academic centers, the personal computer phenomenon was evolving in a parallel universe, with little communication between the two camps. They eventually merged, but not for a while. The former camp relied on DEC PDP's or VAX computers, IBM mainframes, or high-end workstations from companies like SUN, running versions of the Unix operating system and connected to one another by high-speed dedicated communication lines. The latter was teasing out performance from low-powered personal computers running on simple operating systems, connecting via modems to ordinary phone lines.

The social milieu of the two camps was also quite different. Those who built the ARPANET were independent souls who often championed libertarian ideals of freedom and respect for technical competence. But they had to be mindful of their source of funding—would Wisconsin Senator William Proxmire discover one of the more frivolous uses of the 'net and award them a "Golden Fleece" award?[11] Would the military brass at the Pentagon shut the program down if they considered those uses a violation of military communications protocols, which had been long set by the Defense Communications Agency (DCA)? AOL users, who owned their own hardware, had no such fears. AOL chat rooms allowed users to have their say about almost any topic. Once users were able to set up their home computers with genuine Internet and World Wide Web access (not an easy task at first), AOL lost its reason for being. But its legacy lives on.

Among the other online services for PCs that emerged before the Internet, one more deserves mention. "Prodigy" was a joint venture among IBM, Sears, and CBS, the latter dropping out after an initial investment. The service was announced in 1984, at a time when the personal computer phenomenon in the United States had stabilized around the IBM PC architecture. In many ways, Prodigy was a U.S. version of the French "Minitel" system, discussed in Annick Bureaud's paper in this book, but it planned to use personal computers, not specialized terminals, to connect to the telephone network. As with AOL, local phone numbers were established to avoid long-distance charges.

Prodigy offered a wide variety of services, including email and chat, for a monthly fee of $9.95 for a suite of basic features.[12] It had several features, however, which set it apart from the other services mentioned above. The first was its extensive use of graphics. With the exception of ASCII graphics used in early computer games, in BBSs, and by artists, graphics were quite rare among the early personal computers, and the limited speeds of the modems then in use made it difficult to integrate them into the user's online experience. The Prodigy service prestored a lot of graphical data on the user's disk drive and only had to transmit a fraction of the data necessary for a colorful image. This allowed for a much richer experience that could potentially broaden the appeal of online access. A second feature was tied to that. With color graphics, Prodigy was able to attract advertisers, who were thus able to subsidize the service and keep the cost to the consumer low. That was a mixed blessing, and it went against the values of the online communities that had grown up around the ARPANET and its descendants at university laboratories. Like it or not, colorful advertising is a primary method of funding World Wide Web activities today.

Prodigy also offered discussion forums, but it found that users who used these facilities heavily did not generate much ad revenue. Reflecting its corporate parenthood of IBM and Sears, Prodigy monitored the content more closely than AOL. That did not resonate well with users, who much preferred AOL's lighter touch. AOL developed a reputation for the adult-oriented content of its chat rooms, a reputation that Steve Case, a man of conservative personal habits, did nothing to discourage. Case was known to frequent the chat rooms and engage in conversations with their inhabitants. The result was that by the mid-1990s, AOL had, for the general public, far outstripped its rivals to become the preferred "on-ramp" to the information superhighway—a way for personal computer owners to participate in many of the features of the Internet, especially the social features, with little of the difficulties. By 1995, personal computer users could connect directly to the Internet and take advantage of the World Wide Web browser Navigator from Netscape. But in order to do that, a user also had to establish an account with an Internet Service Provider; buy, install, and configure special connections software; install an email reader; and in general do a lot of fiddling to get what AOL was offering in a simple bundle. AOL's pioneering role came to an end with the bursting of the Internet Bubble in 2000, not long after it merged with the traditional media firm Time-Warner. CompuServe, Prodigy, and AOL all transformed themselves into Web-based services, with varying success. World Wide Web "portals" such as Yahoo! offered the same user experience as AOL or Prodigy without tying the user to a particular service.

Throughout the 1990s, Moore's Law and its brethren were in full force. The personal computer began as a primitive device appreciated mainly by hobbyists and electronics enthusiasts. It eventually became a stable, mature product, but only after a period of intense effort, especially in software development. The enthusiasts worked tirelessly to overcome the machine's physical limitations, but in doing so, they also pioneered new methods of social interaction—exploring ways to use the personal computer as a communications device, in addition to its use as a calculator, database, or game platform. They developed predecessors to today's chat rooms, using the metaphor of the Citizens Band radio; and the common discussion space, patterned after campus cork and pin bulletin boards. They explored ways to share programs and data within the constraints of the commercial telephone system and its regulations.

Using the computer as a communications device was precisely what the developers of the ARPANET, the direct ancestor of today's Internet, wanted also to do, although for years the two groups worked independently of one another. With the advent of high-speed modems, faster microprocessors, and communications software that integrated Internet and World Wide Web access with PC operating systems, the two groups found one another. The personal computer became the standard way to access the Internet and engage in social activities, and the flexibility of the PC was augmented by portable devices in the new millennium. The social experience pioneered by the PC hobbyists triumphed. As it did, the personal computer democratized and transformed the Internet into the social platform it is today.

Notes

1. Gordon E. Moore, "Cramming More Components onto Integrated Circuits," *Electronics* (April 19, 1965): 114–117. Grant Compton was the artist who drew the cartoon.

2. Tom Standage, *The Victorian Internet: The Remarkable Story of the Nineteenth Century's On-Line Pioneers* (New York: Bloomsbury, 1998).

3. The so-called "UART," which stood for "universal asynchronous receiver/transmitter." That obscure circuit had an impact far beyond its modest function.

4. Some of these early video terminals had names that suggested their legacy: "Glass TTY" (TTY was a symbol for the Teletype) or "dumb terminal."

5. Alfred Glossbrenner, *The Complete Handbook of Personal Computer Communications, the Bible of the On-line World*, 3rd edition (New York: St. Martin's Press, 1990). It is instructive to compare the successive early editions of this book, first published in 1984, to chart the volatile and rapid changes underway in computer-mediated communications.

6. Joe Corn, *The Winged Gospel: America's Romance with Aviation, 1900–1950* (New York: Oxford University Press, 1983).

7. Perhaps the only place on the Internet today where the level of discourse remains high, and where prominent people log on using their real names, is the site Quora.com.

8. In the movie referenced above, Burt Reynold's character's handle was "Bandit," and Jerry Reed's "Snowman." Pseudonyms on digital media lack some of the color of the CB world, but they have their own charm.

9. Kara Swisher, *AOL.com: How Steve Case beat Bill Gates, Nailed the Netheads, and Made Millions in the War for the Web* (New York: Times Books, 1998).

10. Glossbrenner, *Complete Handbook*, 1983 edition, p. 68.

11. "Sen. William Proxmire uncovers wasteful government spending, 1975-1987." Available on the Wisconsin Historical Society website at http://www.wisconsinhistory.org/turningpoints/search.asp?id=1742. Accessed June 10, 2015.

12. Ibid, p. 167.

3 Daily Life in Cyberspace: How the Computerized Counterculture Built a New Kind of Place

Howard Rheingold

Introduction to "Daily Life in Cyberspace"

Judy Malloy

Reprinted from legendary cyberspace pioneer Howard Rheingold's classic, *The Virtual Community: Homesteading on the Electronic Frontier*,[a] "Daily Life in Cyberspace: How the Computerized Counterculture Built a New Kind of Place" occupies the core position of introducing what early social media platforms were like in the days before the World Wide Web.

Howard Rheingold has served as immersed scribe, experienced guide, and acclaimed sage for the virtual community since 1985, when he first logged on to The WELL.

Always supportive, always ready to help and encourage newcomers, he continues to promote social media–based creativity and scholarship, while at the same time—with his painted shoes and colorful signature outfits—his presence brings vitality to the field.

Here, in a book about social media archeology and poetics, his perspective as an artist/writer is particularly important. As he noted in *The Whole Earth Review* in 1988:

Because I am a writer, I used to spend my days alone in my room with my typewriter, my words, and my thoughts. On occasion, I ventured outside to interview people or to find information. After work, I would reenter the human community, via my neighborhood, my family, my circle of personal and professional acquaintances. But I was isolated and lonely during the working day, and my work did not provide any opportunity to expand my circle of friends and colleagues. For the past two years, however, I have participated in a wide-ranging, intellectually stimulating, professionally rewarding, and often intensely emotional exchange with dozens of new friends and hundreds of colleagues. And I still spend my days in a room, physically isolated. My mind, however, is linked with a worldwide collection of like-minded (and not so like-minded) souls: My virtual community. If you get a computer and a modem, you can join us.[b]

For later generations who seek to understand the roots of social media, Howard Rheingold's "Daily Life in Cyberspace" provides a compelling description of the experience of early social media. It chronicles with whom he shared and built community on The WELL, beginning in the mid-1980s—including memorable spirits, such as beloved Sysop, David

Hawkins, and incisive SRI International futurist, Tom Mandel (both alive at the time but now no longer with us). Along the way, the reader experiences not only the creation of communities of interest in cyberspace but also how people of all kinds—a self-employed productivity consultant who lives in the Marin countryside "overlooking the ocean near Bodega Bay," a music store owner who is also Secretary of Ecological Economics of Alaska, a Japanese writer who is interested in ecology and electronic democracy—gathered in cyberspace and created early social media.

ARPANET, The Community Memory, PLATO and many others created the protocols and platforms, but Howard Rheingold was the first to chronicle social media and advise and caution on its uses. In "Daily Life in Cyberspace," he observes that:

Reading a computer conference transcript in hard copy—on paper—misses the dynamism of the conversation as it is experienced by regulars; the back-and-forth dialogue over a period of time that regular participants or observers experience can be reconstructed by looking at the time stamps of the postings, however. In terms of communication rhythms, e-mail and computer conferencing can be levelers. This is one way in which computer conferencing differs from other communications media. The ability to think and compose a reply and publish it within the structure of a conversation enables a group of people to build the living database of Experts on the WELL, in an enterprise where contributors all work at their own pace. This kind of group thinks together differently from how the same group would think face-to-face or in real time.[c]

Twenty years or so later in *Net Smart*, he observes that:

Every PC as well as smart phone is a printing press, broadcasting station, political organizing tool, and site for growing a community or marketplace. Knowledge, power, advantage, companionship, and influence lie with those who know how to participate, rather than those who just passively consume culture.[d]

Notes

a. Howard Rheingold, *The Virtual Community: Homesteading on the Electronic Frontier*, rev. ed. (Cambridge, MA: MIT Press, 2000).

b. Howard Rheingold, "Virtual Communities," *Whole Earth Review*, no. 57 (Winter 1987), 78–80.

c. Rheingold, 2000.

d. Howard Rheingold, *Net Smart* (Cambridge, MA: MIT Press, 2012), 249.

I was still toting around my 1969 edition of the *Whole Earth Catalog* when I read an article about a new computer service that Whole Earth publisher Stewart Brand and his gang were starting in the spring of 1985. For only $3 an hour, people with computers and modems could have access to the kind of online groups that cost five or ten times that much on other public telecommunication systems. I signed up for an account. I had previously suffered the initiation of figuring out how to plug in a modem and use it to connect to computer bulletin-board systems (BBSs) and the Source (an early public information utility), so I was only a little dismayed that I had to learn a whole new set of commands to find my way through the software to the people. But established WELL users were extraordinarily helpful to newcomers, which more than made up for the bewilderment caused by the software. I started reading the conferences and began to post my own messages. Writing as a performing art! I was hooked in minutes.

Over a period of months, I fell into the habit of spending an hour or two every day gazing in fascination at this window into a community that was creating itself right in front of my eyes. Although the system was only a few months old, the air of camaraderie and pioneer spirit was evident among the regulars. Those $3 hours crept up on me in 10- to 30-minute minivisits during the workday and hour-long chunks in the evening. Still, my daily telecommunicating expenses were less than the price of a couple of drinks or a double cappuccino. The cumulative economic impact of my new habit came home to me when my first month's bill was over $100.

As it happened, a friend of mine had to deliver some artwork to the *Whole Earth Catalog* people at the Sausalito office where the WELL also was located. So I went along for the ride. When we got to the rambling series of ancient offices in one of the last bohemian enclaves of the Sausalito houseboat district, I asked for the WELL. I was led to a small room and the staff of one, Matthew McClure. I talked with Matthew about the possibility of diminishing my monthly bill by starting and hosting a conference about the mind.

Hosts are the people who serve the same role in the WELL that a good host is supposed to serve at a party or salon—to welcome newcomers, introduce people to one another, clean up after the guests, provoke discussion, and break up fights if necessary. In exchange for these services, WELL hosts are given rebates on their bills. I was worried that my hosting duties might take up too much of my time.

Matthew smiled at my question. I know the meaning of that smile now, although it puzzled me then. He recognized what was happening to me. He judged it to be a good thing to happen to me, and to the WELL. He was right. But it was still Mephistophelian. He said, "Some hosts get away with less than an hour a week."

That was the fall of 1985. By the fall of 1986, the WELL was a part of my life I wasn't willing to do without. My wife was concerned, then jealous, then angry. The night we had the climactic argument, she said, referring to the small, peculiar, liberal arts college where we first met, "This is just like Reed. A bunch of intelligent misfits have found each other, and now you're having a high old time." The shock of recognition that came with that statement seemed to resolve the matter between us.

The WELL is rooted in the San Francisco Bay area and in two separate cultural revolutions that took place there in past decades. The *Whole Earth Catalog* originally emerged from the Haight-Ashbury counterculture as Stewart Brand's way of providing access to tools and ideas to all the communards who were exploring alternate ways of life in the forests of Mendocino or the high deserts outside Santa Fe. The *Whole Earth Catalog*s and the magazines they spawned—*Co-Evolution Quarterly* and its successor, *Whole Earth Review*—seem to have outlived the counterculture itself, since the magazine and catalogs still exist after 25 years.

One of *Whole Earth*'s gurus, Buckminster Fuller, was fond of using the analogy of the tiprudder—the small rudder on very big ships that is used to control the larger, main rudder. The tiprudder people who steer the movements and disciplines that steer society—the editors and engineers, scientists and science-fiction writers, freelance programmers and permaculture evangelists, grassroots political activists and congressional aides—continued to need new tools and ideas, even though they were no longer a counterculture but part of the mainstream. These cultural experimenters continued to feed *Co-Evolution Quarterly* and then *Whole Earth Review* through decades when magazines died by the thousands. Even the idea that you could publish books on the West Coast was a revolution when it happened; in 1992, when *Publishers Weekly* ran an article on the history of West Coast publishing, it started with the *Whole Earth Catalog*. The first *Whole Earth Catalog* was the first idealistic enterprise from the counterculture, besides music, that earned the cultural legitimation of financial success.

The *Whole Earth Catalog* crew, riding on the catalog's success, launched a new magazine, *The Whole Earth Software Review*, and, after the WELL was started, received a record-breaking $1.4 million advance for the *Whole Earth Software Catalog*. It was time for the string of successes to take another turn: the WELL was the only one of the three projects to succeed. The *Whole Earth Review* is what survived in print; the WELL did more than survive.

The inexpensive public online service was launched because two comrades from a previous cultural revolution noticed that the technology of computer conferencing had potential far beyond its origins in military, scientific, and government communications. Brand had been part of the faculty at an online institute devoted to stretching the imaginations of business leaders—the Western Behavioral Sciences Institute (WBSI)—which introduced him to the effectiveness of computer conferencing. WBSI was also where he connected with Larry Brilliant.

Brilliant and Brand shared a history at the center of several of the most colorful events of the 1960s: Brand was "on the bus" with Ken Kesey and the Merry Pranksters

(Kesey's pot bust, as described in Tom Wolfe's *Electric Kool-Aid Acid Test*, happened on the roof of Brand's apartment; Brand was one of the organizers of the seminal Trips Festival that gave birth to Bill Graham Presents and the whole rock concert scene). Brilliant had been part of the Prankster-affiliated commune, the Hog Farm (which had organized the security arrangements for Woodstock around the judicious use of cream pies and seltzer bottles and had whipped up "breakfast in bed for 400,000"). After his Hog Farm days, Brilliant became a doctor and an epidemiologist and ended up spearheading the World Health Organization's successful effort to eliminate smallpox.

Brilliant was involved with another health-care effort aimed at curing blindness in Asia, the Seva Foundation, and he found that Seva's far-flung volunteers, medical staff, and organizational directors could meet and solve problems effectively through computer conferencing. When a medical relief helicopter lost an engine in a remote region of Nepal, the organization's online network located the nearest spare parts, gained key information about ways to cut through local bureaucracies, and transported the needed parts to the crippled aircraft. Brilliant became one of the principles of NETI, a business that created and licensed computer conferencing systems. After they met via WBSI's conferencing system, Brilliant offered Brand the license to PicoSpan (the WELL's conferencing software) and the money to lease a minicomputer, in exchange for a half interest in the new enterprise. The new enterprise started out in the *Whole Earth Review*'s charming but ramshackle office, leased a dozen incoming telephone lines, installed what was then a state-of-the-art minicomputer, and set up modems, and in 1985, the WELL was born.

Brand and Brilliant both hoped the WELL would become a vehicle for social change, but instead of trying to mold it in a specific image, they wanted to see the vehicle emerge spontaneously. The WELL was consciously a cultural experiment, and the business was designed to succeed or fail on the basis of the results of the experiment.

The person Stewart Brand chose to be the WELL's first director—technician, manager, innkeeper, and bouncer—was Matthew McClure, not coincidentally a computer-savvy veteran of the Farm, one of the most successful communes that started in the 1960s. Brand and McClure started a low-rules, high-tone discussion, where savvy networkers, futurists, intelligent misfits of several kinds who had learned how to make our outsider status work for us in one way or another could take the technology of CMC to its cultural limits. When McClure left a year and a half later, another Farm veteran, Cliff Figallo, took over. While Figallo managed the business, yet another Farm veteran, John "Tex" Coate, was charged with building the community.

The Farm veterans had tried for more than a decade to create a self-sufficient colony in Tennessee. At the Farm's height, more than 1,000 people worked together to try to create their own agricultural society. It still exists and is still surprisingly self-sufficient. They homebirthed and homeschooled, built laundries for washing hundreds of diapers, grew soybeans, and even extended their efforts to other countries—Cliff Figallo had spent years in Guatemala on behalf of Plenty, the Farm's international development arm, helping Maya villages install hygienic water systems. Matthew and Cliff and John and their families, including eight children, left the Farm after 12 years, partially out of disagreement with the way it was governed, partially out of weariness. Self-sufficiency is very hard work.

Brand thought the Farm alumni were perfect choices for their jobs at the WELL. Matthew was the only one with prior computer experience, but what they knew from the front lines of communal living about the way people reach decisions and create cultures collectively—and the ways people fail to reach decisions and create cultures—more than made up for their lack of computer savvy. By 1992, the WELL staff had grown to 15, the original minicomputer was long gone, and all the Farm veterans had moved on to other enterprises.

By the time I had been ensconced in the WELL for a year, it seemed evident to me that the cultural experiment of a self-sustaining online salon was succeeding very well. At that point, as I was becoming convinced that we were all setting some sort of cultural precedent, I interviewed online both Matthew McClure and Kevin Kelly, who had been part of the original group that founded the WELL.

One of the advantages of computer conferencing is the community memory that preserves key moments in the history of the community. Sure enough, although I had not looked at it in years, the online oral history was still around, in the archives conference. The responses were dated October 1986.

Matthew McClure recalled that "Stewart's vision was very important in the design." The vision that McClure and Brand agreed on involved three goals: to facilitate communications among interesting people in the San Francisco Bay area, to provide sophisticated conferencing at a revolutionary low price, and to bring email to the masses. To reach a critical mass, they knew they would need to start with interesting people having conversations at a somewhat more elevated level than the usual BBS stuff. In Matthew's words, "We needed a collection of shills who could draw the suckers into the tents." So they invited a lot of different people, gave them free accounts, called them "hosts," and encouraged them to re-create the atmosphere of a Paris salon—a bunch of salons. Brand, a biologist, insisted on letting the business grow instead of artificially stimulating it. Instead of spending money on glossy advertising, they gave free accounts to journalists.

McClure recalled two distinct growth spurts. First, the word about the WELL spread among the more adventurous members of the bay area's computer professionals, and the free journalist accounts paid off as WELLites began to write and publish articles about the WELL. Brand went to Cambridge to write a book, and the hosts seemed to have the run of the place.

The next major event, McClure recalled, was the organization of the Deadhead conference and subsequent promotion via interview and occasional remarks on local radio. Suddenly we had an onslaught of new users, many of whom possessed the single characteristic that most endears a user to a sysop [system operator: ratchet jaws [habitual talkativeness. The Deadheads came online and seemed to know instinctively how to use the system to create a community around themselves, for which I think considerable thanks are due to Maddog, Marye, and Rosebody. Not long thereafter we saw the concept of the online superstar taken to new heights with the advent of the True Confessions conference. ... Suddenly our future looked assured.

Kevin Kelly had been editor of *Whole Earth Review* for several years when the WELL was founded. The Hackers' Conference had been his idea. Kelly recalled the original design goals that the WELL's founders had in mind when they opened for business in 1985.

The design goals were:

1. That it be free. This was a goal, not a commitment. We knew it wouldn't be exactly free, but it should be as free (cheap) as we could make it.
2. It should be profit making. After much hard, low-paid work by Matthew and Cliff, this is happening. The WELL is at least one of the few operating large systems going that has a future.
3. It would be an open-ended universe.
4. It would be self-governing.
5. It would be a self-designing experiment. The early users were to design the system for later users. The usage of the system would co-evolve with the system as it was built.
6. It would be a community, one that reflected the nature of Whole Earth publications. I think that worked out fine.
7. Business users would be its meat and potatoes. Wrong.

"The system is the people" is what you see when you log into TWICS, an English-language conferencing system in Tokyo. The same turned out to be true for the WELL, both by design and by happenstance. Matthew McClure understood that he was in the business of selling the customers to each other and letting them work out everything else. This was a fundamental revelation that stood the business in good stead in the years to follow. His successor, Farm alumnus Clifford Figallo, also resisted the temptation to control the culture instead of letting it work out its own system of social governance.

People who were looking for a grand collective project in cyberspace flocked to the WELL. The inmates took over the asylum, and the asylum profited from it. "What it is is up to us" became the motto of the nascent WELL community.

Some kind of map of what "it" is can help you to understand the WELL. Here is a snapshot of the WELL's public conference structure. Keep in mind that each conference can have as many as several hundred different topics going on inside it (like the Parenting conference topic list in chapter 1), and each topic can have several hundred responses. For the sake of space, this listing does not include 16 conferences on social responsibility and politics, 20 conferences on media and communications, 12 conferences about business and livelihood, 18 conferences about body-mind-health, 11 conferences about cultures, 17 conferences about place, and 17 conferences about interactions.

List of Public Conferences on the WELL

ARTS AND LETTERS

Art Com		Photography	(g pho)
Electronic Net	(g acen)	Poetry	(g poetry)
Art and Graphics	(g gra)	Radio	(g rad)
Beatles	(g beat)	Science	(g sf)
Books	(g books)	Fiction	(g sf)
Comics	(g comics)	Songwriters	(g song)
Design	(g design)	Theater	(g theater)
Jazz	(g jazz)	Words	(g words)
MIDI	(g midi)	Writers	(g wri)
Movies	(g movies)	Zines/Fanzine	(g f5)
Muchomedia	(g mucho)	Scene	
NAPLPS	(g naplps)	Scene	

RECREATION

Bicycles	(g bike)	Games	(g games)
Boating	(g boat)	Gardening	(g gard)
Chess	(g chess)	Music	(g music)
Cooking	(g cook)	Motoring	(g car)
Collecting	(g collect)	Pets	(g pets)
Drinks	(g drinks)	Outdoor	(g out)
Flying	(g flying)	Recreation	
		Sports	(g sports)
		Wildlife	(g wild)

ENTERTAINMENT

Audio-videophilia	(g aud)	Movies	(g movies)
Bay Area Tonight	(g bat)	Music	(g music)
CDs	(g cd)	Potato!	(g spud)
Comics	(g comics)	Restaurants	(g rest)
Fun	(g fun)	Star Trek	(g trek)
Jokes	(g jokes)	Television	(g tv)

EDUCATION AND PLANNING

Apple Library		Environment	(g environ)
Users	(g alug)	Earthquake	(g quake)
Brainstorming	(g brain)	Homeowners	(g home)

Biosphere II	(g bio2)	Indexing	(g indexing)
Co-Housing	(g coho)	Network	
Design	(g design)	Integrations	(g origin)
Education	(g ed)	Science	(g science)
Energy	(g power)	Transportation	(g transport)
		Whole Earth Review	(g we)

GRATEFUL DEAD

Grateful Dead	(g gd)	Tapes	(g tapes)
Deadlit	(g deadlit)	Tickets	(g tix)
GD Hour	(g gdh)	Tours	(g tours)
Feedback	(g feedback)	Tours	(g tours)

COMPUTERS

AI/Forth/Realtime	(g real-time)	Mac System7	(g mac7)
		MIDI	(g midi)
Amiga	(g amiga)	NAPLPS	(g naplps)
Apple	(g apple)	NeXt	(g next)
Arts and Graphics	(g gra)	OS/2	(g os2)
Computer Books	(g cbook)	Printers	(g print)
CP/M	(g cpm)	Programmer's Net	(g net)
Desktop Publishing	(g desk)	Scientific computing	(g scicomp)
Hacking	(g hack)	Software Design	(g sdc)
Hypercard	(g hype)		
IBM PC	(g ibm)	Software/Programming	(g software)
Internet	(g internet)		
LANs	(g lan)	Software Support	(g ssc)
Laptop	(g lap)		
Macintosh	(g mac)	Unix	(g unix)
Mactech	(g mactech) (g mactech)	Virtual Reality	(g vr)
Mac Network Admin	(g macadm)	Windows	(g windows)
		Word Processing	(g word)

THE WELL ITSELF

Deeper technical View	(g deeper)	Hosts	(g host)
		Policy	(g policy)
MetaWELL	(g metawell)	System News	(g sysnews)
		Test	(g test)
General technical	(g gentech)	Public programmers	(g public)
WELLcome and help	(g well)		(g public)
Virtual Communities	(g vc)		
	(g vc)		

SOME POPULAR PRIVATE CONFERENCES ON THE WELL—Mail the hosts listed for information on their criteria for admission.

BODY—MIND—HEALTH

Crossroads	(g xroads)	mail rabar for entry
Gay (private)	(g gaypriv)	mail hudu for entry
Men on the WELL	(g mow)	mail flash for entry
Recovery	(g recovery)	mail dhawk for entry
Women on the WELL	(g wow)	mail reva for entry
Sacred Sites Int'l.	(g ssi)	mail rebop or mandala for entry

ARTS, RECREATION

Aliens on the Well	(g aliens)	mail flash for entry
Band (for working musicians)	(g band)	mail tnf or rik for entry
WELL Writer's Workshop	(g www)	mail sonia for entry

GRATEFUL DEAD

Deadplan	(g dp)	mail tnf for entry
Grapevine	(g grape)	mail rebop or phred for entry

COMPUTERS, COMMUNICATIONS

The Matrix	(g mids)	mail estheise for entry
Producers (radio)	(g pro)	mail jwa for entry

Populations

The *Whole Earth* crowd—the granola-eating utopians, the solar-power enthusiasts, the space-station crowd, immortalists, futurists, gadgeteers, commune graduates, environmentalists, social activists—constituted a core population from the beginning. But a couple of other populations of early adopters made the WELL an open system as well as a specific expression of one side of San Francisco culture. One such element was the subculture that had been created by a cultural upheaval 10 years after the counterculture era—the personal computer (PC) revolution.

"The personal computer revolutionaries *were* the counterculture," Brand reminded me when I asked him about the WELL's early cultural amalgam. Apple cofounder Steve Jobs had traveled to India in search of enlightenment; Lotus 1-2-3 designer and founder Mitch Kapor had been a transcendental meditation teacher. They were 5 to 10 years younger than the hippies, but they came out of the zeitgeist of the 1960s and embraced many of the ideas of personal liberation and iconoclasm championed by their slightly older brothers and sisters. The PC was to many of them a talisman of a new kind of war of liberation: when he hired him from Pepsi, Steve Jobs challenged John Sculley, "Do you want to sell sugared water to adolescents, or do you want to change the world?"

Personal computers and the PC industry were created by young iconoclasts who had seen the LSD revolution fizzle, the political revolution fail. Computers for the people was the latest battle in the same campaign. The *Whole Earth* organization, the same Point foundation that owned half the WELL, had honored the PC zealots, including the outlaws among them, with the early Hackers' conferences. Although the word *hacker* has taken on criminal overtones in the popular parlance, restricting it to urchins who break into other people's computer systems, the original hackers were young programmers who flouted conventional wisdom, delighted in finding elegant solutions to vexing technical problems, and liked to create entire new technologies. Without them, the Department of Defense's ARPA research never would have succeeded in creating computer graphics, computer communications, and the antecedents of personal computing.

The young computer wizards and the grizzled old hands who were still messing with mainframes showed up early at the WELL because the guts of the system itself—the Unix operating system and "C" language programming code—were available for tinkering by responsible craftspersons. The original hackers looked around the system for security holes and helped make the WELL secure against the darkside hackers. Making online tools available to the population, rather than breaking into other systems, was their game.

A third cultural element making up the initial mix of the WELL, which otherwise has drifted far from its counterculture origins in many ways, were the Deadheads. Books and theses have been written about the subculture that has grown up around the band the Grateful Dead. They had their origins in the same milieu that included the Merry Pranksters, the Hog Farm, and the *Whole Earth Catalog*. The Deadheads, many of whom weren't born when the band started touring, have a strong feeling of community that they can manifest only in large groups when the band has concerts. Deadheads can spot each other on the road via the semiotics of window decals and bumper stickers or on the streets via tie-dyed uniforms, but Deadheads didn't have a *place*.

Then several technology-savvy Deadheads started a Grateful Dead conference on the WELL. GD, as it came to be known, was so phenomenally successful that for the first several years, Deadheads were by far the single largest source of income for the enterprise. Because of the way the WELL's software allowed users to build their own boundaries, many Deadheads would invest in the technology and the hours needed to learn the WELL's software, solely in order to trade audiotapes or argue about the meaning of lyrics—and remain blithely unaware of the discussions of politics and technology and classical music happening in other conferences. Those Deadheads who did "go over the wall" ended up having strong influence on the WELL at large. But very different kinds of communities began to grow in other parts of the technological-social petri dish that the Deadheads were keeping in business.

Along with the other elements came the first marathon swimmers in the new currents of the online information streams, the professional futurists and writers and journalists. Staff writers and editors for the *New York Times, Business Week*, the *San Francisco Chronicle, Time, Rolling Stone, Byte, Harper's*, and the *Wall Street Journal* use the WELL as a listening post; a few of them are part of the community. Journalists tend to attract other journalists, and the purpose of journalists is to attract everybody else: most people have to use an old medium to hear news about the arrival of a new medium.

Persons

One important social rule was built into the software that the WELL lives inside: Nobody is anonymous. Everybody is required to attach their real userid to their postings. It is possible to use pseudonyms to create alternate identities or to carry meta-messages, but the pseudonyms are always linked in every posting to the real userid. The original PicoSpan software offered to the WELL had an option for allowing users to be anonymous, but one of Stewart Brand's few strong influences on system design was to insist that the anonymity option should *not* be offered.

Two of the first WELLites I met were Dhawk and Mandel. Like new recruits or rookies in any ongoing enterprise, we found ourselves relating to each other as a kind of cohort. A lot of that early fraternization was necessitated by the confusing nature of the WELL's software. The development of human-user interfaces for CMC was in the Pleistocene era when PicoSpan was designed. It isn't easy to find your way around the WELL, and at first there is always the terrifying delusion that everybody else on the WELL can see all the mistakes you make as you learn your way. The WELL's small staff was available to help confused newcomers via telephone, but the more computer-savvy among the newcomers were eager to actively encourage others. David Hawkins had worked as an engineer and electrician, and he found that he quickly learned enough about the WELL's software to act as an unpaid guide for many of us who joined around the same time he did.

David Hawkins was studying to be a Baptist minister, and he was recently married to Corinne, a woman he had met at the seminary. He was from the Deep South. I had never known a Baptist minister or a good old boy. David changed his original career plans to enter the ministry, and Dhawk spent more and more time online, helping the lost, comforting the afflicted. The real people behind the online personae were important to him. I remember that no more than a year after he joined the WELL, David Hawkins drove for nearly an hour, every day for most of a week, to visit an online acquaintance who had undergone minor surgery. Dhawk helped me find my way around, and he visited me face to face early in the game. I was one of many who felt obligated to pass along the favor when we noticed newcomers floundering around, looking for a way to connect with each other through the WELL's software.

Tina Loney, userid Onezie, is another dedicated community-builder. A single mother of two daughters, a public school teacher, and a proud resident of Berkeley, she is a zealous nurturer of the heart elements of the WELL, as the host of the Parenting conference, and one of the people who showed up at the first WELL real-life party and still rarely misses the face-to-face get-togethers. She's a fierce fighter with a temper that comes through her words. I've "watched" her daughters grow up and leave the nest, via Onezie's online reports, and she has watched my daughter grow from toddler to schoolgirl. We've been on the same side of many online battles, and on a few occasions have struggled against each other over one issue or another.

Maddog named himself after something a friend had called him once, in reference to his occasional verbal ferocity. He's a sweet guy if you meet him in person, but online David Gans in the Maddog days did his best to live up to the moniker. He is a dedicated and educated Deadhead, by profession as well as avocation. His book about the band is required reading for hitchhiking tour rats and limousine Deadheads alike. He produced an hourly radio program of Grateful Dead music and lore for a San Francisco station. It was at a Dead concert that he and two of his companions, Mary Eisenhart (marye), a computer journalist-editor, and Bennett Falk (rosebody), a programmer, received the inspiration to start a Grateful Dead conference on the WELL.

The WELL drew me into the Deadhead milieu and real-life contact with David Gans. I watched and participated as the WELL's Deadhead community helped him grow through a major life crisis. The management of the local radio station canceled his show. David was devastated. He loved to do radio, he loved to evangelize about his favorite band, and the cancellation hit him hard. After much commiseration and anguish online, somebody suggested that David syndicate the show to other stations. That way he could have revenge on the station that canceled and reach even more people than he had reached before.

He was skeptical at first, but so many of his online cohorts urged him to do it that he couldn't very well refuse to try. The idea turned out to be a good one; Gans started Truth 'N Fun productions to distribute his weekly programs to scores of public radio stations around the country, and along the way he changed his userid from Maddog to tnf. The Maddog persona is still latent and jumps out snarling online every once in a while, but tnf continues to be one of the people who most visibly reinvents himself on the WELL's center stage.

David Gans is one of several dozen people who seem to influence strongly the WELL's conversational flavor simply because we spend so much time reading and posting. David in particular sometimes seems to have his frontal lobe directly wired into the WELL. It helps to see him at work in real life. Like many of us, he works at home in a custom-designed office-studio. He has a wraparound audio console in front of him—the control center for his radio production. Studio speakers are on either side, focused on the one chair in the middle of all the equipment. A television is mounted at eye level above the audio monitor. A telephone is at his right hand, and directly in front of him is the computer and modem. David Gans marinates himself in media for a living. Not exactly what Peter Drucker envisioned when he coined the term *knowledge worker*, I'd bet. David Gans, like a lot of others these days, is multitasking. The WELL is just part of the information flow.

Then there is Mandel, who appears at first glance to better fit the image of the information-age specialist. He brought some kind of intellectual respectability to the high-tech bull session. He was a professional futurist at a real-live think tank. He had solid research, facts and figures, to back up his assertions. If you wanted to argue with him, you'd better do your homework. Mandel, who joined the WELL the same day as Dhawk, a few weeks before I joined, was another one of the instant online regulars, along with me, the freelance writer; Onezie, the schoolteacher; Dhawk, the Baptist minister turned Unix hacker; Maddog, the Deadhead radio producer; and a dozen others.

Mandel's employer is SRI International (which started out as Stanford Research Institute, where Jacques Vallee and Doug Engelbart did some of the pioneering research in computer conferencing in the 1970s). Municipal and national governments and the biggest corporations in the world pay SRI for a few hours of Tom Mandel's pontifications about the future of publishing or paper or transportation or—and here I can see his wicked grin—communication. Tom not only is paid well for his WELL addiction, he was applauded for it by his clients and the consulting firm that employs him. He doesn't even have to feel guilty. When he is having fun, he is still working.

You can't talk about the WELL as a community without meeting Tex, the innkeeper, bartender, bouncer, matchmaker, mediator, and community-maker, another communard who emerged from 12 years on the Farm with a reality-tempered commitment to community-building and a deep distaste for anything less than democratic governance. I knew all about him long before I set eyes on him. He was a born online autobiographical entertainer, and what he said about the way things ought to be had some bite because he had put more than a decade into living his communal ideals. I knew he had worked as an interstate truck driver, as a carpenter restoring houses in the poorer sections of Washington, DC, and as an activist in the South Bronx. He was working as an automobile mechanic when Matthew McClure hired him to help deepen and broaden the WELL community. He had four children. I had constructed my own mental image of him from what he had disclosed about himself online. When I met him in real life, his boyish looks surprised me. I was prepared for a grizzled, tobacco-chewing, potbellied guy—the kind of guy who would call himself Tex even though he's from California.

One big part of Tex's persona, to everybody who knows him in person, is his friendly disregard for other people's personal space. Online and off, Tex likes to shed formalities and talk person to person, to say what's really on his mind and in his heart. In person, he does this very close up, literally "in your face." He's over six feet tall and big boned. It isn't easy to keep your distance when he gets a grip on your shoulder and talks to you earnestly from less than three inches away. When Tex left the WELL in 1992 to take a position in another online service, the WELL community threw a combination testimonial, bon voyage party, and roast. One by one, people came up to the podium, grabbed Tex, and talked to him from about three inches away.

His "in your face" style and his anecdotes from 12 years of trying to make a real-life intentional community work represented a core value of the WELL that has survived beyond the years of Farm-vet management: a commitment to using the medium to make real human connections, and more—to try to find better and better ways to live with each other in cyberspace. From the beginning, the exchange of information and the sharing of emotional sympathy exemplified by the Experts on the WELL topic and the Parenting conference were accompanied by some of the less attractive attributes of human groups. Whatever community is, it is not necessarily a conflict-free environment. There has always been a lot of conflict in the WELL, breaking out into regular flamefests of interpersonal attacks from time to time. Factionalism. Gossip. Envy. Jealousy. Feuds. Brawls. Hard feelings that carry over from one discussion to another.

When one of those online brouhahas happened and people started choosing sides and unkind words were being said, Tex and I often walked in the hills above Sausalito and talked about how and why online life can become unpleasant and how to make it work. We kept concluding that simple, corny, all-powerful love was the only way to make a community work when it is diverse, thus guaranteeing friction, and at the same time committed to free expression, which can and does get out of hand. A core of people must flat out believe in the possibility of community and keep coming back to that amid the emotional storms in order for the whole loosely coupled group to hold together at all. When you complicate the situation with the real-life courtships and marriages and divorces and affairs and breakups that tend to happen to the same cohort of people when they stay in touch over a number of years, you have an atmosphere that can get overheated at times.

Who are the WELL members in general, and what do they talk about? I can tell you about the individuals I have come to know over seven years, but only a few of them, and the WELL has long since been something larger than the sum of everybody's friends. On the WELL, I subscribe to an automatic service created by another WELLite who likes to build tools for the community; the Blair Newman Memorial Newuser Report collects the biographical descriptions that new WELL members publish in a public file when they join and sends them to me in one long electronic message every day or two. If I'm too busy to bother, I just delete those electronic messages. Email makes it very easy *not* to read something, but it gives me a sense of who is joining the

WELL, and why, when I let those bios scroll by at 2,400 bits per second. Every once in a while I save a few to a file as a methodologically sloppy kind of survey.

The following is a very small random sample of the different user biographies I culled over a few months in 1991 and 1992.

I am a self-employed productivity consultant. I live out in the country overlooking the ocean near Bodega Bay. The phone, fax, and email let me work here, and still be in the business community.

I reside in Seoul Korea where I practice public relations for the U.S. government.

I am a physician, specializing in women's health, including contraception, abortion, and estrogen replacement therapy after menopause. I am the medical director of an abortion clinic. I was a member of the Mid-Peninsula Free University in the 70's and organized concerts, including the Dead, Big Brother, Quicksilver, Jefferson Airplane, etc. I am interested in philosophical/ethical issues surrounding the beginning of life and the end of life and the functional value of rituals and traditions.

I am a 19 year old college student struggling to find myself. I enjoy sitting in a field of dandelions with no socks. I spend too much time playing on my computer. I am an advertising/business major so I will be here for five or more years. I am trying to find the meaning of life ... helpful hints are appreciated. I wish that penguins could have wings that worked (Breathed), solutions are being contemplated ...

I am a student from Prague, Czechoslovakia, studying in San Francisco's Center for Electronic Art computer graphic and design program.

Librarian for USDA, sysop of "ALF" bbs

I am a lawyer, working as a law clerk to 3 state judges in Duluth. I am 31 and single. I graduated from the Naval Academy in 1982, and the University of Minnesota Law School in 1990. I opposed the Gulf War, and I was Paul Wellstone's deputy campaign director in the 8th Congressional District. I like sailing and long hikes in the woods and the shore. My other interests include law and Italy. My biggest issues today are single payer universal health care and proportional representation.

I'm interested in land-use planning (I'm helping to put together the Sacramento County General Plan), and in Management of Information Systems (I'm writing an article about this, and am interested in a career in it). I'm a real estate broker and developer now, with several years experience using and reviewing software for IBM-compatibles.

I am a born-again phreak, at age 33. My modem is my life! OK, the weightlifting, the fast car, they are all fun, but the modem is the biggie! As a matter of fact, I met my husband on a bbs! But, I realize that I have only tapped the surface of what my little Hayes can do, and I want to learn it all!

Daily Life in Cyberspace

I am a self-employed systems and software consultant, primarily on large military command-and-control systems. My newest interests are in neural networks and fuzzy logic, with parallel processing as an enabling technology. I am usually interested in discussing new technologies and new applications of technology, along with the societal implications, with almost anybody, anywhere, anytime.

Music Store owner, Secretary of Ecological Economics of Alaska.

Captain, US Army

I am a Japanese writer who are very much interested in ecology and the electronic democracy. I am going to spend two years with studying (joining?) ecological movement and sharing network as a tool for making the new world here in Berkeley.

I work at the only hospital dedicated to the cure and eradication of leprosy in the United States. I also spent 6 months in Romania after the December, 1989, Revolution.

One of the reasons people value places like the WELL is the intellectual diversity it offers. With a divergent group, you get separate, nonoverlapping personal networks of expertise. If you could use that diversity as a kind of living encyclopedia, you would find that communion, the immeasurable matters of the heart that the Parenting conference provides, is not the only kind of value that people derive from virtual communities. The knowledge-sharing leverage of a large, diverse group of people who are motivated to help one another, and whose differences of place and time are erased by CMC, can be considerable.

Gift Economies and Social Contracts in Cyberspace

No single metaphor completely conveys the nature of cyberspace. Virtual communities are places where people meet, and they also are tools; the place-like aspects and tool-like aspects only partially overlap. Some people come to the WELL only for the community, some come only for the hard-core information, and some want both. The WELL contains places of the heart for me, but it is also a valuable and unemotional information-seeking device that has become an integral part of my professional routine. The Parenting conference might be a sacred circle, but the News conference can be much more like a combination of intellectual marketplace and mind-game parlor. When I first found my way into the WELL, I was looking for information, and I found it. By that time I realized that the people who have the information are more interesting than the information alone; the game-like and tool-like aspects of sharing information online drew me in further. Later, I began meeting some of the knowledge traders in more communitarian places online.

I was hungry for intellectual companionship as well as raw information. While many commuters dream of working at home, telecommuting, I happen to know what it's like to work that way. I never could stand to get out of my pajamas if I didn't have to, so I've always worked at home. It has its advantages and disadvantages. But the occupational hazard of the self-employed, home-based symbolic analyst of the 1990s is isolation. Information-age hunters and gatherers were lone wolves until we found the Net.

The kind of people that Robert Reich, in *The Work of Nations*, called "symbolic analysts" are natural matches for online communities: computer programmers, writers and journalists, freelance artists and designers, independent radio and television producers, editors, researchers, and librarians. For some time now, these early adopters have been joined by the first ranks of the mainstream CMC users. Increasingly, many people who paint houses or build boats or work in an office or a hospital or sell real estate, but who are curious about new cultural phenomena and not afraid of using a computer keyboard to express themselves, are mixing it up with the knowledge workers. People who work for themselves, whether it is with their hands or their symbols, have been plugging into the Net for the kind of tactical and emotional support others get at the office or factory.

Since so many members of virtual communities are workers whose professional standing is based on what they know, virtual communities can be practical instruments. If you need specific information or an expert opinion or a pointer to a resource, a virtual community is like a living encyclopedia. Virtual communities can help their members, whether or not they are information-related workers, to cope with information overload. The problem with the information age, especially for students and knowledge workers who spend their time immersed in the info flow, is that there is too *much* information available and few effective filters for sifting the key data that are useful and interesting to us as individuals.

Programmers are trying to design better and better software agents that can seek and sift, filter and find, and save us from the awful feeling one gets when it turns out that the specific knowledge one needs is buried in fifteen thousand pages of related information. The first software agents are now becoming available (e.g., Archie, Gopher, Knowbots, WAIS, and Rosebud are the names for different programs that search through the vast digital libraries of Internet and the real-time feed from the news services and retrieve items of interest), but we already have far more sophisticated, if informal, social contracts among groups of people that allow us to act as software agents for one another.

If, in my wanderings through information space, I come across items that don't interest me but I know would interest one of my worldwide affinity group of online friends, I send the appropriate friend a pointer or simply forward the entire text (one of the new powers of CMC is the ability to publish and converse via the same medium). In some cases, I can put the information in exactly the right place for ten thousand people I don't know, but who are intensely interested in that specific topic, to find it when they need it. Sometimes one of the ten thousand people I don't know does the same thing for me.

This informal, unwritten social contract is supported by a blend of strong-tie and weak-tie relationships among people who have a mixture of motives and ephemeral affiliations. It requires one to give something and enables one to receive something. I have to keep my friends in mind and send them pointers instead of throwing my informational discards into the virtual scrap heap. It doesn't take much energy to do that since I have to sift that information anyway to find the knowledge I seek for my own purposes; it takes two keystrokes to delete the information, three keystrokes to forward it to someone else. With scores of other people who have an eye out for my interests while they explore sectors of the information space that I normally wouldn't frequent, I find that the help I receive far outweighs the energy I expend helping others: a marriage of altruism and self-interest.

Lee Sproull and Sara Kiesler, two social scientists who have been observing the ways people in organizations use CMC, point out in their book *Connections: New Ways of Working in the Networked World* that this kind of informal lore exchange is a key, if invisible, part of every organization.

"Does anybody know ... ?" is a common phrase in organizations—typically heard in informal encounters in office hallways, before meetings begin, at the water cooler, coffeepot, and in lunchrooms. In the terms of the general information procedure, one person asks a search question that may be vague or ambiguous. Usually the asker is seeking a piece of current or arcane information, not easily found in official documents. The audience for such questions usually knows the asker and is sympathetic or at least tolerant because the behavior is conventional, the questions are not onerous, and answerers themselves may one day need to ask a question.

In the conventional world, if the asker's acquaintances cannot provide an answer, the asker is stymied. But with electronic communication, the asker has access to a much broader pool of information sources. An oceanographer broadcast a message to an electronic network of oceanographers: "Is it safe and reasonable to clamp equipment onto a particular type of insulating wire?" The official instructions said, "Do not clamp." Right away the sender got several messages from other places saying, "Yes, we do it all the time, but you have to use the following type of clamp." The oceanographer did not know the people who responded and would never have encountered them in a face-to-face setting, but through electronic communication, he benefited from their knowledge and experience. Folklore is an important part of science and technology, consisting of idiosyncratic information about how equipment really works and what tricks you have to know to get the experiment to come out right. It never appears in journal articles or manuals, and it is typically conveyed by word of mouth. With electronic communication, folklore can be more broadly accessible.

Early in my history with the WELL, I was invited to join a panel of experts who advise the U.S. Congress's Office of Technology Assessment (OTA) on the subject of communication systems for an information age. I'm not an expert in telecommunications technology or policy, but I do know where to find a group of such experts and how to get them to tell me what they know. Before I went to Washington for my first panel meeting, I opened a conference in the WELL and invited assorted information freaks, technophiles, and communications experts to help me come up with something to say. An amazing collection of minds flocked to that topic, and some of them created whole new communities when they collided.

By the time I sat down with the captains of industry, government advisers, and academic experts at the panel table, I had more than two hundred pages of expert advice from my own panel. I wouldn't have been able to integrate that much knowledge of my subject in an entire academic or industrial career, and it took me (and my virtual community) only a few minutes a day for six weeks. In my profession, I have found the WELL to be an outright magical resource. An editor, or producer, or client can call and ask me if I know much about the Constitution, or fiber optics, or intellectual property. "Let me get back to you in twenty minutes," I say, reaching for the modem.

The same strategy of nurturing and making use of loose information-sharing affiliations across the Net can be applied to an infinite domain of problem areas, from literary criticism to software evaluation. It's a good way for a sufficiently large, sufficiently diverse group of people to multiply their individual degree of expertise, and I think it could be done even if the people aren't involved in a community other than their place of employment or their area of specialization. But I think it works better when the community's conceptual model of its own activities includes a healthy amount of barn raising along with the horse trading.

Reciprocity is a key element of any market-based culture, but the arrangement I'm describing feels to me more like a kind of gift economy in which people do things for one another out of a spirit of building something between them, rather than a spreadsheet-calculated quid pro quo. When that spirit exists, everybody gets a little extra something, a little sparkle, from their more practical transactions; different kinds of things become possible when this mind-set pervades. Conversely, people who have valuable things to add to the mix tend to keep their heads down and their ideas to themselves when a mercenary or hostile zeitgeist dominates an online community.

In the virtual community I know best, elegantly presented knowledge is a valuable currency. Wit and use of language are rewarded in this medium, which is biased toward those who learn how to manipulate attention and emotion with the written word. Sometimes you give one person more information than you would give another person in response to the same query simply because you recognize one of them to be more generous or funny or to the point or agreeable.

I give useful information freely, and I believe my requests for information are met more swiftly, in greater detail, than they would have been otherwise. A sociologist might say that my perceived helpfulness increased my pool of social capital. I can increase your knowledge capital and my social capital at the same time by telling you something that you need to know, and I could diminish the amount of my capital in the estimation of others by transgressing the group's social norms. The person I help might never be in a position to help me, but someone else might be. That's why it is hard to distinguish idle talk from serious context-setting. In a virtual community, idle talk *is* context-setting. Idle talk is where people learn what kind of person you are, why you should be trusted or mistrusted, what interests you. The agora—the ancient Athenian market where the citizens of the first democracy gathered to buy and sell—was more than the site of transactions; it was also a place where people met and sized up one another. It's where the word got around about those who transgress norms, break contracts. Markets and gossip are historically and inextricably connected.

Parents, libertarians, Deadheads, radio producers, writers, homeowners, and sports fans all have particular places to hang out in the WELL. But in the News conference, the WELL's town square, there is a deliberately general topic, Experts on the WELL, that continues to be a paradigm of one of the ways people can spin banter into an unstructured repository of valuable unclassifiable expertise.

The premise of Experts on the WELL is simple. If you have a problem or a question concerning any topic, from plumbing to astrophysics, you pose it. Then you wait seven minutes or a week. Sometimes nothing happens, and sometimes you get exactly what you want. In many instances, the answer already exists elsewhere in the WELL, and the topic serves as a kind of community librarian service that points the query toward the right part of the WELL's collection of information. In some instances, the information requested exists in someone's head, and that person takes the time to type it.

The reward for knowing the answer and taking the time to enter it into the WELL is symbolic but not inconsequential. People who come up with accurate and well-worded answers win prestige in front of the whole virtual stadium. Experts compete to solve problems; the people who harvest solutions become believers. For $2 an hour, you gain access to your own think tank. You just have to know how to prime it and mine it.

Most topics on the WELL are about something specific. A topic in the Pets conference might be about places to board dogs; a topic in the Parenting conference might be about discipline or coping with measles. Experts on the WELL is about several things at the same time, and the topic is expected to change regularly. The topic serves as an intelligent community filter, where people seeking information can be directed to the specific part of the WELL where their area of inquiry is a topic of discussion. This is how social norms of helpfulness to newcomers contend against the ponderous difficulty of the WELL's software that makes it difficult for new people to find their way. In a surprising number of cases, somebody from the WELL's diversity happens to know the definitive answer to a question about angelology or automatic transmissions or celestial navigation or where to find a good martini. Many of us read the topic for amusement and for the odd bits of expertise we can pick up along the way.

Experts on the WELL is about more than simple fact-finding. It is also about the pleasure of making conversation and creating value in the process. Although all these responses were originally typed on a terminal or computer keyboard, and are available for people to read long after they were typed, the postings in a computer conference are experienced by those who read and write them as a form of conversation as well as a form of publication. In the case of the WELL, it's a conversation in which 16% of the people contribute 80% of the words, but many people are listening invisibly, and all are free to join. In that sense, there's a theatrical element to this medium—written conversation as a performing art. One of CMC's distinguishing characteristics is the way it mixes aspects of informal, real-time communication with the more formally composed, write-once-read-forever mode of communication.

Computer conference conversations are dialogues that are situated in a specific place (the conferencing system, the conference, the topic) and time. The place is a cognitive and social one, not a geographic place. The WELL is a kind of place to those who come to it, and within the WELL, the News conference is a more specific kind of place inside the larger place, and within the news conference, the Experts on the WELL topic has its own flavor, its own cast of characters, its own norms and rhythms. The way casual conversation is organized in a hierarchical structure, with descriptive names at every level, enables people to use the record of the conversation as a database in which to search for specific information. The way words and ideas are structured by computer conferencing systems is different from more familiar structures, such as books or face-to-face discussions, so we don't have a default mental model that helps us think about the structure.

An architectural model of the WELL can help you create a mental model of these spaces within spaces. If you think of the WELL as a building, you can walk down the halls and look at the signs on the doors to different rooms of various sizes. The sign on the door tells you about the general subject of the conversations that take place inside—sex or art or politics or sports or literature or childrearing. The building is the conferencing system. The rooms are the conferences. And within each conference room, imagine a number of blackboards covered with writing. Approach one of the blackboards, and you will see a sign at the top that indicates which subtopic of the conference room's specified domain is under discussion. In the health conference, you might have topics about medicines, topics about different diseases, topics about medical discoveries, topics about the politics and economics of health care. Each of those topics has its own blackboard, known in the WELL as the topic level. That's where Experts on the WELL exists, as a topic in the News conference.

At the top of the blackboard, a person begins a new topic of conversation by asserting a proposition,or asking a question, or more generally describing an area for general discussion. Immediately after and under the introduction, somebody writes a response. On the WELL, that is the response level. When you know how to navigate the different levels of such a system and use tools provided by the system to automate that navigation, the sense of place helps you structure the system in your memory.

Reading a computer conference transcript in hard copy—on paper—misses the dynamism of the conversation as it is experienced by regulars; the back-and-forth dialogue over a period of time that regular participants or observers experience can be reconstructed by looking at the time stamps of the postings, however. In terms of communication rhythms, email and computer conferencing can be levelers. This is one way in which computer conferencing differs from other communications media. The ability to think and compose a reply and publish it within the structure of a conversation enables a group of people to build the living database of Experts on the WELL, in an enterprise where contributors all work at their own pace. This kind of group thinks together differently from how the same group would think face to face or in real time.

Sara Kiesler, a social psychologist who studied how email systems changed the nature of organizations, was one of the first to observe businesses systematically and study the impact of CMC on the organization. Dr. Kiesler confirmed and legitimated what CMC pioneers had known from personal experience when she noted in *Harvard Business Review* that "computer-mediated communications can break down hierarchical and departmental barriers, standard operating procedures, and organizational norms." Kiesler's observations supported the theory long popular among online enthusiasts that people who often dominate conversations face to face because of rank or aggressive demeanor are no more visible than those who would remain silent or say little in a face-to-face meeting but say a lot via CMC. Businesses are the next organizations to be subjected to the same new kinds of social forces that were experienced by the research and academic communities when they went online.

Kiesler also offered evidence that people communicate across and around traditional hierarchical organizational boundaries if their mutual interest in a particular subject matter is strong enough; groups make more daring decisions via CMC than they do face to face; work that later turns out to be important is sometimes accomplished in informal conversations as well as in structured online meetings.

Clearly, people in the Parenting conference are enmeshed in a social interaction different from that of people in Experts on the WELL, and a college student indulging in the online role-playing games known as Multi-User Dungeons lives in a different virtual society from a participant in a scholarly electronic mailing list. Point of view, along with identity, is one of the great variables in cyberspace. Different people in cyberspace look at their virtual communities through differently shaped keyholes. In traditional communities, people have a strongly shared mental model of the sense of place—the room or village or city where their interactions occur. In virtual communities, the sense of place requires an individual act of imagination. The different mental models people have of the electronic agora complicate the question of why people seem to want to build societies mediated by computer screens. A question like that leads inexorably to the old fundamental question of what forces hold any society together. The roots of these questions extend farther than the social upheavals triggered by modern communications technologies.

When we say "society," we usually mean citizens of cities in entities known as nations. We take those categories for granted. But the mass-psychological transition that people made to thinking of ourselves as part of modern society and nation-states is historically recent. Could people make the transition from the close collective social groups, the villages and small towns of premodern and precapitalist Europe, to a new form of social solidarity known as society that transcended and encompassed all previous kinds of human association? Emile Durkheim, one of the founders of sociology, called the premodern kind of social group *gemeinschaft*, which is closer to the English word *community*, and the new kind of social group he called *gesellschaft*, which can be translated roughly as *society*. All the questions about community in cyberspace point to a similar kind of transition that might be taking place now, for which we have no technical names.

Sociology student Marc Smith, who has been using the WELL and the Net as the laboratory for his fieldwork, pointed me to Benedict Anderson's work *Imagined Communities*, a study of nation-building that focuses on the ideological labor involved. Anderson points out that nations and, by extension, communities are imagined in the sense that a given nation exists by virtue of a common acceptance in the minds of the population that it exists. Nations must exist in the minds of its citizens in order to exist at all. "Virtual communities require an act of imagination to use," points out Marc Smith, extending Anderson's line of thinking to cyberspace, "and what must be imagined is the idea of the community itself."

It's far too early to tell what the tools of social psychology and sociology will help us make of the raw material of group interaction that proliferates in cyberspace. This is an area where adroit use of the Net by scholars could have a profound effect on the nature of the Net. One of the great problems with the atmosphere of free expression now tolerated on the Net is the fragility of communities and their susceptibility to disruption. The only alternative to imposing potentially dangerous restrictions on freedom of expression is to develop norms, folklore, and ways of acceptable behavior that are widely modeled, taught, and valued, that can give the citizens of cyberspace clear ideas of what they can and cannot do with the medium, how they can gain leverage, and where they must beware of pitfalls inherent in the medium, if we intend to use it for community-building. But all arguments about virtual community values take place in the absence of any base of even roughly quantified systematic observation.

Right now, all we have on the Net is folklore, like the Netiquette that old-timers try to teach the flood of new arrivals, and debates about freedom of expression versus nurturance of community. About two dozen social scientists, working for several years, might produce conclusions that would help inform these debates and furnish a basis of validated observation for all the theories flying around. A science of Net behavior is not going to reshape the way people behave online, but knowledge of the dynamics of how people do behave is an important social feedback loop to install if the Net is to be self-governing at any scale.

II "Opening the Door to Cyberspace"

4 Community Memory: The First Public-Access Social Media System

Lee Felsenstein

The first publicly available social media system, which opened in 1973 near the UC Berkeley campus, surprised its creators with the breadth of creative uses to which it was put by the users. Accessed through walk-up Teletype terminals on a mainframe time-sharing system, the Community Memory system was free, let users enter their own information and search commands directly, and relied on users' imagination to define indexing words and make searches.

Positioned in an entryway of a student owned record store in front of a musician's bulletin board, the system quickly attracted heavy usage by musicians, as might be expected, but also attracted items such as typewriter graphics, advertisements for a poet, and a spontaneous learning exchange item. The system went through three generations over a span of 19 years and was structured as a community information exchange available to noncomputer users. In the last generation, 10 terminals were base-level IBM PCs running text browsers on monochrome displays with coin acceptors attached. The mainframe had reduced to a 386 XT running a version of Unix. Various information retrieval methods were implemented during the three generations, with the final system using both an indexed and a networked database structure so that items could be comments on other items as well as be organized by index words. A local theater critic has reported that he learned to write through the second-generation system as a teenager, trading rhetoric with his friends.

I have said that Community Memory "opened the door to cyberspace and found that it was hospitable territory," due to the creativity it facilitated on the part of the users. Community Memory failed to become self-sufficient in large part because the rising level of online home computer use overshadowed the idea of neighborhood information centers, and no effective marketing work was done to promote that idea.

Figure 4.1
Community Memory terminal in use. 1973. Photo: The Community Memory Project.

August 8, 1973

August 8, 1973, was not a particularly noteworthy day in Berkeley, California. Leopold's Records one block from campus was open for business as usual in its upstairs location at 2518 Durant Avenue, selling records to students and student-age people. Leopold's had been established a few years before by the Associated Students of the University of California—the student government—as an attempt to drive down the prices of music recordings (it was successful there).

Leopold's had no display window. Access was through a front door, up a stairway, and through a corridor. A steady stream of customers made this trip in search of the latest recorded music at reasonable prices. At the end of the corridor, before entering the main sales room of the store, a bulletin board hung on one wall containing hundreds of cards and pieces of paper, all used by people trying to make a living in the music business.

On this day, however, something was new. Before the bulletin board was a table, and on it sat a squat box made of corrugated cardboard with a window on its sloping front made of clear vinyl plastic. Two holes in the front face revealed vinyl sheets with asterisk-shaped cutouts, like flexible cat doors. On the right side of the box was bolted a small cooling fan that purred.

A handmade poster, with psychedelic lettering, read "Community Memory." Inside the box was a teleprinter—a Teletype Model 33 ASR that had gone through three years' service as a commercial time-sharing computer terminal. Urethane foam glued inside of the cardboard muffled the whirr of the teleprinter's motor and the "chunk-chunk-chunk" of its print head.

Standing beside the terminal was a young person, dressed similarly to most of the students and other people entering the store. As they came toward the terminal, this person would say, "Would you like to use our electronic bulletin board? We're using a computer."

"Computer" at the time was a word fraught with implications of social control through technology. No individual owned a computer, it was understood; they were far too expensive, and besides, their uses were for maintaining and extending the regimented order of society, an order viewed with fear and loathing by members of the counterculture for whom Leopold's was a cultural mainstay. The organizers of Community Memory expected to have to contend with shock and outrage at having defiled the record store with this oppressive machinery.

Quite to the contrary, as it turned out—almost 100% of reaction was positive, evincing pleasant anticipation, with the most common comment being, "Oh, wow, can I use it?" The attendant was standing beside the terminal—not between it and its prospective user. The two holes invited users to put their hands through to the keyboard, whose carriage return key was colored green.

Pushing the return key caused a prompt to be printed on the paper visible through the window: "Type a word you're interested in and push the green button." Doing so usually (but not always) resulted in the printer typing out the first lines of a number of bulletin-board items that had earlier been entered by other users who had indexed it using that word among others.

The information was stored on the disc drive of a time-sharing SDS-940 mainframe computer in a warehouse in San Francisco. Under the table in Berkeley sat an early modem capable of no more than 300 baud (bits per second—Teletypes were limited to 110 baud). Cradled in the modem box sat the handset of a Bell model 500 telephone connected to a "foreign board" phone line in Oakland with local dialing rates to San Francisco. Each morning the terminal's attendant would place a single call to the computer's number and place the handset in its cradle. Direct electrical connection to the phone network was still forbidden and the subject of a lawsuit. The data were coupled acoustically to and from the handset, and the free call lasted all day.

The poster explained an example of using this new appliance through the example of someone offering ducks for sale—"Duck eat snails—ducks for sale"—and showing how such an item would be entered with indexing words attached so that users could find it later.

There were no personal computers, no Internet, no "cloud," no wireless. Everyone knew what a paper bulletin board was and how it worked; the information came from a subset of the people who looked at it. Sure enough, within a few months, the musician's bulletin board had shifted over to the Community Memory terminal.

This was the system through which, as I have said, "we opened the door to cyberspace and discovered that it was hospitable territory." Where did it come from, what forms did it take, and where did it go?

History

Community Memory was the result of the work of many people, all of whom identified to different degrees with the counterculture of the time. The group that obtained the computer used for Community Memory was started when three computer science students, Pamela Hardt, Christopher Macie, and Nels Neustrup, left the University of California at Berkeley in 1970 after the Cambodia invasion and the Kent State killings, which impelled them to attempt to put into practice ways of making computers available to people who stood in opposition to the trajectory of American society.

I had spent several years, both while at UC Berkeley and while working in Silicon Valley, exploring forms of media that might be effective in facilitating broad-based community formation, a process I had experienced in 1964 during the Free Speech Movement, an event that was seminal to the development of the counterculture in the Bay Area.

My explorations brought me into contact with Michael Rossman, who was then distributing sections of his manuscript for "On Learning and Social Change" among communal households in Berkeley.[1] Rossman's writings examined possible technological tools involving the telephone system that might be useful in social reorganization, and I wanted to see what I, as an electronic engineer, could do to help realize these devices.

At the same time, I also made contact with a household including Jude Milhon (later "St. Jude" of *Mondo 2000* magazine), Efrem Lipkin, and Mark Szpakowski. Lipkin was a systems programmer who was exploring in the "World Game" events structured around the work of R. Buckminster Fuller and serving to attract like-minded people.

As a result of my experience and analysis, I had concluded that changing the technology and expectations of information technology was fundamental to enabling the ongoing social change required to approach a sustainable and humane society. I became convinced that centralized edited media (such as the press, including the underground press) would never make possible the many-to-many communication that had worked magic at Berkeley. I determined to focus on systems using the telephone network.

Computers did not enter into my thinking until I was assigned to learn the BASIC computer language by my employer in 1970. The course was conducted by young associates at a "service bureau" where access to mainframe computers was available. They proudly demonstrated how they could tell that the computer we were using had been turned off and our jobs shunted to another computer thousands of miles away, and showed how files could be made available to various tiers of users through variations on the file names.

Based on this information, I realized that a network of computers could facilitate formation and re-formation of communities of interest without requiring centralization. This was a radical thought for the time, and I was taken aback, wondering how to implement this, asking myself where I could get access to a computer.

Within a year, I had the outlines of an answer. I was directed to Resource One, Inc., a nonprofit established by Hardt, Macie, and Neustrup with the goal of making computer services available to the counterculture. Through an inspired solicitation effort taking advantage of the San Francisco business establishment's inability to fully grasp the implications of the counterculture, the group was able to secure the long-term loan—in effect the donation—of a time-sharing mainframe computer with enough money to set it up and add a significant disc storage unit. I joined Resource One in 1972, immediately after my delayed graduation from UC, and brought Lipkin and his household into contact with them.

Resource One had cultivated contacts in the computer counterculture, centered as it was around Xerox Palo Alto Research Center and Stanford's Artificial Intelligence (AI) Lab. From former members of the defunct Berkeley Computer Corporation, they obtained an operating system for the SDS-940 along with the help of L. Peter Deutsch to install it and get it running. Fred Wright, a denizen of the Stanford AI Lab, designed and built an interface for the 940 that emulated a minicomputer and would accept a commercially available controller for the new disc storage unit. At the time, Fred was the only person in the group who dressed in "hippie style." None were, so far as I could tell, involved in any significant way with drug usage.

Resource One had taken on the corporate shell left by the defunct San Francisco Switchboard—one of hundreds of tiny volunteer information-referral services usually oriented around an interest area or identifiable community. I had explored among these services and found them to be motivated by their causes and not interested in interconnecting, but the stated intent of Resource One was to provide an information-sharing service for the switchboards.

At one point in 1972, Resource One was visited by Richard Greenblatt, a legendary software hacker from the MIT area. I recall his animated challenge to the group to develop an information retrieval program in 24 hours, a development process that became the point of origin for Resource One Generalized Information Retrieval System (ROGIRS), which incorporated sophisticated features allowing dynamic updating of index pointers. Efrem Lipkin took a leading role in this development and brought in Milhon and Szpakowski to help.

In our discussions, we agreed that the computer would not have the capability to store and retrieve primary materials, the actual communication among people that would be the stuff of community. Rather, the computer could be useful in the exchange of "secondary information"—pointers to other people that would facilitate communication through direct contact, telephone, or even postal media.

We also understood the undesirability of making one computer the hub for all information, as it could be co-opted by the power structure should it ever show signs of becoming a danger. We were aware of the ongoing development of computer networking technology and saw it as a way out of the trap of centralization; although it was not available to us at the time, we designed to be able to use it in the future. We felt that the human networks of the switchboards (then a significant artifact of the counterculture), amplified by the use of ROGIRS, would allow for synergistic exchanges and interactions among young people.

Unfortunately, once we had the system working adequately and went out to reconnect with the switchboards, we ran into a brick wall of incomprehension as well as economics. We were proposing that switchboards, none of whom had a measurable budget, somehow spend $150 per month (in 1973 dollars) to rent a Teletype terminal with modem in order to use a software system that we could not explain to them. We got precisely nowhere with the switchboards.

A series of visits to other possible users ensued, and one conversation was seminal. Librarians at the Bay Area Reference Council—the "library of libraries" that made possible the sharing of books among libraries—heard us out and then commented that we seemed to be a library with no books on the shelf. They suggested that we put some "books on the shelf" and then see where we could go with that.

Lipkin, who had been part of the discussion, took this as a challenge to put one or more terminals in public locations to see what information might accumulate on the system. It was not an idea I would have originated, but it seemed to be a possible route to avoiding irrelevance. A group of us within Resource One began designing the "electronic bulletin board" needed to implement this experiment.

The capability of ROGIRS to handle new indexing words at any time became the enabling technology for what was to follow. Although the designers assumed that this bulletin board would have at most three sections (jobs, housing, and cars), there was no effort expended to limit the topics accordingly, and users were instructed to use their imaginations in proposing new index words and in using them.

Several people, including Milhon and Szpakowski, contributed to "seeding" the Community Memory system with items requesting suggestions or responses, and these items involved a wider range of topics than the three originally assumed to be dominant. Resource One received a friendly welcome and approval from the student senate, who, as owner of Leopold's Records, had to give assent to stationing a terminal there and installing a phone line.

No announcement was made of the opening of the Community Memory terminal. I placed an article in the *Berkeley Barb*, the most prominent underground paper in the area, explaining the theory and practice of Community Memory. The announcement ran a week after the opening. Otherwise we made no further publicity.

Initial Results

"Music" became the largest identifiable topic, given the proximity to the musician's bulletin board and the economic nature of the information exchanged thereon. I myself witnessed more than one example of stereotypically countercultural users advancing to the keyboard without acknowledging my [proffered] assistance and performing well-specified searches to extract lists of prospective clients. I got the strong impression that their consciousnesses were several levels above mine at the time.

Although the consensus among the programmers had been that only three topics (jobs, housing, and cars) would account for the bulk of the traffic, in reality the results were much more broadly spread out, based at least on keyword frequencies. Notable examples included a poet, Jon Thompson, advertising his wares with teaser verses and a phone number for further negotiations. Several versions of typewriter graphics showed up (e.g., a picture of a sailboat). There were many interesting exchanges between people who knew each other only through the user names they posted with each item.

One of the most significant examples was the "bagels dialogue," resulting from an item seeded by Milhon asking where in the Bay Area bagels might be found (this at a time before the proliferation of the bagel as a popular bread). Two responses gave the expected location of stores, but a third response gave a phone number and offered the services of "an ex-bagel maker (who) will teach you how to make bagels."

This was golden because it was a spontaneous validation of an idea posited by the philosopher Ivan Illich in his controversial book *Deschooling Society*.[2] Illich took a radical approach to deconstructing the educational establishment by abolishing institutional education. The replacement would be, according to Illich, the development of informal networks of instruction such as he had observed in Central America. Illich posited that "learning exchanges" could come into existence as marketplaces for skill exchange, and (in a coda) that perhaps computers could be useful in implementing such exchanges.

Unfortunately, we were never able to follow through and discover whether anyone took advantage of this or other learning exchange offers; we considered the system a development prototype rather than a valid research example. In one personal example, however, the concept of the system was validated for me when I made contact in 1974 with a former member of my student residence co-op who was moving into the digital product area and needed help with engineering. This contact resulted in our sharing a workshop and my licensing several significant personal computer designs to his company.

During this implementation of Community Memory under Resource One, Lipkin, Milhon, and Szpakowski turned out several printed summarizations and selections of items. Some of these are in the possession of the Computer History Museum in Mountain View, California.

In January 1974, Resource One made an arrangement with the librarian at the Mission Branch of the San Francisco Public Library to place a terminal on their front desk. This became the second terminal of the first-generation system, joined by a third terminal in the Community Memory satellite office in Berkeley, located in a communal house at 1545 Dwight Way. This terminal was not advertised for walk-in access, nor was the fourth one, located in the Oakland offices of Vocations for Social Change.

Also in early 1974, the terminal at Leopold's was closed down and moved to the Whole Earth Access Store on Shattuck Avenue in downtown Berkeley. This was a display (CRT) terminal leased with a service contract. Because modems at the time cost about $300, there had been an ongoing discussion among the potluck attendees in Menlo Park about a homemade version, and I undertook the design of such a modem, specifically architected for reading data from cassette tape to provide "teaser screens" without loading the computer.

A prototype of this design running at 300 baud was used with the Whole Earth terminal, and the design later became the first kit modem marketed among the personal computer hobbyists (the adaptability needed to read data from cassette tape required a self-adjusting capability that eliminated a major calibration problem). Problems with the maintenance of the terminal led me to consider how to design computer hardware

Figure 4.2
Lee Felsenstein. Community Memory terminal, as installed at the Computer History Museum. 2015. Photo: Anita Medal.

to be survivable in public-access environments, which opened my path to the design of personal computers.

Theory and Background

The beginning of what became known as "the counterculture" stemmed from the Free Speech Movement at Berkeley in 1964. Starting as a reaction by students active in politics and civil rights activity against a clumsy attempt by the University of California administration to suppress organizing opportunities on campus, the ensuing struggle broadened to include thousands of students over a period of two months, climaxing in a massive sit-in with almost 800 arrests, after which the faculty voted to support the student position.[3]

During the process of the struggle, a mass of poorly connected students came together to function as a community, mobilizing distributed resources and creating a bidirectional network of information exchange, with the result that the vaunted administration of the university was fought to a standstill and forced to abandon the principle of "*in loco parentis*," under which student populations were infantilized. One of the participants, Marvin Garson, observed after the dust had settled that "barriers to communication among students had gone down" in the process, with the result that strangers had coalesced into a community capable of concerted action.[4]

The phrase "the dust had settled" should in fact be withdrawn; the dust never settled. In the immediate aftermath of the Free Speech Movement, thousands of students left the university to establish what became the Haight-Ashbury neighborhood, with its effort at community formation. Students established counterculture publications, and others set up the first underground newspapers. Organizing protest activity against the developing Vietnam War started, as did the environmental movement and many other social justice movements.

In 1969, another upheaval occurred in the form of "People's Park," an anarchistic effort to turn unused university land into a park built by members of the community. Crushed eventually with deadly police force and military occupation, it provided powerful experiences and imagery of community formation and action, along with experience in what works and what doesn't, of rhetoric versus experience.

When such popular upheavals happen, they produce secondary and derivative activity that can have long-term importance. A group of architects and architecture students under the name People's Architecture compiled and published (in the underground press in 1970) a plan for a future Berkeley implementing countercultural and ecological principles. Along with many other ideas was posited the concept of the "Life House"—houses whose occupants make a room openly available to neighborhood people as a micro-community center.

This was already happening in the countercultural Bay Area—people who took a particular interest in one or another area would declare themselves to be the "switchboard" on that topic, an information-exchange node using index-card and telephone technology. The "Life House" concept broadened the area of interest, narrowed the physical area of coverage of switchboards, and meshed to some extent with what many collective households were already doing—posting bulletin boards in their front yards and "free boxes" where people could leave surplus clothing and goods for others to take as they saw fit.

At the time, I was making the rounds of various collective households in Berkeley hoping to devise ways to implement a courier-delivered mimeographed ("desktop-published," to use a later term) classified ad publication. My experience working in Silicon Valley had informed me about some basic capabilities of computer networks, and I began to conceptualize a network of Life Houses residing on a computer network able to instantly exchange information of all kinds, allowing users to match needs and resources in order to further community formation and re-formation.

I could see that the renaissance sought by the counterculture could be brought to immanence through a sensitive application of digital technology within a social context. Theodore Roszak, writing in *From Satori to Silicon Valley*, a precursor to *The Cult of Information* (1986), later ridiculed this outlook as a "zany" combination of "reversionary" and "technophilic" worldviews,[5] but I have never been able to understand the supposedly necessary contradiction between these outlooks. They represent two sides of human nature, which often must be brought into reconciliation, but not polar and mutually destructive facets. Thesis and antithesis, in a healthy individual or society, result in synthesis.

Thus, in 1971, when I was introduced to Resource One, I found a group that had done yeoman work in securing the previously unattainable computer, a mainframe designed for time-sharing use together with sufficient money to make a good start in setting up and using it. They pursued a vision of using it to interconnect and share the files of the switchboards, and they had developed a deep network of support among computer experts in the Bay Area with countercultural understandings. Stewart Brand interviewed the group in his research for the book, *II Cybernetic Frontiers*,[6] an outgrowth of an article for *Rolling Stone* magazine that appeared in 1972.

I brought to the group my ideas and observations about community enablement and introduced Michael Rossman to the group (I was able to give Rossman a crude terminal in the form of a "portable" repackaged Teletype 33 teleprinter that required some repair effort; in this way, he was able to access the EIES conferencing system run by Turoff and Hiltz at the New Jersey Institute of Technology).

The concept of Community Memory took shape as a network that was as distributed as the technology would permit (we understood the perils of making centrally vulnerable the pool of information necessary for the continued survival of communities not necessarily supportive of the power structure), with access points distributed around neighborhoods having significant countercultural populations, accessible by anyone who could communicate in writing (the available technology for computer-mediated exchange).

This network of centers would be accessible for commercial and noncommercial purposes because economic activity is essential for community survival, along with cultural and interpersonal exchange. The computer network would not be crafted as a medium of primary communication but rather for handling secondary information—pointers on whom to call, write, or visit in order to develop a relationship for a given purpose.

The important network would be a human network of people who knew and related closely to other people, a network that could function without the computer but whose members would never be able to come together in urban society without the computer's capabilities. It would be an antidote to the mass-media intermediation and in effect begin the reconstitution of the "commons of information."[7]

Practice

One can see why we did not want to concentrate on computer-literate and computer-enabled users in building this network. To simply split off a minority for empowerment would not do; we had to bring everyone along, or at least make it possible. In later system implementations, we would put terminals in co-operative supermarkets, laundromats, cultural centers, and stores, never requiring anyone to learn the use of an editor or file structures.

We did what can only be called leading development work on text-based WYSIWIG display, all using the most basic personal computer architectures as intelligent clients, which could be implemented on what I call "trailing edge technology"—minimal basic IBM PC systems. Carl Farrington architected and implemented a text browser for our third-generation system (1990–1992), which would today make most efficient use of wireless interconnection (Lipkin had foreseen packet-switching, relational databases, and the predominance of the C language in the 1980s). We implemented systems based on indexing databases, relational databases, and a hybrid indexing and network database architecture.

For all this technological acumen and accomplishment, we suffered for want of skills in marketing (which I define as "selling the idea of the system"). We relied too much on the concept of "build it and they will come." At one point late in the process, I had a discussion with someone who was not only computer literate but very connected into the network of Berkeley neighborhood organizations, telling him that we probably had "the cure for the common meeting." He was instantly interested and asked for more details. Alas, I did not follow through, as I now understand I should have, by working with him to establish an experimental use case of our technology that would help relieve the deadly necessity for participants in neighborhood process to tolerate endless meetings. He did not take the lead, and there was no further work in that area, and I could not give it my full attention due to economic circumstances and my lack of confidence in convincing others to do work that I had conceptualized.

The Community Memory Project closed operation in 1992 when funding ran out and most of the members had drifted away. By then we had gone to inputting much information from public sources at our office, although our theory held that a robust system would require people to be as much sources of information as consumers. We donated the contents of our storage locker to the Computer History Museum and were seen no more.

Looking Forward

There is no indication, in my opinion, that the need for a widely networked information-exchange structure tailored for secondary information in a locally centered physical context is any less now than at any time since 1973. In fact, it would seem that its time may just be coming, based on conversations with members of the latest generation of technologically empowered activists.

Community Memory's model was based on understanding behavior in public space—the processes of display, nucleation, interchange, circulation, and reconstitution that has been going on for millennia wherever the public comes together—in agoras, piazzas, market fairs, and town squares. While implementations evolve, the basic habits remain the same because they emerge from cultural evolution. Thus, people in the street transfixed by their smart phone screens have their imitators a century earlier in people lining the streets transfixed by newspapers. What is different is the potential for changing the topology of information flow through radically different media.

As the first public-access social media system, and as one that came to no definitive research conclusion, Community Memory should, I believe, be revived as a participatory implementation with an attached research arm. There are several possible ways in which this could be done, but those are beyond the purview of this chapter.

Epilogue

As we have seen, it is not a stretch to claim that "we opened the door to cyberspace and determined that it was hospitable territory," as written above. Subsequent explorations have remained within fairly tight boundaries, with limited imagination applied. As the sage Ted Nelson reportedly once said, "Everybody wants to be second." In a world for which sustainability has become an ever more urgent issue, it is irresponsible to act as if only the newest, most popular, and temporarily remunerative artifacts are worthy of attention and that old work is irrelevant. I look forward to seeing how new eyes see the artifacts and experiences from generations previous. Fortunately, technology moves fast enough that most of us are still around to watch and possibly advise.

Notes

1. Michael Rossman, *On Learning and Social Change* (New York: Random House, 1972).

2. Ivan Illich, *Deschooling Society* (New York: Harper and Row, 1971).

3. *Free Speech Movement Archives*. Available at http://www.fsm-a.org/.

4. Marvin Garson speaking at a Berkeley Independent Socialist Club forum on the Free Speech Movement in January of 1965 (no recording or transcript exists).

5. Theodore Roszak, *From Satori to Silicon Valley: San Francisco and the American Counter Culture* (San Francisco, CA: Don't Call It Frisco Press, 1986).

6. Stewart Brand, *II Cybernetic Frontiers* (New York: Random House, 1974).

7. Lee Felsenstein, "The Commons of Information," *Dr. Dobbs' Journal* (May 1993), 18–24. Available at http://besser.tsoa.nyu.edu/impact/s94/speakers/felsenstein/felsenstein-article.html.

Bibliography

Felsenstein, Lee. "Community Communications Proposal," Paper for the "Mayday Convention," Atlanta, GA, August 1971. Published in the *Journal of Community Communications*, no 1 (January 1976).

5   PLATO: The Emergence of Online Community

David R. Woolley

The PLATO system was designed for Computer-based Education. But for many people, PLATO's most enduring legacy is the online community spawned by its communication features.

PLATO originated in the early 1960s at the Urbana campus of the University of Illinois.[1] Professor Don Bitzer became interested in using computers for teaching, and with some colleagues he founded the Computer-based Education Research Laboratory (CERL). Bitzer, an electrical engineer, collaborated with a few other engineers to design the PLATO hardware. To write the software, he collected a staff of creative eccentrics ranging from university professors to high school students, few of whom had any computer background. Together they built a system that was at least a decade ahead of its time in many ways.

PLATO was a time-sharing system. It was, in fact, one of the first time-sharing systems to be operated in public. Both courseware authors and their students used the same high-resolution graphics display terminals, which were connected to a central mainframe. A special-purpose programming language called TUTOR was used to write educational software.

Throughout the 1960s, PLATO remained a small system, supporting only a single classroom of terminals. About 1972, PLATO began a transition to a new generation of mainframes that would eventually support up to 1,000 users simultaneously.

PLATO Notes: Original Development

In the summer of 1973, Paul Tenczar asked me to write a program that would let PLATO users report system bugs online. Tenczar was the head of the system software staff, and I was a 17-year-old university student and junior system programmer. I had been with CERL for about a year, learning the ropes and doing minor programming tasks at minimum wage.

We already had a way for users to report bugs, but it was just an open text file called "notes." Anyone could edit the file and add a comment to the end (after something like "+++Fixed-RWB").

This was simple enough, but there were problems. For one thing, only one person could edit the file at a time. For another, there was no security at all. It was impossible to know for sure who had written a note. Most people signed or at least initialed their comments, but there was nothing to enforce this. Occasionally some joker would think it was fun to delete the entire file.

It was just such an incident that prompted Tenczar to ask me to develop a replacement. His idea was a simple refinement of the method we had been using: a user would type a problem report into a special-purpose program, which would automatically tag it with the date and the user's ID and store it safely in a tamper-proof file. The same program would allow convenient viewing of the stored notes. Each would appear on a split screen, with the user's note on the top half and the system staff's response below.

It occurred to me that half a screen might not be enough space for some notes and that some problems might require back-and-forth conversation between a user and the system staff. A limit of one response per note wouldn't permit much dialogue.

I came up with a design that allowed up to 63 responses per note and displayed each response by itself on a separate screen. Responses were chained together in sequence after a note, so that each note could become the starting point of an ongoing conversation. This is what would today be called a linear discussion forum, and PLATO Notes was apparently the first of its kind.

My first prototype kept all notes in one file. Upon entry you would see an index of the most recent notes, listing each note's number, date, title, and number of responses. You could then select a note to read or page back through the index to find older notes.

As I showed this to other members of the system staff, we began to talk about other ways that this program might be used beyond just problem reports. We thought it would be nice to have a separate area where new users could ask questions and get help from more experienced users and another area where the system staff could announce new PLATO features. So I added a top-level menu to let people choose among three notes files: System Announcements, Help Notes, and General Notes.

Notes was released on August 7, 1973. It was named after the text file it replaced, so that people accustomed to typing "notes" would be taken to the right place.

Every note or response appeared on its own screen. Because PLATO was designed for education, its architecture was biased toward carefully crafted full-screen displays. It was easy to place text or graphics at specific locations on the screen, but it was nearly impossible to scroll text. For Notes, this was both an advantage and a drawback. One nice feature was that the note title, date, time, and author's name always appeared in the same place. After using Notes for a while, your eye "knew" exactly where to look for these things.

On the down side, each posting was limited to 20 lines of text so as to fit on one screen. The only way to overcome this was to write a series of responses, but that allowed other responders to slip in and disrupt the flow. Still, the 20-line limit had the virtue of encouraging brevity.

Most options for reading notes required only a single keypress. While reading a response, for example, one keypress could perform any of these functions (among others):

- proceed to the next response
- go back to the previous response
- go back to the base note
- skip to the next base note
- begin writing a new response

There were too many options to list them all on every screen. Most prompts were quite minimal, but a Help key was universally available. It would display a complete list of the options available at any point.

Notes quickly became an indispensable part of the landscape. It appeared just as PLATO was beginning a phenomenal growth spurt made possible by the new mainframe. Although PLATO had been evolving for more than a decade by this time, to the new flood of users coming online, PLATO without Notes was hard to imagine.

The PLATO Architecture

PLATO was designed to be extremely responsive to keys. Every keypress was processed individually by the central mainframe, but the response (or "echo") was usually so fast as to appear instantaneous. An echo time of 100 milliseconds was excellent; anything over 250 was considered unacceptable.

This was vital especially because displays did *not* appear instantaneously. Originally, all PLATO terminals communicated at 1,200 bps. At that speed, a long posting in Notes might take up to 10 seconds to fill the screen. But a single keypress aborted the display and moved on if the first line or two of a note didn't spark your interest.

Talkomatic and "Term-Talk"

Any competent PLATO programmer could quickly hack together a simple chat program that let two users exchange typed one-line messages. PLATO's architecture made this trivial. A few such programs existed on PLATO before 1973, but they did not get much use probably because the user community was quite small, and most terminals were still in a single building.

In the fall of 1973, Doug Brown designed a program that let several users chat as a group. He wrote a simple prototype to demonstrate the concept and called it Talkomatic.

The real magic of Talkomatic was that it transmitted characters instantly as they were typed, instead of waiting for a complete line of text. The screen was divided into several horizontal windows, with one participant in each. This allowed all the participants to type at once without their messages becoming a confusing jumble. Seeing messages appear literally as they were typed made the conversation feel much more alive than in line-by-line chat programs.

I worked with Doug to expand Talkomatic to support multiple channels and add other features. Each channel supported up to five active participants and any number of monitors who could watch but couldn't type anything.

Empty channels were open to anyone, but any active participant in a channel could choose to "protect" it. This prevented anyone from monitoring the channel, and the participants could then decide who else to admit.

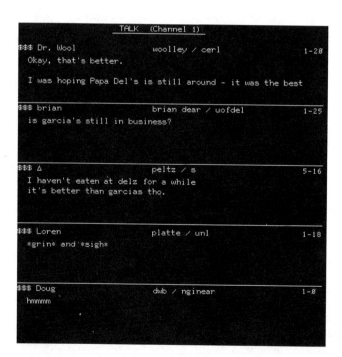

Figure 5.1
PLATO System. A live conversation in a Talkomatic channel.

Talkomatic was an instant hit. Soon it was logging more than 40 hours of use per day. It was not officially part of the PLATO system software, and in fact it was used mostly for what administrators would consider frivolous purposes. There was no way to contact a specific person to let them know you wanted to talk, so it was more like a virtual water cooler than a telephone substitute. People would hang out in a channel and chat or flirt with whoever dropped by.

But Talkomatic was so appealing that it inspired the system staff to create an officially supported chat feature. It became known as "term-talk" because it could be accessed from anywhere on PLATO by pressing the TERM key and typing "talk." The TERM key was originally meant to provide hypertext-like branching to term definitions. In practice, it was rarely used for terms, but it was handy for instant access to features such as "talk."

A "term-talk" conversation was limited to two people, but it had its own advantages: you could page a specific person, and you could use it without exiting from whatever else you were doing. A person receiving a page would see a flashing message at the bottom of the screen identifying the pager and could use "term-talk" to accept. The bottom two lines of the screen then became a miniature Talkomatic. An unwanted page could be rejected with "term-busy" or simply ignored until the pager gave up.

A feature was later added to "term-talk" that allowed the participants to switch to "monitor" mode, in which one person could actually view the other's screen. The person being monitored was free to move about the system normally, editing files, running programs, and so on. This was extremely useful for remote consulting: someone who needed help could literally show an online consultant what he or she was trying to do while maintaining a conversation at the bottom of the screen. To ensure privacy, monitor mode could be initiated only by the person whose screen was to be monitored.

Personal Notes

With Notes and "term-talk" in place, it began to seem natural to use PLATO as a means of communication. What it obviously lacked was a way to send private mail. Kim Mast tackled the job, and in August 1974, Personal Notes was released.

Personal Notes was similar to Notes in many ways: each note appeared on a separate screen, and options such as moving to the next or previous note, deleting a note, or responding were available as single keypresses. There was no index of notes, however. Entering Personal Notes took you immediately to the first note you had not yet read. From there, you could move forward or backward through your notes.

Kim and I worked together to integrate all of PLATO's communication features into a seamless package. For example, while viewing a note, you could copy it to a notes file, forward it as a personal note, send a personal note to the author, or initiate a "term-talk" with the author. All of these options were available.

Notes Categories

The success of Notes led to overcrowding. There were only two notes files that users could write in, and they were used for practically everything. It became a chore to wade through the volume of notes written every day, and people began to ask for a way to filter out notes they weren't interested in.

My initial solution was this: the system staff would define a list of categories, such as "bug reports," "suggestions," "events," "jokes," and so on. Anyone writing a note would assign it to one of these categories. Users could select which categories they wanted to see when reading notes, and their selections would be stored permanently as part of their user records.

In early 1975, I created a version of Notes that supported categories and released it to a limited group of users for testing and comments.

Suggestions from users were vital to PLATO's evolution, and Notes was no exception. Because I had written Notes originally, it was my turf, and I made most of the decisions about what features were implemented. But I had the benefit of lots of ideas from users as well as the rest of the system staff. Often a suggestion would strike me immediately as a great idea, and if it were not too difficult, it might be implemented and released within a day or two. Not all ideas were implemented by any means. But sometimes I would argue against a proposed change, only to be convinced of its merit by cogent arguments or the sheer number of people voicing support for it. Thus, Notes was shaped largely by a consensus of the entire PLATO community.

The notes categories concept was well received at first, but it got bogged down in controversy over features and never made it to general release. A particularly contentious issue was how notes should be presented. One faction wanted to see all notes in chronological order, with the categories serving only as a filter to skip unwanted notes. Others wanted categories to serve an organizing function, as well: all the notes from one category would be shown, then all the notes from the next category, and so on.

Strange as it seems now, I held out against organizing notes by category. I was used to reading notes about many different subjects all jumbled together, and I wanted to be able to see all the new notes listed together in one place. But support for more organization grew, and I began to see that I was in a losing battle.

In the meantime, however, other problems became apparent. First, I realized that as the volume of notes increased, there would be technical problems with keeping everything in one large file. Second, it wasn't clear how many categories would be needed. I had designed in a limit of 60, which seemed like a vast increase over the 3 we had been living with. But if we ever needed more, it would be difficult to increase the limit. After months of wrangling, my concept of notes categories seemed fatally flawed. I really didn't know where to go with it.

About this time, a few people began to ask for private notes files. We had all seen how useful Notes was for discussing development of PLATO. Couldn't the concept be extended to allow any small group of people working on a project to communicate among themselves? In fact, a group in Chicago that was using PLATO to develop pharmaceutical courseware wrote a clone of Notes for their own use.

Suddenly the future clicked into focus. I abandoned the categories project and began to implement Group Notes.

Group Notes

Group Notes was a generalization of the original Notes. Now there could be an unlimited number of notes files, and users would be able to create private notes files for use by their own work groups. Group notes files would serve the same purposes for which notes categories were designed but with none of the inherent problems. The 60-category limit vanished. Distributing notes across many files solved the technical problems of dealing with large volumes of information. The burden of managing notes files would be distributed as well; no longer would the system staff have to oversee everything. Yes, notes would be organized by subject, as so many people had insisted. Group Notes is one of those ideas that, with hindsight, seems glaringly obvious.

Group Notes was released in January 1976, and thereafter the use of Notes skyrocketed. Soon there were public notes files for subjects such as books, movies, religion, music, interpersonal relationships, and science fiction, as well as many private notes files for work groups.

The internal structure of notes files still had not changed much since 1973, and it was beginning to show its weaknesses. In particular, it made it difficult to implement a sorely needed option to read all responses written since a certain date and time. So I rewrote Notes almost from scratch and converted all notes files to a new internal structure in July 1976.

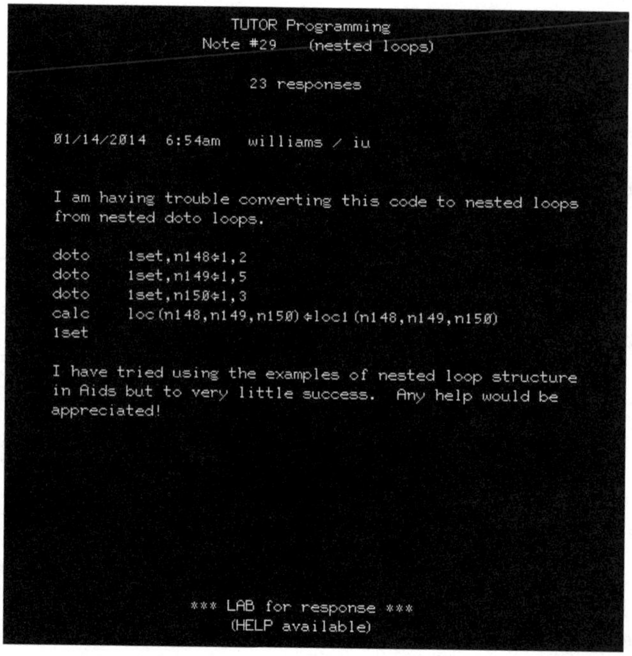

Figure 5.2
PLATO System. An example of how PLATO Notes was used for community technical support and help. Although the image is from the Cyber1 system, a reconstructed PLATO system, aside from the recent date, this is exactly how PLATO Notes looked in the 1970s.

Access Lists

Access lists were the key to Group Notes. A person who created a notes file was automatically registered as a "director" of the file. A director could edit the access list to specify who else could access the notes file and with what privileges. Access could be specified for individual user IDs or entire work groups, and any level of access could be granted to the general public (anyone not specifically listed).

There were six access levels:

- Director
- Read/write
- Read/respond
- Read-only
- Write-only
- No access

Read/write was the most common type of access. It permitted both writing new notes and responding to existing notes.

Read/respond permitted responding but not writing new notes.

Write-only access permitted a user to write new notes but not read or respond. It was sometimes used as a blanket access level for the public, providing a way for someone to request access to a private notes file. It was also useful for collecting comments from the public about some issue while maintaining the privacy of each person's remarks.

Generally, anyone who could read a notes file could also view its access list, although the director can choose to prohibit this.

Reading by Date

Notes offered a way to read all notes and responses written since a certain date and time. This feature was designed so that you could sequence through all new postings using a single key. For every note with new responses, the base note was displayed first to provide context. A keypress then skipped to the first new response. Pressing the same key repeatedly sequenced through the rest of the response chain and then skipped to the next note with new responses.

In 1978, John Matheny implemented the Notes Sequencer, a great boon to habitual notes readers. The Sequencer let you create a personal list of the notes files you read regularly and automatically keep track of the last time you read each one. Using the Sequencer, you could quickly scan all the notes files in your list for new postings with a minimum of keypresses.

Deleting Notes

Someone who had second thoughts after posting a note or response could delete or edit it as long as no responses had been added after it. This restriction was meant to avoid garbling the thread of a conversation. Deleting a response from the middle of a chain could make the following responses seem nonsensical. But an author who desperately wanted to delete a posting anyway could appeal to the notes file director, who could delete any posting without restriction.

A director could delete a response from the middle of a chain without disturbing subsequent responses. However, if a director deleted a base note, all responses disappeared with it. Directors frequently used this power to clean out a notes file, removing old notes that were no longer of interest.

Anonymity

The idea of anonymity in Notes was controversial when first proposed, but the issue was resolved by leaving it to the discretion of each notes file director. If a director chose to allow anonymity, then anyone posting a note or response in that notes file was given the option of making it anonymous.

An anonymous note was truly anonymous. Not even the notes file director nor the system staff could determine who posted it because the user ID was not saved anywhere. The word "anonymous" appeared in the header where the user ID would normally be.

PLATO Notes avoided some of the problems that have plagued experiments with anonymity in other conferencing systems. It was not possible to masquerade as someone else because Notes did not allow the use of pseudonyms. The only identification that could appear in the header is the author's actual user ID or the word "anonymous." The fact that anonymity was the choice of each user was important, too. Someone could post an anonymous note saying, "I'm David Woolley, and I kick my dog every morning," but everyone reading it would know that the author specifically chose to make this note anonymous, so the identity claimed in the text was not to be taken seriously.

Most notes file directors did not permit anonymity, but it was useful in some situations. Anonymity could be abused, but a notes file director could delete offensive postings. Later versions of Notes even allowed a director to review anonymous postings before they became publicly visible.

Director Messages

Another privilege that notes file directors had was to flag their postings with a "director message," a single line of text that appeared above the standard header. Directors often used the message to flag official postings, such as statements about policy or notices that an inappropriate note was deleted. The director could specify what the message should say, but a single message had to suffice for all situations because there was only one director message per notes file. Typical messages ranged from the serious ("OFFICIAL MESSAGE") to the humorous ("Not Operating With a Full Deck").

A director could toggle the message on or off for any posting, even those written by other people. For example, in a notes file used to report problems, a director might set the message to "FIXED" and use it to flag problems that have been resolved.

"Term-Comments"

One of the ways that Notes supported PLATO's educational purposes was through a feature called "term-comments." While running a program, a user could press the TERM key, type "comments," and then type a note to the program author. Such comments were collected in a notes file that the author had associated with the program. Each note was tagged with a header indicating the exact point in the program where the comment was made, so if a note reported that "entorpy is misspelled on this page," the author knew exactly where to look for the error.

Linked Notes

Around 1975, Control Data Corporation set up its own PLATO system in Minneapolis and began turning PLATO into a product. By 1985, more than 100 PLATO systems were operating at sites around the world, about 60% of them running full time. Some of them were linked together with dedicated lines so that files and notes could be exchanged easily. Both Group Notes and Personal Notes were modified to support intersystem links in 1978.

A notes file could be linked between any number of systems. From a user's viewpoint, a linked notes file was exactly like any other, except that the notice "Linked Notesfile" appeared on the index display, and in the headers of some postings, a system identifier appeared after the author's user ID.

When a note or response was posted in a linked notes file, it appeared immediately in the local copy of the file and was put in a queue to be broadcast to all systems which share that notes file. The Notes software did its best to keep the file identical on all systems, although it couldn't guarantee that responses in a given chain appeared in exactly the same order everywhere. There could be a delay of several minutes to an hour before a response was posted on linked systems (or even longer if one of the systems was down for an extended period).

Linear versus Tree Structured Discussions

Although Notes evolved in many ways over the years, one thing that never changed was the linear structure of its notes files. One or two PLATO users wrote experimental versions of Notes using tree-structured (or "threaded") notes files, but most people who tried them found them hard to use, and the idea did not catch on.

My own feeling is that a linear structure is much more conducive to ongoing discussion. Human conversation is inherently disorganized, and a tree structure attempts to impose too much discipline. Conversations often tend to fragment and dissipate quickly in a tree. Some people seem at home with a tree structure, but in my experience, more people find it rather foreign and overly complex.

With a linear structure, each base note and its chain of responses resemble a conversation that we might have with a group of people gathered around a table. The conversation might drift or develop multiple threads, but if that becomes a problem, it is easily dealt with by simply starting new base notes to carry on divergent threads.

Multiplayer Games

There were myriad games on PLATO. Some were for single players, but the most popular ones involved two or more players at separate terminals.

Games were certainly not a priority when PLATO was designed, but it turned out that its architecture supported multiplayer games superbly. The crucial features were:

- shared memory areas
- standardized terminal
- high-resolution graphics display
- central computer processing of every key
- fast key response
- ability to abort display output

Rick Blomme wrote PLATO's first two-player game in the late 1960s, a simple version of MIT's Spacewar. Possibly the most popular game in PLATO history was *Avatar*, one of several dungeons'n'dragons games. *Empire*, a multiplayer game based on *Star Trek*, was another favorite. Other multiplayer games range from *Airfight* (a precursor to Microsoft Flight Simulator) to *Wordwar* (a spelling and speed-typing game) and card games such as contract bridge.

Most games were written by unpaid programmers. The only reward they could hope for was the prestige of having written a popular game. A number of games that originated on PLATO have been re-created commercially as video arcade or personal computer games.

The Online Community

The sense of an online community began to emerge on PLATO in 1973–1974, as Notes, Talkomatic, "term-talk," and Personal Notes were introduced in quick succession. People met and got acquainted in Talkomatic and carried on romances via "term-talk" and Personal Notes. The release of Group Notes in 1976 gave the community fertile new ground for growth, but by that time, it was already well established. The community had been building its own additions to the software infrastructure in the form of multiplayer games and alternative online communications. One such program was Pad, an online bulletin board where people could post graffiti or random musings. Another was Newsreport, a light-hearted online newspaper published periodically by Bruce Parello, aka The Red Sweater.

With the abundance of special interest notes files made possible by Group Notes, a lot of creativity was unleashed. Someone started a series of short-lived notes files simply for fun. One week it would be "Nice Notes," the next week it would be "Nasty Notes," then "Neurotic Notes," then "Psychotic Notes." Some PLATO users developed their own unique online personas. One of the best known was Dr. Graper (actually a student at the University of Delaware named David J. Graper). He began posting wild, surrealistic stories in a public notes file where they were not exactly appropriate, but they were so hilariously entertaining that people clamored for more, and eventually someone created a notes file called Grapenotes as a platform for his ravings.

The early PLATO community was concentrated in Illinois and consisted mostly of people in academia: educators turned instructional designers, and students hired as programmers. Later it grew to include more people from business, government, and the military as Control Data marketed PLATO as a general-purpose tool for training. It also grew geographically, spreading across the United States and around the world. The building that housed CERL became something of a Mecca to the far-flung PLATO community. Many people traveled to Urbana to see the lab and meet those of us who worked there. It was odd to meet people face to face after getting to know them online. My images of people based on their postings in Notes sometimes turned out to diverge wildly from reality.

The growing PLATO community also developed all of the problems that are now well known in online communities, such as flaming, men impersonating women as a prank, and so on. Free speech was the general rule, but there were a few much-discussed incidents in which political postings in notes files were officially quashed for fear of jeopardizing PLATO's funding. Nobody on PLATO had ever experienced an online community before, so there was a lot of fumbling in the dark as social norms were established.

Over the years, PLATO has affected many lives in profound ways. So many real-life marriages have resulted from online encounters that such stories no longer seem remarkable.

Usage Statistics

The CERL PLATO system logged 10 million hours of use between September 1978 and May 1985 (a period for which the most complete statistics are available).

About 3.35 million of those hours (more than one third) were spent in Notes. About 3.3 million messages were posted. By the end of this period, there were about 2,000 notes files.

No figures are available for time spent in Personal Notes, "term-talk," or Talkomatic. But some numbers are known for games. Avatar alone accounted for about 600,000 hours, and Empire claimed another 300,000 or so. All told, games probably accounted for about 20% of PLATO usage during this period.

Few statistics are available for the many Control Data systems, but none was as large as the CERL system. An educated guess is that CERL accounted for about 25% of all PLATO usage worldwide.

The numbers are incomplete, but it is probable that people interacting with other people represented at least half of all PLATO usage. This is remarkable considering that the designers of PLATO never envisioned that communication between people would play more than an incidental role.

The PLATO Diaspora

Control Data ran into trouble in the late 1980s and sold or closed many of its businesses. Simultaneously, microcomputers were becoming a more cost-effective platform for education than PLATO with its mainframe-based architecture, and over time, the Control Data systems were shut down. Control Data's PLATO technology was eventually acquired by Minneapolis-based Edmentum, Inc., which sells PLATO-branded courseware and instructional management systems.

At the University of Illinois, where it all began, PLATO was renamed NovaNET and spun off as a for-profit business, now owned by Pearson Education, Inc. On NovaNET, a team headed by Dale Sinder rewrote Notes in 1991. Among the new features are multipage notes and better search capabilities. But all of the key features of PLATO Notes, including linear structure of its notes files, were preserved. NovaNET's Personal Notes was also redesigned to make each user's mailbox look and work much like a group notes file, with the user as its director and write-only access for everyone else. It was also enhanced with the ability to send and receive Internet email.

In 2004, a group of enthusiasts revived PLATO using software that emulates the old Control Data mainframes and developed a terminal program for Windows and Macintosh that emulates the old PLATO terminals. The revived system is called Cyber1. Logging in today, you can experience PLATO much as it looked in the early 1980s. Information and access are available at cyber1.org.

Lotus Notes and Other PLATO Progeny

As an educational/multimedia system, PLATO has many direct and indirect descendants. One of the best known is Adobe's Authorware, an authoring system for the Macintosh and Windows.

As a communication system, PLATO has numerous other offspring. Many people who experienced the online PLATO community were inspired to replicate it on other platforms.

IBM Notes, originally known as Lotus Notes, is the best-known example. It was developed by Ray Ozzie, Tim Halvorsen, and Len Kawell, all of whom had worked at CERL in the late 1970s. It would be an exaggeration to call Lotus Notes a clone of PLATO Notes because Ozzie expanded the concept to include powerful capabilities that were never contemplated for PLATO. But many of its basic features were modeled after PLATO Notes.

Here are a few other direct descendants of PLATO Notes:

- Notesfiles, a public domain UNIX version of Notes written by Ray Essick and Rob Kolstad. In the early 1980s, it contributed significantly to the rise of USENET newsgroups.
- News readers **tin** and **tass**. The **tass** reader, written by Rich Skrenta, was modeled after the Notesfiles software mentioned above. Iain Lea's **tin** then evolved from **tass**.
- DEC Notes (originally called VAX Notes), a product of Digital Equipment Corporation originally written by Len Kawell.
- NetNotes, a client-server conferencing system created by OS TECHnologies Corp. as an improvement on DEC Notes. An add-on product called WebNotes made a NetNotes server accessible through the World Wide Web.
- COCONET's "Discussion" feature. COCONET was a UNIX-based software platform for running interactive multimedia online services, written by Brian Dear and largely modeled after PLATO.
- Notefile, a Notes clone written in ALGOL for the Burroughs B6700 by John Eisenberg at the University of Delaware.
- FORA, a messaging and chat system for DOS written by Jim Bowery.
- The Connection, a XENIX-based BBS program written by Greg Corson.
- READ, a conferencing system based on the PDP-10 computer written by Rich Braun.

In the spring of 2014, Doug Brown and I revived Talkomatic, this time as a Web application. The underlying technology is completely different from the original, but the basic functionality is the same. Each participant still has his or her own section of the screen, and messages are still transmitted a character at a time as they are typed. This sets Talkomatic apart from the plethora of other chat rooms that exist today. The new Talkomatic is free for anyone to use at http://talko.cc/.

In September 2014, Ray Ozzie followed up on his tremendously successful Lotus Notes with a new product: a voice-based smart phone app for group communication. He dubbed the new app "Talko" in homage to PLATO's Talkomatic.

Over time, ideas spread, evolve, mingle, and diverge. The social media landscape of today includes giants such as Facebook, Twitter, and YouTube, alongside countless other platforms with an incredible variety of features and user communities. After decades have passed, it becomes difficult to trace the tangled roots of this phenomenon.

But the places we gather online today are all intentional communities. PLATO was an accidental one that emerged spontaneously in an environment created for other purposes. In 1970, few suspected that a human community could grow and thrive within the electronic circuitry of a computer. PLATO demonstrated that this is not only possible but inevitable.

Acknowledgments

Many thanks to Al Avner, a veritable fount of statistics. Additional information was provided by Rick Blomme, Jim Bowery, Rich Braun, Greg Corson, Brian Dear, Sherwin Gooch, Mark Goodrich, Rob Kolstad, Dave LePage, Kim Mast, John Matheny, Dale Sinder, Joe Sneddon, Dan Tripp, and John S. Quarterman's book, *The Matrix* (John S. Quarterman, *The Matrix: Computer Networks and Conferencing Systems Worldwide* [Bedford, MA: Digital Press], 1990).

An earlier version of this article, published online in 1994, is available at http://just.thinkofit.com/plato-the-emergence-of-online-community.

Note

1. University of Illinois at Urbana-Champaign, Computer-based Education Research Laboratory PLATO Reports, PLATO Documents and CERL Progress Reports, 1958–1993. Available at http://special.lib.umn.edu/findaid/xml/cbi00133.xml.

6 alt.hypertext: An Early Social Medium

James Blustein and Ann-Barbara Graff

Before there was an "easy" and a straightforward way to locate people with similar interests to one's own, to find collections of online resources about topics—in other words before WWW search engines, hyperlinked directories (even before gopher), and blogs with trackback links—the Usenet newsgroup alt.hypertext was an important place to find out what topics in hypertext were and who was interested in them. In the disparaged[1] (and somewhat anarchic) *alt* hierarchy, alt.hypertext was a "place" for the discussion of many kinds of hypertext and announcements of meetings and systems. alt.hypertext hosted discussion of hypertext in general— its possibilities and promise. Some of the discussion was of how to best use and develop what were becoming the dominant conventions of hypertextual documents, especially using links.

Because it was in the disparaged *alt* hierarchy, alt.hypertext was not propagated through the entire 'net, and it was often inaccurately documented. The other popular hierarchies had strict rules about newsgroup creation that involved formal proposals to the Usenet community and votes. Administrators would not make newsgroups for those hierarchies unless the rules were followed. Posts that were addressed to groups that had not been formally approved were not circulated or were relegated to the *junk* newsgroup. In the *alt* hierarchy, anyone could create a newsgroup simply by posting a newsgroup message in the control newsgroup. At some Usenet sites, administrators would create groups in the *alt* hierarchy simply because there were messages addressed to that group. Furthermore, what began as an informal list of one-line descriptions of newsgroups taken from the formal proposals had great authority. The descriptions of *alt* groups are not based on proposals but seemingly composed whimsically by the compiler, David C. Lawrence, himself. The inaccurate descriptions make it more difficult for potential participants to find the appropriate newsgroup, for example, the description of alt.hypnosis is "When you awaken, you will forget about this newsgroup," and alt.personals is described as "Do you really want to meet someone this way?" In the case of alt.hypertext, the one-liner describes the group's purpose as "Discussion of hypertext—uses, transport, etc."

Like much of Usenet, the purpose of alt.hypertext came from the loose community of people who regularly posted there. Members of the self-organizing community created frequently asked questions (FAQs) lists in 1993 and again in 1997. The lists included definitions of relevant terms (what is hypertext anyway?), pointers to systems (e.g., HyperCard and the WWW), and other resources. Meeting announcements were a regular feature of the postings.

By 1991, alt.hypertext was such a well-recognized forum for people with an interest in hypertext that the WWW was first announced there by Tim Berners-Lee. Likewise the Mosaic browser (precursor to Firefox and IE) was announced there. Presumably alt.hypertext was chosen for the announcement of the WWW because its community was recognized as the most likely to understand and appreciate the importance and potential of the nascent WWW.

By the mid-2000s, many people had ceased posting, and presumably reading, alt.hypertext. This exodus was the result of a confluence of circumstances: The September That Never Ended (a phenomenon explained later in the text), the rise of blogging, and personality conflicts that can arise in any social media. The few curated archives that survive record interesting and serious discussions of somewhat esoteric concepts within hypertext.

The lifecycle of alt.hypertext provides an emblematic case study of a social medium. For this chapter, we have reviewed many years of postings to alt.hypertext that Henry Spencer collected.[2] We focus more on the social media aspects than the specific topics discussed in the group.

The group seems to have been created (in 1987) as an ad hoc extension of a discussion in the comp.society.futures newsgroup about how best to make a hypertext version of (or user interface to) Usenet. It is clear that many of the early posters (users who posted messages, as opposed to the lurkers who read but did not publicly comment) were inspired by the first international conference about hypertext (ACM Hypertext '87).[3]

Discussion quickly moved to specific practical aspects of how Usenet functioned (e.g., How do we get this group propagated? Is this the best namespace/place for such a group?), and then to more general aspects of hypertext.

As more people became members of this community, it was more difficult to maintain a narrow focus.[4] Part of the reason for the topic drift was that alt.hypertext was not moderated (i.e., there was neither an official charter defining what topics and messages were appropriate [or not] nor anyone to approve [or withhold approval from] postings). However, even moderated newsgroups tended to be, in Bob Webber's words, "self-archiving"[5] because topics tend to be discussed repeatedly. The repetition is largely due to the nature of Usenet: messages were not kept online indefinitely, and even after DejaNews (and now Google Groups) made vast collections of old postings available for reading, it is the rare person who reviews more than a few weeks of postings before asking a question or commenting on a topic.

Every September there would be a large influx of new posters to Usenet corresponding to the college admissions cycle in the United States. Many of the new posters would not be aware of the etiquette and mores of the fora or past discussions. Usually, in a month or two, Usenet would settle down again.

However, 1993 was an especially bad year. Although there were likely other factors,[6] many blame America Online (AOL) for the severity of the problem because AOL users did not receive the common warning message[7] that their postings should not be made for trivial reasons. Subscribers to AOL, as paying users (unlike most of the existing population), seemed to feel an entitlement to post what they liked when they liked and wherever they liked. Usenet has never quite recovered from the onslaught that is known as The September That Never Ended.[8,9]

Clearly posters thought of hypertext through the lens of their own interest. Some posters wanted namespaces that clearly separated hypertext for information (à la Robert Horn's Information Mapping)[10] from experiments with and studies of the rhetorical structures possible with so-called nonlinear text, whereas others were interested in hypertext for literature. See the appendix for examples of posts. In the first five years (1987–1992), there were many requests for software and postings about experiments with software. There was, however, little discussion of the two earliest hypertext systems on the Internet: Microcosm and Hyper-G.

Postings (or articles) in Usenet were not available indefinitely, and as there were many posts over the years, it was impossible for anyone who was not reading the postings from the beginning of the group to know all that had gone before. Thus, the readership and focus of the group shifted over time but always in a fluid way. The "place" known as alt.hypertext was constant, but the nature and type of discussions changed over time, and there was little collective memory. By way of example, 10 years after its creation, the origins of the group and its ostensible purpose were unknown, and discussion of its name and purpose continued.[11]

As the WWW became entrenched, much of the discussion was about how to achieve various effects using the tools of the WWW. There was also a substantial number of posts complaining that there was too little discussion of conceptual aspects of hypertext and too much "noise" (inappropriate postings by people who were not interested in working with an existing community).

As the environment became more toxic, members of the community went elsewhere, continuing only to post announcements (of meetings and software releases). In a posting from 2003 regarding the FAQ, one of us (Blustein) wrote,

The world has changed and many more people use blogs and wikis for discussion of issues, and dissemination of results, that this group used to be suitable for. I don't wonder why former posters sought alternative forums [sic] or why newbies wouldn't all immediately think of Usenet as the place to hold discussions.

The implication was that because the content of blogs (postings and comments) is controlled by their authors, the so-called signal-to-noise ratio of a blog is much higher than in an unmoderated newsgroup. The major disadvantage of a blog compared to a working newsgroup is the range of distribution—newsgroups were much easier to find than individual blogs, and every author's posts could be found in one location.

When a newsgroup is functioning well, when there are vibrant discussions that connect people and projects, it is more effective than a blog as a social medium. Even in a functioning newsgroup, topics of discussion might repeat, but they are either introduced to new participants (like an introductory course or a hobbyist magazine) or the repetition is only part of the discussion. Discussions of hypertext (and closely related topics) continue but primarily outside of Usenet (sometimes in the comments and trackbacks of blogs) because Usenet is currently so spam-filled. But the archival record reveals how earlier generations of posters thought and wrote seriously about hypertext and hypertext systems. Today, the aggregation of posts can be mined to find names of experts to seek for further discussions in other places (real or virtual).

Appendix: Two Examples of Posts on alt.hypertext

Posting #1 (1998)

```
From: jill.walker@lili.uib.no (Jill Walker)
Newsgroups: alt.hypertext
Subject: Re: Anti-hypertext
Date: Wed, 25 Feb 1998 09:49:21 GMT
Organization: UiB
Lines: 35
Message-ID: <34f3e632.59518895@nntp>
References: <6csu2g$an$1@nnrp2.dejanews.com> <34f2b947.16013777@nntp>
<bernstein-2402981546290001@mfs-05-252.port.shore.net>
Reply-To: jill.walker@lili.uib.no
NNTP-Posting-Host: pc-173.hf.uib.no
On Tue, 24 Feb 1998 15:46:29-0500, bernstein@eastgate.com (Mark
Bernstein) wrote:>
In article <34f2b947.16013777@nntp>, jill.walker@lili.uib.no wrote:
>
>> On Mon, 23 Feb 1998 16:43:26-0600, harmony_4f@hotmail.com wrote:
>>
>> >I am currently invovled in a class that has strong opinions about
hypertext,
```

alt.hypertext

```
>> >any anti-hypertext, "published" articles would be greatly appriciated. I am
>> >also interested in the other side of the issue.
>>
>> Try Sven Birkerts _The Gutenberg Elegies_ Boston: Faber and Faber
>> 1994 for nostalgia and pessimism or Johndan Johnson-Eilola _Nostalgic
>> Angels_
>
>_Nostalgic Angels_ is published by Ablex. I'm not at all sure that I'd
>characterise Johndan Johnson-Eilola as "anti-hypertext," however. Outright
>opposition to hypertext now seems pointless, but much good work has been
>done in elucidating its strengths and weaknesses.
```

I agree with you, Johnson-Eilola isn't "anti-hypertext," but he certainly points out a lot of problems or perhaps rather possible dangers to be aware of I particularily remember his ideological objections: how hypertext is not "automatically" liberating. Instead of giving the reader freedom, it can for instance increase the effiency "post-capitalists" adore (especially functional hypertext), or dissolve resistance to the text.

One important question Johnson-Eilola got me wondering about is whether it is at all possible to criticise the medium hypertext in hypertext? Isn't Bolter's question in Writing Space rather scary: "why do we need any critics when the text in effect criticizes itself?" (p 165)

Jill

Posting #2 (1994)

```
From: Venanzio@hookup.net (Venanzio Jelenic)
Newsgroups: alt.hypertext
Subject: Re: Hypermedia premises
Followup-To: alt.hypertext
Date: Thu, 21 Jul 1994 01:21:24-0500
Organization: SimulNet Learning Concepts
Lines: 63
Distribution: world
Message-ID: <Venanzio-210794012124@venanzio.hookup.net>
References: <30jjqtINNpmb@uwm.edu>
NNTP-Posting-Host: venanzio.hookup.net
```

In article <30jjqtINNpmb@uwm.edu>, tch2@csd4.csd.uwm.edu (Thomas C Hughes) wrote:
> > Hypertext is a medium of communication; it has nothing to do with the
> > organization of brains.
>
> Actually hypertext (or any non-linear media) may change how we think. I
> think we're changing from a print-based culture with all that implies to an
> electronic "oral" culture—and people in oral societies think differently
> than folks in print cultures.
>
> > The point is that links let writers create
> > structures that communicate more deeply and more effectively to a
> > broader and more diverse audience than linear media.
>
> Hypertext is just a different means of communication, but "more effective"
> than print, uh, that's like saying TV is better than radio.
Hypertext, as a medium of communication hardly exists yet.
True hypertext is a two-way communication process, one which allows information to flow in, and out, of the system. An author can easily annotate a picture, text, object, at will. Another author can pick up on
this, or head off in another lateral direction at will. The result is, can be, a real jumble of info, one hardly worth reading.
Isn't this what Xanadu was aiming for?
I believe the issues surrounding hypertext contribution/annotation are only
now beginning. The "Oral" culture has had evolutionary time-scales to develop, the hypertext one is an infant in comparison.
I developed an experimental hypertext system that might be of interest to
others here. It was built on the "oral" tradition, built from graphics up
(hypertext as originating from a will to "see" what a picture leads to,
rather than the standard "click here to continue" instructions …). Try taking a look, it might help change the way we think about "hypertext."(sorry, works only on Macintosh computers. …)
I would be happy to discuss this with any willing participants.

alt.hypertext

```
The files are available via ftp from: sumex-aim or umich.edu or their
mirrors:
examples:
link2-simulnet-10-demo-hc.hqx 1040k 3/14/94 sumex-aim.stanford.edu
/info-mac/app/link2-simulnet-10-demo-hc.hqx
link2-simulnet-10-demo-hc.hqx 1040k 3/14/94 sics.se /pub/info-
mac/app/link2-simulnet-10-demo-hc.hqx
link2-simulnet-demo-hc.hqx 1077k 10/8/93 sumex-aim.stanford.edu/info-
mac/app/link2-simulnet-demo-hc.hqx
link2-simulnet-demo-hc.hqx 1077k 10/8/93 metten.fenk.wau.nl
/pub/mac/info-mac/app/link2-simulnet-demo-hc.hqx
link2-simulnet-demo-hc.hqx 1077k 10/11/93 lth.se /mac/info-
mac/app/link2-simulnet-demo-hc.hqx
--
Venanzio@hookup.net
Simulnet Learning Concepts
Cambridge, Ontario, Canada
(Current email: venanzio@greenspotantiques.com)
```

Notes

1. Unlike in the other major newsgroup hierarchies, no votes are needed to create a newsgroup in *alt*, and newsgroups in *alt* cannot be deleted from the entire network. Because of this seeming lack of regulation, *alt* is sometimes disparaged.

2. Katharine Mieszkowski, "The Geeks Who Saved Usenet," *Salon*, January 2002. Available at http://www.salon.com/2002/01/08/saving_usenet/.

3. *ACM HYPERTEXT '87: Proceedings of the ACM Conference on Hypertext* (New York: ACM, 1987).

4. Larry N. Osborne, "Topic Development in USENET Newsgroups," *Journal of the American Society for Information Science*, 49 (1998):1010–1016. Osborne has categorized types of topic drift and speculated on their causes.

5. Robert Webber (personal communication).

6. One of the stumbling blocks placed before new users was that groups were presented in alphabetical order, which made the *alt* groups appear to the primary groups from which to choose (much like many of the groups' names used to begin with *net*). This may have been done to avoid any appearance of prejudice. A better way that could have guided users would have been to start with two (moderated) groups: news.announce.newusers and news.answers, and then other groups in random order.

7. Many people who posted messages to Usenet would have to agree to (or at least acknowledge with a "clickthrough") the following message: "This program posts news to thousands of machines throughout the entire civilized world. Your message will cost the net hundreds if not

thousands of dollars to send everywhere. Please be sure you know what you are doing. Are you absolutely sure that you want to do this? [y/n]"

8. Eric S. Raymond, "September that Never Ended," in Eric S. Raymond, ed. *The Jargon File* (version 4.4.7), 2003. Available at http://www.catb.org/jargon/html/S/September-that-never-ended.html.

9. Wendy Grossman, "The September That Never Ended," in Wendy Grossman, *net.wars* (New York: NYU Press, 1998), 4–17.

10. Robert E. Horn, *Mapping Hypertext Analysis Linkage, and Display of Knowledge for the Next Generation of On-Line Text and Graphics* (New York: Lexington Institute, 1989).

11. Mark Bernstein, "Re: Why alt.hypertext?" Usenet Message ID: bernstein2104971417540001@wllsma-01-211.port.shore.net, April 21, 1997.

7 DictatiOn: A Canadian Perspective on the History of Telematic Art

Hank Bull

In the beginning was the word, and the word was: Hello.
—Brion Gysin[1]

Canada is a vast country with a small population, sparsely scattered across four and a half time zones. Time zones are, in fact, a Canadian invention, like, some would argue, the telephone (Bell) and broadcast radio (Fessenden). It is therefore not surprising that Canadian artists would be among the first to interrogate the implications of mass media and communications cultures.

Beginning in 1961, Michael Snow's "Walking Woman" series was one of the first attempts to adapt the corporate use of a trademark to art. The ideas of Marshall McLuhan, based at the University of Toronto, exercised a strong influence on Canadian artists. Correspondence art, video, electronic sculpture, and artist-run production and presentation centers all found fertile ground in Canada. General Idea started FILE, a magazine as art, in 1972. In Vancouver, the Intermedia movement brought together musicians, theater artists, poets, filmmakers, and visual artists to collaborate on large-scale performances and environments. In this context, one of the first uses of telex and fax to create works of art took place, realized in 1969 by N. E.Thing Co., the duo of Iain and Ingrid Baxter.[2]

Global Consciousness and Communication

In 1971, Bill Bartlett, living in nearby Victoria, formed a group called the "For Continuous Use League" to promote exchange among artists through the "regeneration, transformation, transportation, communication, and transmutation of images." This group would eventually devote its full energies to artists' telecommunications projects.

Bartlett's credo was that communication is by definition two-way and "interactive" and that technological media should operate in this way, in opposition to the centralized model of the broadcast panoctopus. This concept was first articulated by Bertolt Brecht in "Radio as an Apparatus of Communication" (1932). Bartlett sought to translate Brecht's ideas into a consistent practice.³ He recognized the telephone as a horizontal medium, well suited for two-way, reciprocal, and multidirectional correspondence. He coupled the telephone with a device, pioneered by ham radio visionary Copthorne MacDonald, that would allow the exchange of "slowscan" video pictures over telephone lines. The Robot 400, manufactured by Robot Research of San Diego, could transmit video images at a rate of 8 seconds per frame.

Meanwhile, Toronto artist Norman White obtained the permission of I. P. Sharp Associates (Toronto), a computer time-sharing company, to use its international network as the vehicle for a series of artistic experiments. White and Bartlett were thus able to communicate with artists in other countries exploring similar strategies.

Slowscan and electronic mail became Bartlett's basic tools. In 1978, he formed the "Direct Media Association" and the next year produced "Pacific Rim Identity," an exhibition in the form of a communications lab at the Vancouver Art Gallery. Links were established among various members of the Peacesat users group using the ATS-1 NASA satellite. Correspondents included contacts in Raratonga, Santa Cruz, Wellington, and Vancouver. I was the gallery technician for that exhibition, and I became immediately hooked by the potential of telecommunications art.

The simultaneity and immersiveness of global communication could already be felt in mail art, where each artist could be at the center of the network. Everyone has a front row seat, John Cage used to say. The virtuality of mediated, distance communication favored the construction of fictional identities such as Genesis P. Orridge, Anna Banana, Mr. Peanut, and Doctor Brute. The Western Front, founded in Vancouver in 1973, was a hotbed of mail art, connected to peers in other countries. In that same year, French artist Robert Filliou, visiting Dick Higgins in Vermont, called up various friends and correspondents in North America on the telephone, producing an audio cassette work called "Research on the Eternal Network." Talking with Kate Craig at the Western Front, you can hear him deciding to visit this newly formed organization on the West Coast.⁴

Filliou visited Vancouver several times and had an important influence on the development of artist networks in Canada. One of his concepts, *The Principle of Equivalence*—that any thing or gesture can be either Well Made, Badly Made, or Not Made—opened a field of collective creative research in which any contribution could have value. Filliou said, "The most important work one can do as an artist is to support the valid work of another artist."⁵ This was an encouragement to build mutually supportive practices that include artists and non-artists, art and non-art. His notion of the *Eternal Network* conceived of all walks of life, and all human destinies, as being connected. The term was taken up by the artist-run centers and lent itself equally to the emerging telecommunications network. Filliou emphasized the interactive conversation, in particular through an overarching series of interventions he called *Teaching and Learning as Performing Arts*.

My first involvement with media art was a weekly radio show, first broadcast on CFRO-FM, Vancouver Co-operative Radio in 1976. Produced by myself and Patrick Ready, the *HP Radio Show* (Hank and Patrick) was a peripatetic mix of oddball banter, games, inventions, and audience participation. Featured weekly were messages from a spaceman with the oddly Welsh-sounding name of Llalla. His voice had first appeared to us on an audio cassette. Someone had left a tape recorder on record by mistake, with no microphone plugged in, and when they played back the tape, they heard a voice from beyond. These tapes were circulated by post, and one came to us. The lectures of Llalla, delivered in a slow, incantatory style, were riveting. His most memorable spiel went something like this:

"Greetings, Earth beings. I am Llalla. ... We have been in communication with your leaders for some time. Why is it that they have not informed you of our presence? It has to do with power. What is power? Power is not money. Money is only the symbol of power. Real power is hydro-electric, oil, nuclear power, electricity. Whoever controls real power controls the world. ... Now, imagine that someone comes along from another planet, equipped with the knowledge to give ordinary individuals access to as much real power as they need, for free, simply by setting up a certain antenna-like apparatus on the roof. What would be the result? The power company would be out of business. Power would devolve to the people. Don't you think that your leaders, who represent the interests of those who control power, would do whatever they could to prevent this information from being delivered to you? They would lose control of power. That is why they have kept entirely silent on the subject of their extra-terrestrial communications."

The message was as convincing as it was curious, and it placed us clearly in the domain of poetry not politics. Indeed, there was something unearthly and telepathic about the early days of electronic communication, something suggesting a larger set of political values.

Significant Events in the Early Development of Network Art

One of the first two-way satellite video transmissions by artists took place between East and West Coast United States in 1977. The Center for New Art Activities and the Franklin Street Arts Center (Willoughby Sharp and Liza Bear with artists including Keith Sonnier) in New York, collaborated with ArtCom/ La Mamelle (Carl Loeffler, Sharon Grace) in San Francisco, in a three-day NASA-sponsored demonstration, using a US/Canadian Hermes CTS satellite. The exchange, called "Send/Receive," was simulcast on cable TV in San Francisco and New York.

In the same year, Kit Galloway and Sherrie Rabinowitz created *Satellite Arts*, and Douglas Davis, working with Nam June Paik and Joseph Beuys, created a live telecast via satellite to more than 30 countries, including the USSR, as part of the opening of

Dokumenta 6. This was the precursor to Paik's "Good Morning Mister Orwell" (1984) and other television broadcast events employing an international network of live studios.

In January 1980, Kit and Sherrie presented "Hole in Space." This seminal piece involved setting up cameras, monitors, and sound connections linking department store windows in New York and Los Angeles so that casual passersby could see and talk to each other coast to coast. The artists left the animation of the piece to the people who discovered it by accident. The results were astounding and sometimes very moving, as the video document attests. People sang songs together, played games, and even made contact with long lost relatives. Later, during the Los Angeles Olympics of 1984, Galloway and Rabinowitz connected sites in different ethnic communities around Los Angeles with a variety of interactive hardware. Called "Electronic Café," the piece pioneered the use of several new technologies and proved the power of the new media to transcend cultural barriers in building meaningful relations over distance. Its success led directly to the creation of a permanent center in Santa Monica, the Electronic Café, devoted entirely to electronic network art.

Bill Bartlett produced a number of important events during this initial period. In February 1980, he set up "The Conference on the Artist's Use of Telecommunications," hosted by the San Francisco Museum of Modern Art. Nine cities were connected on two parallel telephone conferences. Each took turns transmitting images while the other nodes received. At the same time, a text-based conference took place via the I. P. Sharp network. The network nodes were San Francisco (SFMOMA, Art Com), Vancouver (Western Front), Toronto (Norman White, Trinity Square Video), Boston (Aldo Tambellini, MIT), New York (Willoughby Sharp), Bristol (Roy Ascott), Vienna (Robert Adrian), Honolulu (John Southworth), Sydney (Eric Gidney), and Tokyo (Michael Goldberg, Tsukuba University). Under the direction of Carl Loeffler, the co-producer and host of this telematic arts conference was Art Com, one of the first galleries to move into cyberspace.

Robert Adrian, a Canadian living in Vienna, was quick to foresee the pervasive impact of telecommunications and consistently articulate this challenging problem in artistic terms.[6] For Ars Electronica 1982, he organized "The World in 24 Hours," which connected Linz to consecutive time zones around the planet using slowscan, computer mail, and the newly available fax technology. In "Telephone Music," organized by Adrian and Helmut Mark in 1983, symbolic links were made across the Iron Curtain with György Galántai (Artpool) in Budapest and artists in Warsaw and Berlin. For several years, Adrian filled the tireless role of steward to the network. He worked with Gottfried Bach of I. P. Sharp's Vienna office to design an email program, eventually called ARTEX, which was tailored especially to the needs of artists. I. P. Sharp was a generous sponsor. ARTEX had about 35 users worldwide and remained in use until the service was discontinued in 1991 by the company's new owner, Reuters.

Toronto Community Videotex opened in 1983 with a focus on networks, robotics, and computer graphics. Changing its name to Inter-Access in 1987, it developed an artists' dial-up computer network called Matrix.[7] Inter-Access was the first permanent space in Canada to be fully devoted to computer and network arts, and it is still going strong.

The Western Front, with its background in mail art, radio, performance, and new music, was a natural center for telecommunications. Kate Craig developed a video production program that hosted residencies and maintained good equipment at a time when video was still a relatively expensive medium. As part of this program, a series of exchanges, called WIENCOUVER, was developed between Vancouver and Vienna, where the team of Heidi Grundmann, a radio producer at the ORF (Austrian National Radio), and artist Robert Adrian assembled a formidable pool of talent. During his residency at the Western Front in 1983, Adrian and I organized a well-documented slowscan event, WIENCOUVER IV, with the theme of "Telephone Music." Over the coming years, WIENCOUVER would evolve, hosted by Kunstradio, Grundmann's weekly broadcast of radio art. In 1984, we collaborated to produce a book, *Art Telecommunication*, with essays by Eric Gidney, Roy Ascott, Tom Sherman, and Robert Adrian.[8]

Roy Ascott entered the world of telecommunications art in 1978 with a work called "Terminal Art." In 1980, "Terminal Consciousness" linked eight computer terminals in England, Wales, and the United States to a host computer in California. Ascott had the vision of "a global creative network, a cybernetic matrix." As an educator, theorist, and organizer of events, in the mid-1980s, Ascott designed an art college in Gwent, Wales, fully wired and connected to similar campuses in France, Germany, and Austria. He adopted the word "telematics" to describe the convergence of telecommunications and computers. For the exhibition "Electra" held at the Musée d'art moderne de la Ville de Paris in 1983, Ascott conceived a project called "La Plissure du Texte," which gave 12 cities two weeks to write a collaborative novel. In this important work, each city played the part of a different character in a "fairy tale," responding daily to each other's contributions. Participants joined in from Vancouver, Sydney, Vienna, Amsterdam, Gwent, Bristol, Boston, Pittsburgh, New York, Alma, San Francisco, Santa Monica, and Honolulu. The result of this intense exchange was a bilingual English-French printout of Joycean proportions, including a lot of ASCII art.[9]

In the early 1980s, Bruce Breland, teaching at the Studio for Creative Enquiry at Carnegie Mellon University in Pittsburgh, formed the Digital Art Exchange (DAX). Breland was an enthusiastic and regular contributor to network projects, hosting a number of successful events and exhibitions, notably with Artur Matuck at the University of São Paulo, Brazil, and artists in Senegal ("DAX Dakar D'Accord," 1990).

In Paris, there were various nodes, too many to acknowledge here. Thanks to Minitel, a precuser of the World Wide Web, and *Leonardo* Magazine, which provided theoretical support, a number of initiatives took flight. Fred Forest's "La Bourse de l'imaginaire" brought email, radio, television, and telephone to the Centre Georges Pompidou in 1982. Don Foresta, a regular correspondent on I. P. Sharp, became consultant to the new science and culture museum at La Villette. "Art *Réseaux*," a communications lab, was formed at the Sorbonne, under the direction of Karen O'Rourke.

The Experience of Mediated Engagement

The most important element in the network was not the technology but the relationships among people, and it was an odd and diverse assortment. Some were thinkers, perhaps associated with a university, some were independent hackers, and others were social critics, visual artists, or writers. Long-term friendships were spawned with people you might never get to meet. The rules of mediated engagement were somehow quite new. The disembodied, the virtual, and the ephemeral made sense in the era of the simulacrum, the collapse of meaning, and "death of the author."

A watershed year for me was 1986. Curator Germano Celant made telecommunications a central theme of the Venice Biennale. Roy Ascott developed the concept for a collective work to be called *Planetary Network*, and Robert Adrian organized production, with Tom Sherman and Don Foresta as consulting curators. The idea was to produce a global discussion about world news and information systems, with each node broadcasting its own local news to the network. Audio, slowscan, and email were displayed on monumental screens in Venice. There were many participants in the email discussion, and things became quite heated. Issues of cultural and technical imperialism, censorship, gender, and language took the fore. One debate centered on methodology. Electronic Café argued for an end to special events in favor of regular, even daily, connection. Carl Loeffler and Art Com in San Francisco insisted that the future of email lay on the Internet, just then becoming accessible via The WELL. The movement was bursting at the seams. *Planetary Network* was a pivotal event that marked the transformation of explorations by a relatively small group of artists into a larger cultural movement. The late 1980s saw rapid growth in artist-controlled television, radio, and communications projects.

Small, independent groups often worked outside official structures, using consumer-level equipment and home-made electronics. Certainly for the events I was involved in, most of the effort was spent just getting the connection to work. To hear a voice, read a message, or see a face on the screen, beamed in from afar, seemed like some kind of miracle. To have both sound and picture at once was exhilarating. An event involving several points on a global network, all in touch with each other at the same time, with simultaneous sound, image, and text, was a complex, relatively rare phenomenon that took a great deal of planning and preparation. Months of administrative discussion were required to ensure that such a production would run smoothly.

Long before the appearance of the personal computer, the terminal was our basic communications tool. It looked like a portable typewriter, equipped with a roll of heat-sensitive paper and a rubber coupler to hold the handset of a telephone. Incoming text printed at a rate of 15 baud, about the speed of a good typist. The text would auto-wrap as you typed, but if you forgot and hit return (as one would on a typewriter), it opened a gate allowing any incoming message to print immediately, interrupting your flow. The result was a cut-up, full of surprises and non-sequiturs. Every point on the network had its own print-out of the collective text. The terminal was used for both planning and production.

To connect a slowscan Robot, you had to take the handset of the telephone apart and attach alligator clips to the loudspeaker and microphone. This was, technically speaking, illegal, akin to what the "phone freaks" had been doing to hack their way into the global telephone network. It came with a sense of subverting the system, taking the power of multinational communications into one's own hands. Hooking up a telephone to a set of loudspeakers was a tricky business because the wiring of a telephone is hot. There were failed transmissions, dead lines, and snowy pictures.

It was such an effort to get all this gear working properly, and so exciting when it did, that the matter of content became an afterthought. Imagine Alexander Graham Bell: "Mr. Watson, come here. I want to see you." Thinking of what to say was the very last thing to be considered. It was all about the metadata.

Any poetics of telecommunications art must focus, therefore, not on the message but on the medium itself. It is not so much a question of interpreting the meaning of the so-called art works being transmitted as it is a matter of considering the ways that social structures and political relationships among people are affected by technology.

How did the network effect the audience? First of all, there was no audience; to put it another way, the audience consisted of other performers, the readers were also writers. As we see with the Internet today, there was a collapse of the proscenium arch separating actor from audience, producer from consumer. Access to the tools of telecommunication, like the access to video that preceded it, enurtured hopes for democracy, a sense of being able to "talk back to the media," and spawned an alternative economy of symbolic exchange outside the market. Just as video practitioners before them, telecommunication artists were criticized as seduced by technological utopianism. While this may in part be true, personally, I enjoyed being part of a collective production, of contibuting to a larger whole. Anything that could serve to promote global consciousness and a sense of shared human destiny was fine with me. How better to to critique the society of spectacle than with the tools of spectacle themselves?

A case in point was *Infermental*, the "magazine on video cassette" conceived by Hungarian filmmaker Gábor Bódy in 1980. The transportability of the video cassette enabled

Figure 7.1
Hank Bull, *Lithophone*, fax art, Renshaw Gallery, McMinnville, Oregon, 1996.

the construction of a media bridge across the Iron Curtain. Short clips, trailers, interviews, and information about art could now be shared as a kind of video *samizdat*. In the case of *Infermental*, every year a new edition would appear, compiled by new editors based in a different city—Budapest, Berlin, Lyon, Vancouver, Buffalo, Tokyo, and so on.[10] I produced the Vancouver edition, which aspired to be "a video map of the world" and featured an editorial called *Cross Cultural Television*, in which Antoni Muntadas and I made a montage of television samples sent in by correspondents in 26 countries. Focusing on broadcast news, we considered the ways in which compound editorial decisions—the angles chosen by the camera operator, the cuts mades by the censor, the ideology of the broadcaster, and even the whims of the viewer—affect the way societies are shaped and individual subjectivities are constructed.

Figure 7.2
Karel Dudasek, videophone transmission, January 17, 1991. Photo: Hank Bull.

Taking the media into one's own hands was a way of reclaiming agency, of being able to tell the story your own way. *Art's Birthday*, originated in 1963 by Robert Filliou, had become by the late 1980s a telecom "party," held annually on the 17th of January. Today a multipoint Internet festival with a considerable web presence,[11] *Art's Birthday* in the early 1990s was celebrated with fax and videophone. In 1991, we faced the imminent possibility of war in Iraq. Karel Dudasek, of Ponton Media, had stationed himself in a hotel room in Amman, Jordan, equipped with a videophone. By a tragic twist of fate, just as everything was ready for our January 17th transmission, the bombing of Baghdad started. While millions sat stunned before their TVs, mesmerized by the spectacular and highly theatricalized representation of horrific events, the *Art's Birthday* network was immediately transformed from party to war protest. We found ourselves—some 20 nodes, from Tucson to Tokyo—operating our own independent global media network, with live reporting from Amman relayed and a street protest in Pittsburgh beaming back.[12]

Independent Media and the Interpersonal Communication Network

By the mid-1990s, China's booming art scene was beginning to have global repercussions, but the the country was still fairly closed. The first *Shanghai Biennale* took place in 1996. While the organizers were cautious, provocative work was shown, especially by Chen Zhen. Meanwhile, across town, in the gallery of the Hua Shan College of Art, Shi Yong, Shen Fan, Ding Yi, and I staged *Shanghai Fax—Let's Talk About Money*, China's first international group exhibition. It was forbidden to discuss politics, but the Deng Xiaoping had said, "It's glorious to be rich" and "It doesn't matter what colour the cat, as long as it catches mice." So, under the banner of a grinning Cheshire cat, we invited people to talk economies. Faxes poured in from around the world to a number in Shanghai. This was an early use of telecommunications technology to climb "the Great Firewall."

What is at stake is the role of the artist/poet in the body politic. Attempts to connect with artists in Africa, Latin America, Asia, and Eastern Europe were part of this imagined program. Art may not be able to change the world, but it can construct models of possible futures.

Ezra Pound, writing in Paris in 1913, a city enthralled at the time by electricity, described poets as writing "in new wave-lengths" and would declare that "artists are the antennae of the race."[13] The metaphor of radio, and the notion of the artist as communicator, an agent with the power to tap into the essence of the age and predict the future, a kind of *avant garde* of culture, would dominate late Modernism. The idea of art as transmission, reaching thousands, or even millions of people, is developed throughout the 20th century by theories such as Malraux's "museum without walls," movements such as Pop Art, and has only been more deeply embedded by the dramatic proliferation of electronic image and sound technologies.

Antennae are not only used to transmit; they also receive, a kind of hearing. This receptive side is developed by Jack Spicer in his theory of poetic "dictation"—that the poet has to "tune in," wait for the message to be delivered by those sprites and spooks who inhabit what Gaston Bachelard named the *logosphere*.

Radio is an absolutely cosmic problem: the whole world is talking about it. But we must define a concept. It is this: the Bergsonians have spoken of the biosphere, that is to say a living stratum containing forests, animals and man himself. The idealists have spoken of a *noosphere*, a sphere of thought. Others have spoken of a stratosphere, ionosphere; radio, fortunately profits from an ionized layer. What term could be better suited to this domain of world speech than *logosphere*? We all speak in the logosphere. We are citizens of the logosphere.[14]

Poetry, it was explained to me by the poet Billy Little,[15] is very much like radio—it moves through us all the time, even when it is turned off. To get the message, you have to turn on and tune in.

In the early 1980s, Tetsuo Kogawa, a philosopher and media critic in Tokyo, developed a Japanese version of "micro-politics." His research had put him in touch with Félix Guattari, an active proponent of "pirate radio" in Italy and France. Kogawa promoted the use of a simple low-power radio transmitter, something anyone could make quite cheaply from readily available parts and use to transmit over a small range, of couple of kilometers at most. This sparked a spontaneous "mini-fm boom" across Tokyo, as students and musicians started their own radio stations. The extreme local posited itself an alternative to mass revolution.[16] At issue was freedom of speech; why shouldn't individual producers have the right to broadcast over the airwaves, just as anyone has the right to operate a printing press? Freedom of the press is one of the basic rights established by the democratic revolutions of the late 18th century. Kogawa was connected to Paper Tiger TV in New York. We met in 1992, and he has come several times to Canada. In 1994, he developed a way to amplify the UHF transmitter on a VHS video deck, making it powerful enough to cover a city block or two. We made a project called *Neighborhood Television*, NTV. Our slogan was, "If you like the show, come on over." Kogawa was invited to do an interview with a mainstream radio reporter interested in "pirate television" and Kogawa's idea of "polymorphous media."

"How far does your signal go?" was the first question.

"About 100 metres," replied Kogawa.

The reporter hesitated. "Only 100 metres? Not so far, what sort of thing would you broadcast on a transmitter of that strength?"

"Talking is good."

"Yes, but what would you say?"

"You can say whatever you like."

You can say whatever you like. Since Y2K, the *logosphere* has become an open text. Systems of linguistic and social control are taken apart as quickly as they evolve. Listening is the order of the day, whether by surveillance above or hackers below. We bear witness to a continual deconstruction and rearranging of the rules of syntax and grammar.

The interpersonal communication network differs from the market. It is a gift economy in which the currency is an infinitely renewable resource of ideas and imagination. To the achievement of meaningful gender equity, the response to climate change, or intercultural understanding, sympathetic listening is the key. This is new territory, speaking unfamiliar languages, full of disagreement and uncertainty. Unfamiliar noises clash as new rhythms sound.

Notes

1. In conversation with the author, 1983.

2. A project of the Morris and Helen Belkin Art Gallery, University of British Columbia. Available at http://vancouverartinthesixties.com

3. Brecht's text, along with other key documents of media theory, are collected in John Handardt, ed., *Video Culture* (Layton, UT: Peregrone Smith, 1986).

4. The conversation is recorded in *Robert Filliou: From Political to Poetical Economy*, exhibition catalog, curators: Sharla Sava, Hank Bull, Scott Watson (Morris and Helen Belkin Art Galley, University of British Columbia, Vancouver, 1995).

5. Robert Filliou, Clive Robertson, and Marcella Bienvenue, *Porta Filliou*, videotape produced by Arton's Video Publishing, Calgary, Canada, 1977.

6. A useful list of telecommunications events is available on Adrian's cv at http://alien.mur.at/rax/BIO/telecom.html.

7. Jeff Mann, "The Matrix Artists' Network: An Electronic Community," *Leonardo*, 24:2 (1991): 230–231.

8. Heidi Grundmann *Art Telecommunication* (Western Front and BLIX, 1984).

9. Roy Ascott, "La plissure du texte," available at http://www.medienkunstnetz.de/works/la-plissure-du-texte/.

10. The entire 10 years of INFERMENTAL is now archived and available at the Zentrum für Medienkunst in Karlsruhe, Germany.

11. At the time of writing, *Art's Birthday* has more than 300 members on its Facebook page and more than 30 production nodes around the world.

12. Email discussion and faxes are preserved in FRONT Magazine, March 1991 issue. A video document exists, *Telecom '91*, a Western Front Production.

13. Ezra Pound, "approach to Paris," 1913, quoted in Linda Dalrymple-Henderson, *Duchamp in Context, Science and Technology in the Large Glass and Related Works* (Princeton, NJ: Princeton University Press, 2005), 100, 204. She makes a convincing argument for the determining role played by science and electronic technologies in the formation of 20th-century modernism.

14. Gaston Bachelard, "Reverie and Radio," in the revue La Nef, Paris, February–March 1951, English translation in *RadioText(e)*, Neil Strauss, Dave Mandl, eds. (New York: Semiotext(e) #16, 1993).

15. Billy Little, aka Zonko, had a regular feature on the HP Radio Show called "The Poetry Centre of the Universe."

16. Tetsuo Kogawa, "Free Radio in Japan: The Mini-FM Boom," in Strauss, p. 199.

8 Art and Minitel in France in the 1980s

Annick Bureaud

In June 2012, after 30 years of "good and loyal services," the French Telecom terminated its videotex system, famous worldwide under the name Minitel. From the very beginning, the Minitel was the territory of art projects and art experiments. With a few exceptions, what remains of those artworks, created mostly between 1982 and 1988, is traces, residue, and documentation. The process to discover, collect, and analyze those fragments in order to elaborate their history has just started. At the time of writing (November 2014), I have identified 73 projects created under the umbrella of the online magazine-gallery *ART ACCES Revue* and 33 artworks carried independently by 8 artists or groups of artists. In both cases, it should be taken into consideration that these figures are not definitive. Based on this initial cartography, some of the ideas and aesthetics that informed these works and the contexts in which they were created are presented and discussed here.

The Minitel has often been considered as a "pre-Internet" platform and now as a social media *avant la lettre*. The word "telematic," coined in 1978 by Simon Nora and Alain Minc,[1] refers in French only to the Minitel, but it has been used more widely in English to describe early network art, which precisely couples telecommunication and informatics.[2] To attempt to avoid misinterpretations and anachronisms, and also because this artistic and aesthetic history is intimately linked to the socioeconomic context and its technological history, it is important to briefly summarize this history.

Minitel: A Centralized, Hierarchical, Proprietary System—Socioeconomic, Technological, and Cultural Contexts

Like the Internet, the Minitel was born out of a governmental decision but with the idea that it would be an economic development driving force. To achieve this goal, France privileged a large, general audience instead of a professional one.

The Minitel was a centralized, hierarchical, proprietary system designed to allow users to access information and services from administrations and companies. There was one unique "provider" (the French Telecom). The "services" (content) were technologically developed and hosted by specialized companies. The Minitel terminal was given for free to every household, and both "provider" and "services" were paid via the phone bill, according to the time spent.[3] The telematic network was designed as a vertical structure rather than for horizontal communication. With a primary use of horizontal exchange, social media today are also proprietary systems where you can input only where (and sometimes what) you are allowed to.

The Minitel, which gave its name to the whole system, was a crude computer terminal[4] with a 9-inch screen of 25 lines by 40 columns.[5] An alpha mosaic coding allowed the use of letters, numbers, punctuation signs, and graphic or mosaic characters, all of which could also be flickering or in video inversion. The time necessary to display the information as well as the dynamics between screen pages through the "choices," "next," and "return" functions, activated by the users, were elements in the "composition" or "writing" on the Minitel. If, on the programming side, the display was in color, it was in black, white, and gray on the user's side.[6] The keyboard was small, and the keys were particularly hard to use (far from the comfort of today's smart phones). Additionally, with a 1,200 baud modem, it was extremely slow!

Art with videotext occurred in several countries, including Canada, where Nell Tenhaaf produced animations, and Brazil with Eduardo Kac's animated poems,[7] but it was primarily in France that this creative activity was the most lively due to a convergence of elements.

The first factor was the key role played by two major exhibitions that supported and exhibited this work—*Electra*[8] at the Museum of Modern Art of the City of Paris and *Les Immatériaux*[9] at the Pompidou Center. The second factor was a propitious artistic ecology, including the "art of communication" movement that centered around Fred Forest; the digital and experimental literature and poetry movements; groups of innovative young graphic designers; and connections between these different circles.[10] To this must be added some public support for art and technology at that time; the interest of engineers from the R&D department of the French Telecom; and, not the least important, the business model of the Minitel that allowed for the possibility, or should I say the potential, of reaching a (paying) audience directly at home.

Neither in its forms nor in its aesthetics was this creative activity homogenous. The Minitel has been used as a tool, a material, a medium, a means of creation, as well as for dissemination.

Art of Communication: The Message Is the Medium

Although the Minitel had not been designed for people to communicate among themselves, it is what Fred Forest first used it for in *L'Espace Communicant* ("Communicating Space") exhibited in the *Electra* exhibition at the Museum of Modern Art of the City of Paris in 1983. In the museum, he installed 40 phone lines, 10 minitels, and answering

Art and Minitel in France in the 1980s

Figure 8.1
Fred Forest, *L'Espace communicant*, 1983, picture from the installation in the museum. Photo courtesy Fred Forest.

machines. Their call numbers and access codes were advertised in newspapers and on radio. On site, the public could answer and use the phones, leave and answer "posts" on the minitels, either live or with a delay through "mailboxes." All the communications were amplified.

In the early 1980s, following his work in the "sociological art" movement, influenced by McLuhan's ideas[11] but also in dialogue with philosophers Mario Costa and Vilém Flusser, Forest developed a series of reflections and actions around the concept of "aesthetics of communication." Alongside mass media and other communication technologies, the Minitel took its inherent place in his media space and ecology, and Forest included it in several projects. He never used the Minitel by itself but, as noted, in conjunction with other media in an ephemeral communication apparatus that was not used to deliver information but rather as a means to connect people.

The content of the work is the act of communication itself, people talking to each other, highlighting the virtual plaza of the emerging global village, where strangers meet and don't ask "Where are you?", as they do today when the device is portable, but "Who are you?"

Participatory and Collaborative Art

Audience participation and the creation of collaborative artworks have been key artistic concepts that found fertile soil in electronic communication and digital technologies. The artist was no longer considered a content provider but rather a context provider. In Roy Ascott's words, the authorship would become distributed, or, simply, the public would be described as co-author of the work. The Minitel offered a perfect platform for such experiments. Jean-Marc Philippe collected people's contributions to *Messages des hommes à l'univers* (Messages from Humans to the Universe, 1986–1987) in this way. The answers to the question "If an extraterrestrial intelligence were existing, what would you tell it?" were then beamed to the center of our galaxy via the Nançay radiotelescope.

Two other projects dealt more specifically with collective creation: *La Vallée aux images* (The Valley of Images, 1987–1989) by Jean-Claude Anglade[12] and *Le Générateur poïétique* by Olivier Auber (Poetic Generator, 1986).[13] With *La Vallée aux images*, Anglade proposed to the inhabitants of Marne-la-Vallée, a suburban area near Paris,[14] to create a stained-glass window for the City of Noisiel water tower. The grid of the Minitel was used to collectively draw abstract geometrical images that were then transferred on tarpaulins fixed on the grid that covered the water tower.

Just as the water tower is a symbol of the community (the shared water), the Minitel materializes as common territory of a dematerialized sociability for a collective appropriation of a public physical space. It is interesting to notice that for both Philippe and Anglade, the Minitel is a tool-medium for artworks that are realized in other spaces, immaterial in the cosmos for one, tangible in a suburban social and physical space for the other.

Olivier Auber's *Poetic Generator* is a fully immaterial work that exists in cyberspace during the time of the interaction among the connected audience members. Dealing with issues of self-organization and crowd behavior, it is the collective creation of a global image for which each person has available a square of 20 × 20 pixels, the size of the drawing automatically adjusted itself to the number of participants. Created on the Minitel and then upgraded and remediated from platforms to platforms and in this era to smart phones, it is one of the very few still existing works from the Minitel era.

Figure 8.2
ART ACCES Revue. 1984–1986. Online journal/gallery founded by ORLAN and Frédéric Develay. Home page, screen image. Courtesy archives *ART ACCES Revue*-Frédéric Develay.

Ten years later, similar projects would be created with the then-nascent web—such as Bonnie Mitchell's *ChainReaction* (1995), in which she proposed the creation of collaborative digital images, or *Je suis ton ami(e) ... tu peux me dire tes secrets* by Nicolas Frespech (1997–2001), an invitation to share one's secrets on an online platform, and the still active *Mouchette Suicide Kit* (1996),[15] to name but a few.

Textimage, Interactivity, Nonlinearity, and E-literature

The Minitel was primarily a text-based system. However, despite its graphic limitations, it has been used to create image-based works such as those of Anglade or Auber. It has also been the terrain for digital and experimental interactive and nonlinear writing, where text melted into images and vice versa. For instance, under the label *Toi et Moi Pour Toujours*, a group of young graphic designers, Jacques-Elie Chabert, Camille Philibert, and Jean-René Bader, and the journalist, Jean-Paul Martin, created three "telematic novels" where the screen page displayed the text in such a way that it could also be perceived as an image in what Françoise Holtz-Bonneau called a "textimage." *ASCOO* was a hypertext detective novel. During the *Electra* exhibition, where it was presented, visitors could leave messages to the characters via the minitel; one third of the people who interacted with the piece did. *L'Objet perdu*, shown at *Les Immatériaux*, also a hypertext novel, invited the audience to re-create and complete part of the story that supposedly had been lost. In 1984, the group proposed *Vertiges*, which took the form of an installation in which the stories of seven characters took place on seven minitels spread out in the room on a *Carte du Tendre* revisited. The audience could follow each of the characters and compose his or her story by building paths among the minitels. Experimental nonlinear narratives that utilized interaction and/or physical space (installation) in the case of *Vertiges* were at the core of those works.

The Minitel welcomed not only hypertext graphic stories but also a myriad of forms of experimental and digital literature and poetry. For the most part, these creative works took place in the framework of *ART ACCES Revue* with works by Frédéric Develay, Tibor Papp, Philippe Bootz, Julien Blaine, Henri Chopin, Isidore Isou, and Jean-François Bory, to name but a few.[16]

***ART ACCES Revue* (1984–1986): An Online Art Journal and Gallery**

With the Internet, boundaries between creation (the work), exhibition (showing the work), and publication (writing about the work), as well as public relations (which has turned into viral auto-promotion in the social networks), have become increasingly blurred and fuzzy. Thirty years ago, *ART ACCES Revue* was one of the first, if not the first, online platform at the intersection of creation, curation, and publication. Cofounded by ORLAN and Frédéric Develay[17] in 1984 and presented at *Les Immatériaux* a year later, it included three kinds of creative art: visual arts (the majority), literature, and music. This last category was particularly interesting as the Minitel had no sound. Musicians, such as Franck Royon Le Mée, proposed graphical scores.[18] The structure was the same for everyone: alongside the work were published a text by the artist and one by a critic chosen by the artist.

With a medium that was considered, in today's words, as "innovative" and a vector of "future for society," for ORLAN and Develay, it was crucial to propose a cultural and an artistic alternative to its purely utilitarian and economical use and to raise and explore conceptual issues by challenging its limitations, confronting the then-shiny bland computer images associated with progress, as well as established art forms. ORLAN staged her Sainte-ORLAN character: from screen to screen, one could zoom on the bared breast to reveal first the word "art" at the teat and then the words "new" (in English) and "vieux" (meaning "old"). Fluxus artist Ben Vautier proposed several of his sentences, such as, "Action pour un minitel: le recouvrir d'un drap blanc."[19] Interestingly, in his accompanying text, he wrote, "I am dreaming of a Minitel with which we would send a message in French and it would be received in Bantu at Tombouctou and in Basque in Bayonne."[20]

Obviously, *ART ACCES Revue* also had the same hope, carried on the Internet later on and with video at the time, of a democratization of art, by by-passing the cultural institutions and intermediaries in order to reach the audience directly at home—and based on a new economic model that was considered sustainable.

Within its three years of existence, *ART ACCES Revue* proposed an amazing number of works created by an impressive list of artists, including John Cage, Vera Molnar, Paul-Armand Gette, and Daniel Buren.

Minitel: An In-between Space

The Minitel has been used by artists as a stand-alone medium but also with other media to reach people's homes and/or in installations in public venues, linking,/connecting various spaces, the privacy of the home, public physical space, and the virtual inter-space.

Marc Denjean used the metaphors of the labyrinth and the mandala to highlight the nature of that new space—what we would later called cyberspace[21]—that was proposed by the Minitel and its constitutive elements of interaction, interactivity, and combinatory. For instance, with *Déambulatoire/combinatoire* (1984), he created a maze on the floor of the exhibition venue, on the model of the Chartres Cathedral labyrinth[22]—at the center of which was a Minitel surrounded by a circle of 60 telephones connected to 8 tape recorders playing tales and poems. The audience had to walk the labyrinth to reach the Minitel, on which mandalas were displayed. For Denjean, the Minitel is a "terminal," that is an "end," a "door" to the network within which we can enter and navigate[23] (later we would say "surf"). For him, what is interesting is not what is at "this" end but what happens "in-between" the ends. Furthermore, if we can conduct actions on the Minitel, the "terminal," there is nothing "here," everything is "there," at the other end, in another machine.

In this regard, there is some resonance with today's social networks, where nothing is now stored on our hard drives but rather is stored in an unspecified "there." Turning our devices into mere terminals might be where we are slowly heading to with cloud computing.

The Minitel became a "dead media" in art long before it was deactivated. There are several reasons for this: user costs, which were too high to attract a large enough audience; the complexity and costs of production for artists; the availability of other dissemination tools (e.g., e-literature was published on floppy disks and then CD-roms); and the emergence of more flexible systems.

However, not only a "pre-Internet" platform, the Minitel was a medium among the media used at that time to experiment with, develop, and implement cutting-edge artistic and aesthetic concepts and ideas to create artworks that were part of a larger digital, new media, and contemporary art scene.

Notes

1. In their report to the French Government on the "Computerization of Society," Nora Simon, Minc Alain, *L'informatisation de la société*, Paris, La Documentation française, 1978.

2. For instance, by Roy Ascott in his emblematic essay, "Is There Love in the Telematic Embrace?" in *Art Journal*, New York, College Arts Association of America 49:3 (1990): 241–247.

3. A percentage was going to the "service provider" and another to the French Telecom. Hence, both the Telecom and the content provider would earn money that was in direct relation to the use of the service. Different rates were available.

4. Available at http://en.wikipedia.org/wiki/Minitel. The French page is much more detailed and is worth looking at for the graphics that explain quite clearly how the system worked. Available at http://fr.wikipedia.org/wiki/Vidéotex/.

5. ASCII had 80 columns.

6. At the beginning. Later, color minitels were available.

7. "Videotext works by Eduardo Kac (1985-1986)." Available at http://www.ekac.org/VDTminitel.html.

8. *Electra* was held from December 10, 1983 to February 5, 1984. Curator Frank Popper.

9. *Les Immatériaux* was held from March 28 to July 15, 1985. Curators: Jean-François Lyotard and Thierry Chaput.

10. For instance, at the Gallery Donguy. There were also strong theoretical debates produced within some university circles, as well as hosted in places such as the Canadian Cultural Centre in Paris.

11. Marshall McLuhan, *Understanding Media: The Extensions of Man* (New York: McGraw-Hill, 1964).

12. Jean-Claude Anglade. Available at http://jean.claude.anglade.free.fr/.

13. Olivier Auber, *Poietic Generator*. Available at http://poietic-generator.net/.

14. That area is now where Disneyland Paris is located.

15. *The Mouchette Suicide Kit*. Available at http://www.mouchette.org/suicide/xmas.html.

16. So far I have identified 26 projects, but there is no doubt that there have been even more.

17. Frédéric Martin also took part in the definition of the project.

18. Some of them turned into performances.

19. The English translation is, "An action for a minitel: cover it with a white sheet."

20. "*Personnellement je rêve d'un minitel qui enverrait un message en français et qui serait reçu en bantou à Tombouctou et en basque à Bayonne,*" read on a photograph of the screens of this work in the archives of Frédéric Develay.

21. The word "cyberspace" was coined by William Gibson at the same period. However, it took it some years to be translated and to cross the Atlantic and, more generally, become widespread and the shared denomination.

22. Which is a "path labyrinth and not a "choice" one. Available at http://www.cathedrale-chartres.fr/labyrinthe.php.

23. Although the different services were not hyperlinked in the way that websites on the World Wide Web are hyperlinked.

9   Rescension and Precedential Media

Steve Dietz

In an age of pervasive social media, a look at its precursors raises important issues. What is the intrinsic value of the new? What is the importance of precedent? How critical is content in the context of invention? How do we appropriately memorialize early technology before it becomes commodified and quotidian? When is a poetics impoverished and when is it simply "primitive"? Is art even the right, best, most useful context to think about technology-driven creative practice? How and why do utopian impulses become dyspeptic reality? And, a bit more meta-ly, what is a useful role for the exhibition and the institution in relation to participatory social media?

Around the turn of the millennium, I curated five exhibitions of what might be most broadly termed network-based art: *Beyond Interface: net art and art on the net* (1998), *Art Entertainment Network* (2000), *Telematic Connections: The Virtual Embrace* (2001), *Open_Source_Art_Hack* (2002), and *Translocations* (2003). At the time, many of the artists probably would not have self-identified their work as "social media." Nevertheless, there was an undeniable social network underlying their practice, and this "connectiveness"— of the makers and within the works themselves—coming out of early social media practices described elsewhere in this volume and turning the corner to a wider world web, so to speak, is where these projects can be situated. An inflection point.

This inflection point happened in a much larger, shared, transnational space. Hank Bull's pioneering efforts in Canada and beyond are documented in this volume. New Zealand and Australia had vibrant telematic scenes. The UK and Europe were host to dozens of informal networks, major programs, exhibitions and festivals, and hundreds of artists. Japan's NTT ICC was founded in 1997, reflecting a significant embrace of telematic culture. Artists in South America were also active as early as Eduardo Kac's experiments with Minitel (1984). In the United States, most notably, the independent "digital atelier," ada'web (1994) was a breathtaking experimental online space, The Thing (1991), Rhizome (1996), artnetweb (1993), and many other "lists" were creating an online culture independent of commercial media. Experimentation was rife: The Palace's early 3D chat rooms (1995), Pseudo's streaming content (1993), and Eyebeam Atelier artist residencies (1997), to name just a few. The commercial scene was also paying attention to galleries such as Postmasters (1984) and Sandra Gering Gallery (1991). Even museums were getting involved—with Jon Ippolito at the Solomon R. Guggenheim Museum, Christiane Paul at the Whitney Museum of American Art, Benjamin Weil at San Francisco Museum of Modern Art, and myself at the Walker Art Center attempting to engage institutional contemporary art in the emerging practices of the network.

This chapter does not purport to represent the wider social media context of the time except to the extent that the exhibitions and artists' work are representative of contemporaneous practice.

Beyond Interface

The effect of concept-driven revolution is to explain old things in new ways. The effect of tool driven revolution is to discover new things that have to be explored. —Freeman Dyson

I founded the new media program at the Walker Art Center in 1996, and Dyson's dictum from *Imagined Worlds*[1] was one of the primary intellectual wedges that we used to lever net art and net artists into the Walker's self-image as a uniquely multidisciplinary organization. I was advocating that the network was a new disciplinary area, which any institution of contemporary art should want to explore. One of the first orders of business, as simplistic as it seems now, was to clearly make a distinction between "art on the net"—providing access to images of artworks in the collection, for example—and net art, as complicated and un-unified as that notion remains.

One of the first things I did for *Beyond Interface*[2] was to pull together a steering committee,[3] a practice of collaboration and co-curation that has remained constant. While this may be a personal preference, I think it also reflects the social nature of network-based art—at least some of its utopian impulses at the time—the networks of sharing that are both a precedent for and a rebuke to contemporary commercial platforms such as Facebook.

Practice as Practice

More germane and far more instructive, however, is a work such as *Homework* "by" Natalie Bookchin and Alexei Shulgin. It was initiated by them with the participation of Bookchin's students in an "Intro to Computing in the Arts" course at UC San Diego (UCSD) as well as Shulgin, Heath Bunting, Rachel Baker, Keiko Suzuki, Garnet Hertz, and others. Essentially, Bunting and Shulgin posted Bookchin's assignment to her UCSD students to make net art to a number of lists: 7–11, nettime, Rhizome, and elsewhere.[4] Bookchin then graded each work.[5]

There is a tongue-in-cheek element throughout the project. The headline for Shulgin's post "Attention uncertified [net][internet][web] artists !" and Bookchin's grading comments, along with various back channel and "view source" conversations captured in the digital trail, suggest, however, that the field of [net][Internet][web] art was being explored by example—tool driven. It was a social process, available on the surface to anyone, but at the same time deeply layered, rewarding one's level of engagement.

Beyond Interface was posted the same year as *Homework*, and I think it's plausible that Bookchin and Shulgin allowed their work to be part of a more formal institutional narrative as further experimental exploration. There was a commitment to the idea of open networks that allowed for intersection with mainstream, "broadcast" topologies, such as a museum, even if some of their core principles were inimical. Perhaps the experiment would infect the host. Perhaps it would be rejected. Certainly it wouldn't be fully understood. But the process itself is what mattered and would lead to the next iteration/instantiation/mutation.

Pre-Commodification Culture

Robbin Murphy's *Project Tumbleweed*[6] was even more explicit about the relationship of social media to the museum, but the exemplary aspect of this project was how his work at artnetweb preceded the commodification of social media technology—and culture—in the form of blogging sites such as Blogger and WordPress.

While I'm intellectually interested in the "what's next" of technology and what happens in the research lab, this technology is often expensive, surprisingly primitive despite its sophistication, and of limited access and reach. Such technology usually has its most explosive effect when it becomes a commodity item that anyone can use with relative ease for a reasonable cost. Moore's law is a huge driver of this effect, making heretofore military-grade computational capability as available as a game controller or arduino kit, for example.

The commodification of social media practices such as Murphy's *Project Tumbleweed* is driven partly by advances in computation. The ability to have a computer on/as your phone arguably enables Facebook and Twitter. But the parallel driver is the size of the network and the asymmetrical reach possible for an individual. None of our early network experiments had this scale.

What they did have, however, is ideas, passion, and even compassion. The scale hadn't overwhelmed the practice. Many people preceded the contemporaneous practice of "like, link and share,"[7] of course—many of them represented in this volume—but the interesting part about precedence is not who was first but how a community of practice through that practice helps define a field of activity. Defining, or at least thinking about what you are doing and why you are doing it and how best to do it, is the critical driver of what is actually desired and needed—not just which blogger service has the largest network or the cheapest price or the best enhancements.

Project Tumbleweed was "an evolving investigation of the possibilities of a personal multidimensional on-line ecology based on potentiality rather than simulation." Murphy was interested in this because he saw the virtual as a kind of everything. Online virtually anything is possible, which is a different metaphysical relation to the (online) world than creating ever more immersive representations of the physical world. The two are not mutually exclusive, but they are different. Projects such as *Project Tumbleweed* haunt the world of blogging, tweeting, and contemporary social media, even if only as precedential zeitgeist. For example, Murphy's prescient discussion of "where" an online museum is situated and what its boundaries are or should be is a direct precedent for the Walker Art Center to make the "unprecedented" decision that its latest homepage should function as a general portal to the world of art and culture, not simply the Walker's events and collection.[8]

Art Entertainment Network

If *Beyond Interface* was a survey to help me and my colleagues think about net art, including its relation to the institutional, then *Art Entertainment Network*[9] was an attempt to do the same using a "born digital" exhibition format, which also had a physical presence in the institution.

Working with the brilliant designer Vivian Selbo, who was previously the art director for ada'web, we decided to create the *Art Entertainment Network* exhibition interface in a format native to the web—the portal.

We made two critical decisions early on: create a web-appropriate context for the selected works, and introduce a variability to the context so that the site always looked the same but constantly changed. This dynamism ranged from intentional send ups of mass customization to making each functional element of the portal as an artist project. The goal was to both valorize and satirize the mutability of the web while still making it available in the staid white cubes of the museum.

ICS—It's the Context Stupid

As I have suggested elsewhere, exhibiting net art can be like leipidoptery.[10] How do you keep from killing off the subject of interest? For *Art Entertainment Network*, we created an online Mediatheque[11] for the exhibition, which presented curated videos, sound, and texts from around the web that related to either the exhibited artists specifically or topics of their work in general. The idea was that it's impossible to fully appreciate net art when it is divorced from its natural environment.

Rescension and Precedential Media

Figure 9.1
Walker Art Center, *Art Entertainment Network*, opened February 12, 2000, online interface.

We organized a 12-week online symposium, Entertainment, Art, Technology (EAT).[12] The goal of EAT was to provide a social context for experts and interested audience to come together to discuss various topics triggered by the artwork in both *Art Entertainment Network* and *Let's Entertain*, a parallel gallery exhibition curated by Philippe Vergne. The guest "speakers" included artists from the exhibition as well as theorists from around the world, such as Sara Diamond, Geert Lovink, Mckenzie Wark, and Mark C. Taylor, but also, of course, the general public. EAT was modeled on another online symposium, which I curated and produced at the Walker, *The Shock of the View*.[13]

Figure 9.2
Walker Art Center, *Art Entertainment Network* opened February 12, 2000, in-gallery, portal. Photo: Steve Dietz.

Both were attempts to meaningfully engage the public as co-creators of meaning in relation to their topics.

We also created a marketplace, Artwarez,[14] where you could download artist projects and ssoftware such an RTmark screensaver, C5's softsub, Maciej Wisniewski's netomat(™), Jeff Gates' demographics, and much, much more! The marketplace remains a contested site for much of the net art world, and rightly so. But artwarez was perhaps precognition of a Kickstarter or Indiegogo "community," and the idea of being able to garner support outside of the constraints of traditional institutional funders.

We also created WebWalker, an email newsletter/blog,[15] with Robbin Murphy as the guest editor of at least one issue. Like the marketplace, the e-newsletter occupies uncomfortable territory between a purely promotional, marketing device and an informative publication. In some ways, it has been eclipsed by one's Facebook timeline, which has been commodified, technologically, by aggregators such as Feedly. But the question remains whether such algorithmic publications can reach beyond one's network of mirrors on the basis of something other than commercialism.

At one level, none of this contextual material was particularly earth shattering. Versions of each had been done before, but we still had to invent, more or less from scratch, the capabilities. There was no MailChimp, no Kickstarter, no "Instagram YouTube Complex." This is one leading edge of artist projects: to help imagine the kinds of things that could be done so that the structural technology can be commodified—and in turn subverted by a new generation of artists.

Mutant Bridges

The second defining aspect of *Art Entertainment Network* was how we took many of the standard aspects of a commercial portal, randomized them, and made them into artist projects. Each time the home page—the only page, really—refreshed, the webcam view would shift among different artist projects; the banner ad would alternate between an actual sponsor and Jennifer and Kevin McCoy's fake business, *Airworld*; the date format would change; you could change the background color; you could select different artist soundtracks to browse by; a different artist-designed daily meme was displayed; the order of the title words changed; a different default search engine was used; the privacy policy linked to a site that revealed "Here's some of what I (and every site you've ever accessed) know about you." There was even an artist personals ad service.

Angel Borrego Cubero of the Office for Strategic Spaces has promoted the notion, to which I subscribe, of mutant bridges, writing, "OSS defends the idea that, once having decided to build a bridge, it doesn't cost much extra to add one or more uses or options to its design. Part of the project for a bridge should therefore consist in thinking how the bridge could be used by other trans-species users."[16] If the formative role of pre-social media is to point to a direction of use that is user-driven and not a way to enhance stockholder value, perhaps in the contemporary climate there are ways that economic functions operating as and in public space, including virtual public space, can be recuperated as mutant bridges.

IRL—In Real Life

The museum is a physical site. How do you represent the virtual space and interaction of the network at such a site? Answers generally range from you don't (*Beyond Interface*) to the café/lounge (too many examples to list) to the hard(wired) copy (Documenta X).[17] For *Art Entertainment Network*, we worked with Antenna Design, which proposed a revolving door—a literal portal to the virtual world on the nonexistent other side of the door. This was a one-person interface, which didn't reward long-term engagement, but in the context of a museum exhibition where the average time spent in front of an artwork is measured in seconds, perhaps that's not a critical issue. In any case, as long as it is the human sensorium interacting with social networks, the physical interface is an interesting opportunity that is too seldom explored except as measured by the optimal-size touchscreen.

Telematic Connections

Telematic Connections: The Virtual Embrace, like many of the works in it, is a hybrid affair.[18] Part history, part speculation, partly onsite, partly online, it crosses boundaries among art, communications, and popular culture. Its four sections include installation works, past and recent film clips, online projects, and a "telematics timeline." Through these various media, the exhibition presents the ways in which artists use technology—and the Internet—to explore both the utopian desire for an expanded, global consciousness and the dystopian consequences of our collective embrace, willing or not, of computer-mediated human communications. At the same time, *Telematic Connections* places this emergent work within a historical framework.[19]

The focus on the physical distinguishes *Telematic Connections* from *Beyond Interface* and *Art Entertainment Network*. What effect does the network have in physical space and how does the physical world affect and intersect with the network? The network is not a closed system running only in the ether. Even if it's only the human sensorium, there is almost always an intersection. The works in *Telematic Connections* were explicit about this.

Attention Economy

Victoria Vesna's *n 0 time*,[20] subtitled "a community of people with no time," investigated the sensorial boundaries of online social community. How many people can we really know? How do we know that we know them? These questions have become increasingly urgent in the Facebook universe of friends and "liking" corporations. She writes:

Computer technology promised to save us time and provide a renewed sense of community. Instead we are collectively suffering from information overflow and lack of time, and we have to reconsider the established notions of "community." When thinking through issues around development and design of networked multiuser environments, time is the most critical element to consider. Information demands time; relationships demand time. How do we approach social environments in which relationships are built on information exchange and where physical presence is not necessary?[21]

One of the ways that Vesna controlled for her community of people with no time was to construct a physical interface based on a tetrahedron. This related to some of its formal properties as outlined by R. Buckminster Fuller, specifically its "insideness and outsideness," which parallels the inside/outside of online/physical interactions.[22] The physical installation was also important because the interface was activated by whole-body interactions, which can be more responsive/intuitive as well as a limitation. Our bodies can only do so much at any given moment. For Vesna, this is not intended as a separation or an incompatibility. She states, "there is no question that the most important consequence of this combination [telematic connectivity] is the shift to acceptance of ourselves as collective and distributed entities." The physical in the form of attention is an important control or reminder of the number of connections we can benefit from.

Where Is the Love?

The core question of *Telematic Connections* isn't the physical per se; however, it is the echo of Roy Ascott's notoriously appropriate question, "Apart from all the particulars of personal histories, of dreams, desires, and anxieties that inform the content of art's rich repertoire, the question, in essence, is asking: Is there love in the telematic embrace?"[23]

Eduardo Kac's *Teleporting an Unknown State* is an interesting case in point. A seedling sits in a dark room and a telematic community of viewer-participants must care enough to remotely activate a light source for it to live and grow. When the project was first exhibited in 1996, a seed grew into an 18-inch plant.[24] Over the course of the five installations of *Telematic Connections*, sometimes a seedling grew and sometimes it died and had to be replaced. At times, the distributed community did not care enough to ensure the seed's survival. There are many possible reasons for this, of course, but conceptually, at least, it points to the tension between scaling and intimacy in social media. The algorithmic solution is better targeting of ads; the artist practice is something to care about more. The techniques are not incompatible, but too often they are a parody of each other.

It has become almost a staple of sci-fi fiction to have dynamically generated, immersive environments. You're sitting in your home, on the phone, and as you converse, the walls display images generated by the conversation. Maciej Wisniewski's *netomatheque* was an actual instantiation of this idea.[25] A microphone translates your voice into Internet image searches, which are displayed on the walls of the installation. Over the past decade, everything about this project has become an order of magnitude easier to accomplish technically, from effective voice-recognition software to sophisticated image searching on the web to tiny, short-throw projectors that can easily cover a wall with imagery while remaining unobtrusive to its occupants. On the face of it, Wisniewski tackled the difficult end of the problem, creating, in essence, his own software to accomplish what today an undergrad student could easily connect up. Such design fictions, however, undergird many of the directions that our fundamental technologies might take. As Jon Ippolito writes in his essay "The Art of Misuse,"

What is the ultimate effect of creative misuse? Sometimes misuse becomes the norm. In a 1976 report to the Rockefeller Foundation, Paik coined the provocative term "electronic superhighway"—a phase that Bill Clinton paraphrased in his campaign rhetoric as "information superhighway" and has since permeated public consciousness. More often, however, misuse is of no direct practical value, but does what art is supposed to do: stretch our minds to accommodate not only the box, but what's outside it as well.[26]

Tina LaPorta's *Re:mote_corp@REALities*[27] is a perfect example of a project that drew on webcams and CU-see me sites to create a powerful narrative about disembodied connectivity. Like Wisniewski and others, her project predates contemporary social media, which would have made the "gathering" part of the project infinitely easier, but the story she tells remains powerfully evocative, more a harbinger of *Her* than Skype. In relation to Ascott's question, there is indeed powerful content in the virtual embrace, which connects viscerally to each individual viewer, but it may beg the question of how it is generated. What is the appropriate mixture of algorithmic, curated/edited/created, and user-generated content for there to be love?

Open_Source_Art_Hack

Somewhat ironically, the exhibition *Open_Source_Art_Hack*, co-curated with Jenny Marketou, took place in the media "lounge" designed by Lot-ek at the old New Museum.[28] I say ironically because arguably the "lounge" is an institutional mechanism to domesticate the wild environment of net art, intentionally or not. In the case of *Open_Source_Art_Hack*, the art also exposed the institution's inability to adapt to a social and arguably moral imperative due to its commercial relationships.

One of the projects in the exhibition, Knowbotic Research's *Minds of Concern::Breaking News*, triggered a set of network processes that investigated the security conditions of any nongovernmental organization (NGO) group's server and identify whether it was secure or open to hacking attacks. The results were made available on a news ticker, which visualized the strength or vulnerability of the server for a worldwide audience.

The aforementioned "network processes" were a form of port scanning, which hackers do use to take control of various servers. Knowbotic, however, as they analogized, was "just looking," not entering the organization's domain. Knowbotic had competent legal counsel about this effort, and no one questioned that it was in fact legal at the time under U.S. laws. The kicker, however, was that the New Museum had a commercial contract with its ISP, which specifically prohibited port scanning. The museum was not willing to have its network shut down or find another ISP, and at the last minute, The Thing stepped in and hosted Knowbotic's project for the New Museum.

This is where the commodification of technology intersects with its commercialization and "shrink wrap" licenses take precedence over other legal rights, subverting the very social networks that they claim to support.

Translocations

Translocations was a series of platforms: the physical, networked exhibition installation of "Architecture for Temporary Autonomous Sarai" by collaborators Raqs Media Collective and Atelier Bow Wow; the streaming media platform of the Translocal Channel, which was programmed by a number of artist groups from around the world; and the platforms of individual artworks such as *OPUS* and *Translation Map*, which require the participation of viewers to establish the possibility of translocal communities over the network. These projects and others in *Translocations* envision and promote an open, participatory culture that is translocal, interconnected, hybrid, and in flux.[29]

Five years after *Beyond Interface*, *Translocations* attempted to take what had been learned from the preceding exhibitions at the Walker Art Center and elsewhere and achieve two things in relation to net art and the institution: create an appropriate physical installation, and understand and instantiate the Walker as a network node, not a broadcast center.

Architecture for Temporary Autonomous Sarai

For us, the creation of a sarai was to create a "home for nomads" and a resting place for practices of new media nomadism. Traditionally, sarais were also nodes in the communications system (horse-mail!) and spaces where theatrical entertainments, music, dervish dancing, and philosophical disputes could all be staged. They were hospitable to a wide variety of journeys—physical, cultural, and intellectual. In medieval Central and South Asia, sarais were the typical spaces for a concrete translocality, with their own culture of custodial care, conviviality, and refuge. They also contributed to syncretic languages and ways of being. We would do well to emulate even in part aspects of this tradition in the new media culture of today. ... This might create oases of locatedness along the global trade routes of new media culture. —Raqs Media Collective[30]

Rather than creating a portal to somewhere else, as we did for *Art Entertainment Network*, we worked with Raqs Media Collective and Atelier Bow Wow to create a "Temporary Autonomous Sarai" to host the exhibition node *Translocations* as part of the exhibition *How Latitudes Become Forms*. The idea was inspired by ancient sarais along the Silk Road, modeled in some ways on Dan Graham's *New Space for Showing Videos*[31] and constructed with the experience of Atelier Bow Wow's investigation of pet architecture.[32]

The idea of TAS was that it packed into itself—it was built out of its own crates—and when unpacked, it created a networked set of viewing stations from the individual to the small group to the crowd. Central to TAS was Raqs Media Collective's project OPUS or Open Platform for Unlimited Significance, which, simplistically put, was a way to think about the transmission of culture as recensions rather than precedents and derivatives,[33] and a way for the audience to participate in the making of the meaning of the piece by rearranging it physically.

Nodal Networks

Hosted on TAS were a number of projects designed to orient the platform as a network node, not a center from which to broadcast.

Fran Ilich (Mexico City) organized a blog, "Big(b)Other," with contributors around the world, which was written in both Spanish and English.[34] Big(b)Other was, in part, Ilich writes, "a reaction to the supposed 'reality TV' epitomized by shows in the United States such as *Big Brother*, *Survivor*, *Fear Factor* and any number of other programs that are, in fact, slickly produced and heavily manipulated narratives that have little in common with 'real life.'" It was intended as a transnational narrative of the everyday. Most every day. Sounds exactly like and completely different than Facebook and Twitter.

There was also a "Translocal Channel," which provided regular audio programming from around the world.[35]

Precedence, Recension and Ubiquity

A re-telling, a word taken to signify the simultaneous existence of different versions of a narrative within oral, and from now onwards, digital cultures. … The concept of rescension is contraindicative of the notion of hierarchy. A rescension cannot be an improvement, nor can it connote a diminishing of value. A rescension is that version which does not act as a replacement for any other configuration of its constitutive materials. The existence of multiple rescensions is a guarantor of an idea or a work's ubiquity.[36]

There is indeed a rich pre-history to our contemporary social media universe. This is important to know and remember. Understood as recensions, not deriviatives or improvements, however, and the ubiquity of social media is indicative of an idea that may be as old as human history. It is our choice how to "rescend" it. Nothing is inevitable.

Notes

1. Freeman Dyson, *Imagined Worlds* (Cambridge, MA: Harvard University Press, 1998), 50.

2. Organized on the occasion of the 1998 Museums and the Web Conference, Toronto, Ontario, Canada, April 22–25. Available at http://www.walkerart.org/gallery9/beyondinterface/bi_fr.html.

3. Remo Campopiano, Craig Harris, Susan Hazan, Greg Lam-Niemeyer, Chris Locke, Pedro Meyer, Randall Packer, Paul Vanouse, and Martha Wilson. Available at http://www.walkerart.org/gallery9/dasc/g9_dasc_bifr.html.

4. Available at http://www.easylife.org/homework/. Accessed January 31, 2015.

5. The process is described in greater detail at http://www.walkerart.org/gallery9/beyondinterface/bookchin_fr.html. The Homework page via Wayback Machine available at https://web.archive.org/web/20070103070121/http://jupiter.ucsd.edu/~bookchin/finalProject.html. Accessed January 31, 2015.

6. Project Tumbleweed, Beyond Interface. Available at http://www.walkerart.org/gallery9/beyondinterface/murphy_fr.html.

7. Lutman & Associates, "Like, Link, Share: How Cultural Institutions Are Embracing Digital Technology." Available at http://www.lutmanassociates.com/blog/2015/1/5/like-link-share-how-cultural-institutions-are-embracing-digital-technology/. Accessed February 2, 2015.

8. "Redesign expands walkerart.org from an Information and Promotion Hub to a First-of-its-kind Content Provider," Walker Art Center press release. Available at http://www.walkerart.org/press/browse/press-releases/2011/walker-art-center-launches-newly-redesigned-w/. Note this redesign took place after my tenure at the Walker.

9. *Art Entertainment Network*. Available at http://aen.walkerart.org/, opened February 12, 2000, in conjunction with the exhibition *Let's Entertain*. Available at: http://www.walkerart.org/archive/B/9E13C5FA142230B2616E.htm. Accessed January 31, 2015.

10. Steve Dietz, "Curating Net Art: A Field Guide," in Christiane Paul, ed. *New Media in the White Cube and Beyond: Curatorial Models for Digital Art* (Berkeley: University of California Press, 2008), 76–84.

11. Available at http://aen.walkerart.org/mediatheque/. Accessed January 31, 2015.

12. EAT. Available at http://www.walkerart.org/archive/5/A353659586FB138F6164.htm. EAT was also a nod to the pioneering program Experiments in Art and Technology known as E.A.T., launched in 1967 by the engineers Billy Klüver and Fred Waldhauer and the artists Robert Rauschenberg and Robert Whitman. Available at http://en.wikipedia.org/wiki/Experiments_in_Art_and_Technology/. Accessed January 31, 2015.

13. Available at http://www.walkerart.org/archive/7/B153919DF735B615616F.htm.

14. Available at http://aen.walkerart.org/artwarez/. Accessed January 31, 2015.

15. Available at http://www.walkerart.org/gallery9/webwalker/. Accessed January 21, 2015.

16. Angel Borrego Cubero, "Mutant Bridges / Puentes Mutantes." Available at https://vimeo.com/album/1740556/video/31660455/. Accessed January 31, 2015.

17. As Kathy Rae Huffman notes in her review of the net art program presented at Documenta X, "A point of real concern to the 'Net' community, the blocking off of the Internet was seen as a

total disregard for the unique environment that all artists realize once they are online, the 'network.'" Available at http://www.heise.de/tp/artikel/4/4079/1.html.

18. *Telematic Connections: The Virtual Embrace* was a traveling exhibition organized and circulated by Independent Curators International (ICI), New York, and curated by Steve Dietz. It opened at the San Francisco Art Institute on February 7, 2001, and traveled to four other sites. The website is hosted by the Walker Art Center. Available at http://telematic.walkerart.org/index.html.

19. From the introduction to *Telematic Connections*. Available at http://telematic.walkerart.org/overview/index.html.

20. Available at http://telematic.walkerart.org/telereal/notime_index.html.

21. Available at http://telematic.walkerart.org/telereal/notime_identity.html.

22. Available at http://notime.arts.ucla.edu/notime3/. Accessed January 31, 2015.

23. Available at http://telematic.walkerart.org/overview/overview_ascott.html.

24. Available at http://www.ekac.org/teleporting.html.

25. Available at http://telematic.walkerart.org/telereal/wisniewski_index.html.

26. Available at http://telematic.walkerart.org/overview/overview_ippolito.html.

27. Available at http://telematic.walkerart.org/datasphere/laporta_index.html.

28. Available at http://netartcommons.walkerart.org/index.html.

29. *Tranlocations* opened February 9, 2003, at the Walker Art Center in conjunction with *How Latitudes Become Form*. Available at http://latitudes.walkerart.org/translocations/index.html.

30. Available at http://latitudes.walkerart.org/artists/index2815.html.

31. Available at http://www.walkerart.org/collections/artworks/new-space-for-showing-videos

32. Available at http://en.wikipedia.org/wiki/Atelier_Bow-Wow#Pet_Architecture/.

33. Raqs Media Collective, "A Concise Lexicon Of / For the Digital Commons." Available at http://www.raqsmediacollective.net/images/pdf/e65c6ffd-dbe1-4e57-a8b1-9fb087731ef4.pdf.

34. Available at: http://latitudes.walkerart.org/artists/index79ee.html.

35. Available at http://latitudes.walkerart.org/artists/index3cd8.html.

36. Raqs Media Collective, *op cit.*

III "See you online!"

10 Defining the Image as Place: A Conversation with Kit Galloway, Sherrie Rabinowitz, and Gene Youngblood

Steven Durland

Since meeting in 1975, Kit Galloway and Sherrie Rabinowitz (figure 10.1) have focused their collaborative art career on developing new and alternative structures for video as an interactive communication form. Under their organizational moniker of Mobile Image, the pair has created three major works: Satellite Arts Project (1977) (figure 10.2), *Hole-In-Space* (1980) (figure 10.3), and *Electronic Café* (1984) (figure 10.4), as well as numerous smaller projects. Their sophisticated knowledge of satellite telecommunications has made them sought-after consultants in the field, and their research has resulted in numerous contributions to the technology. They combine the technological and sociological possibilities of two-way communications with artistic sensibility to create elegant models of the way things could be. These "models," a term they prefer to "artworks," serve not only as a vision of how telecommunications could serve humanity but also put forward some provocative notions of the future and function of the artist.

Gene Youngblood has been a writer, lecturer, and teacher on the subject of art and new technology for 17 years. In 1970, he authored *Expanded Cinema*, the first book about video as an artistic medium. Today he is considered one of the most informed and articulate theorists of media art and politics. His theory of the "creative conversation" is an inspiring vision of the role of the artist in society.[1]

Youngblood is a faculty member at California Institute of the Arts, where he teaches the history and theory of experimental film and video. He is currently collaborating with Rabinowitz and Galloway on a new book titled *Virtual Space: The Challenge to Create at the Same Scale as We Can Destroy*. According to Youngblood, *Virtual Space* will examine the political, philosophic, and aesthetic implications of the communications revolution.

In addition to their contributions to *Virtual Space*, Rabinowitz and Galloway are developing a composite-image performance between dancers in the Soviet Union and the United States. A second project-in-progress, *Light Transition*, will use satellite television images to sync contemporary technology with natural systems and ancient technology such as Stonehenge. Critical moments of sun/moon intersection, etc. will

Figure 10.1
Kit Galloway and Sherrie Rabinowitz, 1984.

appear for 20 seconds every half-hour on a cable TV superstation. "This project will be more poetic than our others," said Galloway. "It's basically a celebration of earth's systems and humanmade technological systems. We hope to create a project in which both systems reflect the elegance of the other."

In early March, I visited the Mobile Image studio in Santa Monica and listened in as Rabinowitz, Galloway, and Youngblood discussed the work of Mobile Image and its implications for the future of art and communication.

SHERRIE RABINOWITZ: When Kit and I first met, it was in Paris. I'd been invited over, and I was introduced to him as the person who knows everything about ze video in France, in Paris.

Figure 10.2
Satellite Arts Project: *A Space with No Geographical Boundaries*, 1977. In collaboration with NASA, the world's first interactive composite-image satellite dance performance. Using the U.S.-Canadian CTS satellite, dancers located 3,000 miles apart at NASA Goddard Space Flight Center, Maryland, and the Educational TV Center, Menlo Park, California, were electronically composited into a single image that was displayed on monitors at each location, creating a "space with no-geographical boundaries" or virtual space in which the live performance took place. Performance includes the first satellite time-delay feedback dance; three location live-feed composite performance; flutist Paul Horn playing with his time echo. July performance and three-day performance in November. Pictured: The center dancer, Mitsuko Mitsueda, is in Maryland, keyed into the West Coast location to dance with her partners Keija Kimura and Soto Hoffman in California.

KIT GALLOWAY: Bit of an exaggeration.

SR: Before we met, I was working in San Francisco and helped start Optic Nerve, a group there, and Kit was working in Europe with the Video Heads. Both of us through our experiences had come to two realizations. One is that the power of television is its ability to be live, to support real-time conversation independent of geography. Number two was a sense of the way television is experienced, the way it's taken in, this wash of images with nobody really remembering the context of any particular image. Television is an image environment, and that's how you have to understand it. Making tapes didn't make sense because it wasn't affecting the context or the environment. Kit had become interested in satellites, in 1973, 1974. When we met, we put our ideas together and developed the track that we've been on ever since.

KG: Living in Europe I could see what effect television was having on different countries, reading all the material at UNESCO back when satellites were beginning to appear and were seen as a weapon against illiteracy. I was seeing how most of

Figure 10.3

Hole-In-Space: A Public Communication Sculpture, 1980. A three-day, life-size, unannounced, live satellite link allowing spontaneous interaction between the public on two coasts Video cameras and rear-projection screens were installed in display windows at Lincoln Center for the Performance Arts in New York and The Broadway department store, Century City, Los Angeles. Each screen displayed life-size, full-figure images of people on the opposite coasts. There were no signs or instructions. Passersby drawn to the windows discovered an open channel through which they could see, hear, and talk with people on the other coast almost as if they were standing on the same street corner. Pictured: A woman on screen from New York City leans forward to visit with silhouetted people in Los Angeles.

the world apart from the United States had an international policy for telecommunications. But the United States, under the guise of "free flow of information," was putting forward a policy of first-come, first-served, screw you if you can't get up there and park a satellite.

I was aware of the imbalance, and I got more and more interested in television and its technology as a communications medium. We started looking at ways of using international satellite transmissions. There were no interesting ideas in the mountains of UNESCO documentation. We looked at the idea of collaborative performances between artists in different countries meeting in this composite-image space we had conceived, mixing the live images from remote locations and presenting that mix at each location so that performers could see themselves on the same screen with their partners. That became the premise of the work and experimentation we wanted to do.

In 1975, NASA announced that they were accepting proposals from public organizations to experiment with their U.S./Canadian satellite. They were trying to drum up some public support by providing access. So we hopped on a Russian liner and landed

Defining the Image as Place

Figure 10.4
The Electronic Cafe table at Ana Maria's Restaurant in East Los Angeles. Electronic Cafe, July–September 1984. Called "One of the most innovative projects of the Los Angeles Olympic Arts Festival," officially commissioned as an Olympic Arts Festival Project by the Museum of Contemporary Art (MOCA), Los Angeles. Electronic Cafe linked MOCA and five ethnically diverse communities of Los Angeles through a state-of-the-art telecommunications computer database and dial-up image bank designed as a cross-cultural, multilingual network of "creative conversation." From MOCA downtown and the real cafes located in the Korean community, Hispanic community, black community, and beach communities of Los Angeles, people separated by distance could send and receive slow-scan video images, draw or write together with an electronic writing tablet, print hard-copy pictures with the video printer, enter information or ideas in the computer database and retrieve it with Community Memory (TM) keyword search, and store or retrieve images on a videodisc recorder that held 20,000 images. Electronic Cafe ran six hours a day, six days a week for seven weeks.

in New York, and in a couple of months, we had secured NASA underwriting for a project that was going to last for about a year, a project that to date is probably the most intensive look at the interactive potential of human communications and satellite time-delay problems.

Kit and Sherrie create context rather than content. An artist can enter the context they create and make content, which will now be empowered and revitalized in a way that it could never have been empowered without the context that these people set up. —Gene Youngblood

SR: It's still the most sophisticated. And it's only been in the past few years that people have come to appreciate what it is. It's still not been duplicated.

HIGH PERFORMANCE: How did your projects develop?

KG: We came back here with the idea that we would work domestically, doing model projects, and then, when we had the credentials, to move out internationally.

SR: We did a number of projects—large-scale ones like *Satellite Arts*, *Hole-In-Space*, and *Electronic Cafe* and smaller ones that had a different intensity, laboratory things.

KG: The first project was called the *Satellite Arts Project*.

SR: When we first did *Satellite Arts*, nobody was interested in satellites, everybody was interested in, I don't know, video art. It took a long time. And now others have caught up. We always approached the image as a place. To our way of thinking, the essence, the magic is this ability to carry a living event and then interconnect with satellites to connect places over vast distances. When we started, and even now, there was no aesthetic approach to this. Businesses used teleconferencing, but the aesthetics of what that connection is is a whole new reality that hasn't been explored. In a sense it's a meta-design. It's looking at the live image in telecommunications as part of a grander structure.

KG: You just don't go out, which is relatively easy to do, and rent satellite time and do something that would be of any significant contribution outside of the context of ignorance. There's a great context of ignorance both in the industry and in the art world as to the intrinsic nature of this medium. We focus on the living event, not being too concerned with whether it's artlike or not. We don't produce artifacts, we produce living events that take place over a period of time, to facilitate a quality of human-to-human interaction.

GENE YOUNGBLOOD: When video first started, there were three directions that it took, all simultaneously. One was "artist's video," that is, people like Bruce Nauman and Vito Acconci, who came from the art world, using video to document their art practices. Then there was what is called, for lack of a better term, "electronic video," that is, the exploration of the essence of the medium, represented by people like the

Defining the Image as Place

Vasulkas. Then there was "political video," guerrilla television, and so on, which was essentially a documentary tradition. But that excluded this whole other world of live interconnection, which is what Kit and Sherrie represent. It's an important direction.

KG: Other artists have accessed satellite technology but look at what Douglas Davis has done with his access, or Nam June Paik more recently. Doug's pieces have been very sort of "artist using a satellite with a written scenario," like a theater piece that is all scripted out and some interaction is then portrayed during the access of this technology. It's still held within the context of control.

GY: Spectacle.

SR: One of the essences of our work is scale. When you galvanize that much technology and that many resources, it's not like an artist working in a garrett. As soon as you work with telecommunications, a satellite's part of your structure, as is the society around you. You have to deal with NASA and Western Union to access your satellite time. You have to deal with where the satellite comes in, you have to deal with the real thing, and it's expensive. So the idea of doing something that's self-focused just doesn't seem to be a very ecological political use of the medium. You can't deal with this technology without dealing with it politically.

GY: It really pushes up against a question of how far an artist is willing to go in the direction of not being an artist, giving up the ego identification with the product. That's a central issue with all this. Everybody knows that people in power can use this technology to put on a spectacle, we see it every day. So the fact that all artists could raise the money to do this is not really a revelation. It doesn't constitute a revolutionary use of the medium. But if somebody were to set up a system and then turn it over to people, like Kit and Sherrie do, nobody else does that, nobody. So who in our society is going to do that? The artist as traditionally understood won't do it.

So we need a new practitioner who does what I call "meta-design." They create context rather than content. An artist can enter the context they create and make content, which will now be empowered and revitalized in a way that it could never have been empowered before without the context that these people set up. To me, this is the new avant-garde: the collaboration of the meta-designer and the artist. One not being enough without the other, each needing the other and together constituting a whole new force. A context is created that can be controlled by the people who constitute it. Those people might be artists whose work would then be given an autonomy of context, which it dearly needs, which the whole modern history of art is screaming for. So this is where it gets important.

HP: I'm very intrigued with this idea of using art to empower other people instead of using it to empower yourself.

SR: I don't see the way we create our pieces as based solely on the fact that you have to empower people. The way we embrace the issue is pretty classically art. If you define the aesthetic of the medium by defining what the essence and integrity of that medium is, then good art in the sense of telecommunications means that you create a situation that has to be some kind of communication between people in order to maximize what that technology can do. If you're just scripting from one side to the other side, you don't need a satellite, you can run two tapes. There has to be that quality of tension that defines what communication is, that higher level, which would be, as Gene points out, the conversation. And unless you create that tension in the work, then you're not really looking at the qualities of the medium or the qualities of the art.

HP: Were you aware of a television program called "The People's Summit" that featured a live studio audience in Seattle hosted by Phil Donahue meeting with a live studio audience in Leningrad hosted by Vladimir Posner?

KG: The Phil and Vlad show.

SR: That's who we're starting to work with now, those and others. Not Donahue.

KG: We believe very much in the principle of informal networking, which is aligned with the new phenomenon of citizen diplomacy. Now when you put that in the context of a nationally rated syndicated program like Donahue, then again it becomes a spectacle, and it's not really an informal network. When that happened, I felt that it was very much a disaster, ill-conceived, ill-executed. The consequences were not all that good. Yet in a context of ignorance, even a gesture in that direction is somewhat healing or an improvement over what existed before.

SR: Everybody, including the Russians, is ready for something more interesting, more cultural integration between the United States and the Soviet Union, and also cultural integration between all of us and the electronic culture.

HP: On the order of *Hole-in-Space*?

KG: There's interest in that. Our work with Electronic Cafe was carried to the Soviet Union. They looked at what we did here in Los Angeles as an international model for cross-cultural communications systems. I could show you newspaper articles from Pravda where they took the concept and totally embraced it. They made an Electronic Cafe in Moscow, did some slow-scan and voice-only connections with San Francisco.

SR: One of the things we're working toward now is the composite-image performance similar to what we did in 1977 (*Satellite Arts*). We'll have performers in the Soviet Union and performers in the United States meet in this composite-image space, this virtual space with no geographical boundaries, and in that space they'll perform together and dance together. Dance and performance do not need to be translated.

Defining the Image as Place

When you start communicating and being able to touch and join bodies, it creates a whole other context.

KG: All this sounds very strange, very experiential, but if you do this, you realize communication's power of being able to mix space or exchange spaces.

We focus on the living event, not being too concerned with whether it's artlike or not. We don't produce artifacts, we produce living events that take place over a period of time, to facilitate a quality of human to human interaction. —Kit Galloway

GY: People have kind of a phantom limb sensation, it's actually visceral.

SR: The video image becomes the real architecture for the performance because the image is a place. It's a real place, and your image is your ambassador, and your two ambassadors meet in the image. If you have a split screen, that defines the kind of relationship that can take place. If you have an image mix or a key, other relationships are possible. So it incorporates all the video effects that are used in traditional video art, but it's a live place. It becomes visual architecture.

KG: A lot of work was done discovering things like reversing the scan on the image so that when the dancer moved to the right, the image moved to the right instead of the left, and all the special technology that surrounded the performers to facilitate this interaction. A lot of research and development went into how fast movement or activities could take place with the satellite time delay being present.

GY: It needs to be pointed out that they're doing the same thing that other people have gotten international recognition for in video, the Vasulkas, for example. Most of their life has been this kind of research and development. They enter the digital domain and find out what's possible, and they're internationally renowned for doing so. Kit and Sherrie are doing the same thing. These are new frontiers, and you can't just step in and make art automatically; first you have to research and find out what's there, what's possible, what are the consequences. I think it's interesting that everyone recognized that as a value in video, but it has not been recognized so much in telecommunications. Nam June and Doug Davis just step in and make precious works of art immediately, and there's no sense of exploration or research and development.

KG: Nam June did *Good Morning Mr. Orwell*. He took a lot of the work from *Satellite Arts* without discussing it with us and gave us a credit on the end. He had Merce Cunningham in some studio in New York stumbling in front of a TV monitor that hadn't had the scan reversed, making a fool out of himself by his standards, certainly.

GY: This addresses the postmodern notion that what artists do now is not attach themselves to any particular medium. They just float among whatever mediums are appropriate to whatever they want to say. However, when it comes to creating on this kind of scale, you can't do that. There is only a limited set of technologies that operate at that scale, and if you want to do something meaningful, something new, you have to know that tech, you have to get access to it over extended periods of time, you have to devote your life to it. You cannot come in as a dilettante or an aesthete.

KG: We're not about the whole "access" mentality, which often doesn't really cultivate work. It's like running out, driving a stake in uncharted territory, and saying I was here first.

SR: The art world in general is pretty impotent. I think part of the reason is the whole sense of scale of contemporary society, the scale at which we can destroy ourselves, the little chips that hold how many pieces of information we don't even know, genetic engineering. Really, all the new technological developments are out of human scale. The more that we explore space, the more we feel lonely. It seems to us that the only real power is to work at the same scale that contemporary society is working on. If you don't create on the same scale that you can destroy, then art is rendered impotent.

GY: This is what I meant earlier about the meta-designer and the artist together. The artist is the one who can make the most powerful, the most moving representations of our life situation. Yet the forms that have traditionally been available to the art world just don't meet the scale of the problem. This raises the question of what is political art. My own opinion is that there is no such thing as political art. There is art about political issues. But only situations are political, only circumstances are political. So if you set up a space bridge or a hole in space or an Electronic Cafe as a situation, as a circumstance that spans boundaries of people, and then you put those poignant, powerful representations of the artist in there, then you've got both: art that addresses a political issue within a political situation. The whole thing becomes highly political and powerful.

HP: Do you grapple with the issue that some people bring up, such as Godfrey Reggio, the director of *Koyaanisqatsi*, that technology is inherently evil?

GY: To me it's beneath an answer actually. It's like this book, *Four Arguments for the Elimination of Television*[2]—remember that book? These views are what is called "vitalism." It's like saying in protein there is life. It's voodoo. It doesn't contribute anything constructive to what we have to do.

KG: It doesn't contribute anything, in fact, it's backpedaling. The fact is, if we don't learn how to use this technology to manage the human and material resources of this planet we're screwed. End of story. Fade to black.

HP: What are your realistic goals with this technology?

KG: We're not going to realize our vision in our lifetime, I don't have that much expectation, but somebody has to be creating models to liberate people's imaginations so they can apply them to hope and the possibility of redefining these technologies.

Defining the Image as Place

We've gotten to a point where we've realized the limits of models. We want to take the revolution into the marketplace. We've designed a cost-effective, kickass, multimedia, cross-cultural teleconferencing terminal that will allow communities of common concern to link up and evolve collectively.

HP: Are you talking about providing a set of instructions? Are you talking about providing actual hardware?

KG: We're looking at turn-key hardware solutions. There have been all these attempts at networking cross-culturally, but all of them take place using a different set of technologies. Some are cost effective, but most are made by the teleconferencing industry that's marketing to the Fortune 500, so the markup is like tens of thousands of dollars over the value that's really in the box. We see the possibility of creating a turn-key system that would create a compatibility standard for a multimedia teleconferencing terminal that provides fax, slow-scan, full-motion video, written annotations on pictures, pictorial data management, text conferencing. We can see putting that together at a price that would fall into the small organization price range. That's what we're looking at right now. The creation of a pilot network is our first priority because people must have the opportunity to experience systems like this or like Electronic Cafe to fully realize the indispensable value of technology such as this.

SR: The whole nature of this so-called information and communications society is really dependent on people synthesizing and being creative. It's almost as if capitalism and communism are turning in on themselves and possibly meeting in a new place. It's determined in part by the progression of the technology, which is becoming decentralized, which is becoming dependent on creating this information economy. How do you create information? When you get right down to it, information is based at some point on somebody's discovery or somebody's synthesis or somebody's research and development.

GY: I'd like to address this. We've been talking about the communications revolution. People take this in one of two ways: either it's some kind of 1960s, hippie, utopian idealism or it's a marketing scam by industry—you know, "The communication revolution is here, and our product. ..." There has been no middle-ground discourse between those extremes. I would just like to point out that any interesting thing that people like Kit and Sherrie would do with this technology would by definition have to be a model of what a communications revolution would be like if there were one. McLuhan said, "The medium is the message." What's the medium? Depending on who you talk to, the medium is television, the medium is this and that. But I always understood it this way: the medium is a principle, it's not a piece of hardware. The medium is the principle of centralized, one-way, mass-audience communication. We happen to do that through broadcast TV, but you can do it through cable TV, you can do it through the telephone lines. That's the medium, and that medium is the message. It determines that what will be said over a centralized, one-way, mass-audience communications system will have to be said within a very narrow framework of what is possible, what will be accepted by such a mass audience. That is the medium. A revolution would be to invert that principle through whatever technologies permit you to invert it: a decentralized, two-way, special-audience system. Art-world theorists who criticize anyone who talks about this would have you think that this all was attempted in the 1960s and failed! This is bullshit! The only thing that happened in the 1960s is that we got the vaguest notion that this was even something important to think about. We woke up in the 1960s, and now we're starting to take the first steps to see what direction that inversion might lie in. Do we think it'll happen? That's completely beside the point! The point is that this is the only meaningful thing to do with our lives because we know that no other institutions are capable of addressing the problem.

KG: The trouble is there haven't been enough participants to make a major shift that would land on the cover of *Time* magazine. What we've been trying to do is to get it out into public spaces so that people participate in these environments. We create the context and invite people to come in and do their laundry, hang up their clothes, live there for a while, and see what it's like. To begin to recognize the value of it and to acculturate it for a period of time. Redefine themselves through it.

SR: In an art context, what we've been doing is perfectly logical. If the art world has problems with it, it's because that logic challenges the validity of the art institutions for this new practice in contemporary society.

First you look at communication and then you look at the aesthetic quality of the communication. As soon as you do that, and you're true to your art form and your art logic, not worrying about whether it fits in a gallery or on a shelf, then you very naturally put one foot in front of the other and get to these places. The art logic just marches you right out of the art institutions into life.

If you look at the aesthetic quality of the communication and you're true to your art form and your art logic, then you very naturally put one foot in front of the other and get to these places. The art logic marches you right out of the art institutions into life. —Sherrie Rabinowitz

HP: What about the communications possibilities of videotapes now that everybody's got a VCR?

KG: The great thing about tape rentals is that they're just knocking the side out of broadcast television and cable. It's just knocking them for a loop.

GY: Mailing around personal videotapes and the home VCR network is revolutionary, but the real issue is to have an alternative social world that doesn't stop, that is continuous the way television is. Television is a social world, it's there 24 hours a day, it has a history. They're called series, you know. The news develops a history: "You know what happened yesterday, here's the update," and so on. So it's a world. The point is to have this alternative social world that's always there and never stops and is always validating itself as a possible social world. And then alongside of that are all these other supplementary "periodic" media. I can go to the store and get a tape just the same as I can go to the store and get *Mother Jones* or *Newsweek*. But if it were only these periodic media, this would not constitute an alternative social world of any significant power. So they're complementary.

HP: The Electronic Cafe ran for over seven weeks. Why couldn't it have just kept running indefinitely?

KG: Nobody wanted it to come down, but we couldn't perpetuate it because it was not cost-effective. We put it together with available technology, and it wasn't the system to perpetuate. Now we've got a system that's cost-effective.

SR: One obvious idea that we've thought about is a new electronic museum. It's a way to link people, places, and art works in this new environment.

GY: Electronic Cafe created this new inversion of the art and life situation. The longer it ran, the more it just became life, right? In a sense, you could say the less it became art, the more it became life. Or the shorter it ran, the more it became art, but the less it would be doing what it ought to really be doing, which is becoming life.

KG: Just look at this as interaction with a system. It's looking at creativity applied across the boards and at different levels. Even though Electronic Cafe had to go away, it's successful in that it empowered people in those communities with enough experience to describe what is desirable or what they would want as a system. It's politically hot, culturally hot. It created a lot of travel, an exchange between these communities, and used Los Angeles as a global model. When you look at the archiving aspect of it, this is important because the face of Los Angeles, the demographics, the dynamics of it are so wild that the face of history is going to change so fast that there's not going to be much of a record of it. But when you have environments where people can come and register their opinions and ideas and show their stuff and accomplishments—little kids breakdancing—whatever it happened to be, all that can be there to be looked at under the context of a social space, it's not private.

The other aspect of Electronic Cafe that was very important was that it created a public space in which one could participate in telecommunications anonymously. You could be among people without anyone knowing how many kids you have, how many points you have, what your income is. It was like a public telephone booth. It wasn't the privacy of your own home, where there's a wire right up your consumer tract. It was a place to present your ideas, register your opinions anonymously. You didn't have to sign your name. The artifacts you created—pictures, drawings, writing, computer text—either independently or collaboratively could be, if you desired, permanently stored in the community-accessible archive. People could have access to opinions without being monitored. There always exists the possibility of being monitored when it's in your home. A "commons" was created that was very important in terms of the freedom and what gets to define our personal freedom in this electronic space.

GY: If this isn't political, I don't know what is. They gave people a living experience of one of the hottest political issues of our time—how can we move into electronic space and still be anonymous? Are we going to be anonymous? Is anyone even talking about that? Has the issue even come up? No. You gotta join The Source, you got to give all your data to Compuserve. Anonymity is a possibility that could just vanish, except for those people in East L.A. now who've had that experience, who are therefore much hipper than probably most of the consultants to AT&T who never thought ...

The fact is if we don't learn how to use this technology to manage the human and material resources of this planet we're screwed. End of story. Fade to black. —Kit Galloway

A Selective Resume of Mobile Image (Galloway/Rabinowitz)

Biennale of Venice, Italy, 1986. Part of the "Art, Technology and Informatics" Exhibition. Three evenings of slow-scan video transmitted from Mobile Image's studio in Santa Monica, California, and projected live on large screens in the exhibition hall in Venice, Italy.
Electronic Café, July–September 1984.
Professor/Instructors, University of California Los Angeles (UCLA), 1983. School of Motion Picture/Television. Graduate-level course: "Experimental TV."
Aesthetic Research in Tele-communications (ART-COM), Loyola Marymount University, 1982. Designed and taught full semester multidisciplinary laboratory examining the effects, potentials, and future of interactive video.
Hole-In-Space, 1981. 30-minute award-winning documentary.
Hole-In-Space: A Public Communications Sculpture, 1980.
Satellite Arts Project, 1978. 30-minute award-winning documentary.
Satellite Arts Project, 1977.
Exhibitions and lectures include: Venice Biennale, Italy; Museum of Contemporary Art, Los Angeles; Museum of Modern Art, New York; San Francisco Museum of Modern Art; Long Beach Museum of Art; The Kitchen, New York; American Film Institute, Los Angeles; Tokyo Video Festival; American Center, Paris; Avignon International Arts Festival.
Grants and support include: Times Mirror Corp., Museum of Contemporary Art, Honeywell Corp., National Endowment for the Arts, American Film Institute, Sony Corp., NASA, Corporation for Public Broadcasting, Western Union, General Electric.

Notes

1. Youngblood discusses the creative conversation thusly: "To create new realities, we must create new contexts, new domains of consensus. That can't be done through communication. You can't step out of the context that defines communication by communicating; it will only lead to trivial permutations within the same consensus, repeatedly validating the same reality. Instead we need a creative conversation that might lead to new consensus and hence to new realities, but which is not itself a process of communication. I say something you don't understand, and we begin turning around together: 'Do you mean this or this?' 'No, I mean thus and such.' During this nontrivial process, we gradually approximate the possibility of communications, which will follow as a trivial necessary consequence once we've constructed a new consensus and woven together a new context. Communication, as a domain of stabilized noncreative relations, can occur only after the creative (but noncommunicative) conversation that makes it possible: communication is always noncreative and creativity is always noncommunicative. Conversation, the paradigm of all generative phenomena, the prerequisite for all creativity, requires a two-way channel of interaction. That doesn't guarantee creativity, but without it there will be no conversation at all, and creativity will be diminished accordingly." Excerpted from "Virtual Space: The Electronic Environments of Mobile Image," by Gene Youngblood, IS Journal #1, 1986.

2. Jerry Mander, *Four Arguments for the Elimination of Television* (New York: William Morrow, 1978). Reprinted with permission from *High Performance* #37, 1987, 52–59.

11 IN.S.OMNIA, 1983–1993

Rob Wittig

For a decade before the World Wide Web, a loose collection of creatives associated with the literary performance group Invisible Seattle conducted a series of collaborative literature projects on the bulletin board system (BBS) called IN.S.OMNIA, or Invisible Seattle's Omnia. The Invisibles, as they called themselves, used the visual and procedural estrangement of reading and composing glowing text on a cathode ray tube (CRT) to explore alternatives to the literary conventions of their day. Inspired by European avant-garde frame breakers such as James Joyce, Gertrude Stein, the Surrealists, and the Oulipo, by poststructuralist theorists such as Derrida and Lyotard, as well as by pop-culture absurdists such as TV comedians Monty Python, radio satirists Firesign Theater, and the participatory, collaborative parody Church of the Subgenius, their projects radically remixed writing and were some of the earliest explorations of identity and collaborative creativity in social media.

As an Invisible myself, and a featured player on IN.S.OMNIA, these experiments of the 1980s profoundly shaped my subsequent electronic literature projects such as the fictional website *Fall of the Site of Marsha*,[1] the novel in e-mail *Blue Company*,[2] and the current networked improv narrative (netprov) projects I do with Mark C. Marino and others. The Invisibles threw ourselves into issues that are still being investigated in the mid 20-teens. Among these are: (1) the materiality of writing; (2) pseudonymity, identity, and virtual community; (3) born-digital literary styles and genres; and (4) collaboration and crowdsourced creativity.

The Materiality of Writing

Changes in the material support of writing always reawaken awareness of the materiality of visible language. A random alphabetic shape corresponds to a phoneme; text flows left to right, the codex is bound on … oh, let's say, the left. It's easy to forget that these conventions were once fresh, arbitrary, contingent, and only gradually came to seem inevitable and natural. The glowing CRTs of the 1980s announced a transition that would reveal how many hidden cultural rules were bound up in the technology of print.

The Song of the Modem

In its earliest years, users—sometimes equipped with nothing more than a so-called "dumb terminal" consisting of a modem, a CRT screen, and a teaspoonful of hard-wired text-display software—would hand-dial in to the single dedicated phone line where a single dedicated desktop computer in Gyda Fossland's house answered calls one at a time. You could be all ready to write, only to hear the disappointing beep-beep-beep of the telephone busy signal, meaning some other user was on the BBS, and you'd have to wait. Or the finicky BBS software would be frozen and you'd hear only the ringing of the phone, which in this era before widespread use of answering machines and no voicemail, would last as long as you could stand it. Once you connected, you'd get the thrill of seeing text, in a brutally simple sans serif, begin to flow at 300 bps in a slow, stately, green or blue character-by-character sweep across the black screen.

WELCOME TO INVISIBLE SEATTLE'S ONLINE

LITERARY MAGAZINE, IN.S.OMNIA

ENTER YOUR NAME:

The cursor prompt would sit and blink at you, indicating it was your turn to type a command. No touch screen, no mouse. Just a keyboard. Your chance to play occurred at the end of a text, in the form of that blinking prompt. To hit return was to "send."

ENTER YOUR NAME: Curious User

ENTER A 4 TO 6 CHARACTER PASSWORD: elucubrations

CAN YOUR TERMINAL DISPLAY LOWER CASE?

(Y/N) YES

DO YOU WANT INSTRUCTIONS (Y/N) NO

ENTER NUMBER OF NULLS (USUALLY 0): 0

ENTER SCREEN WIDTH: 80

Curious User logged on at 11:24:17 PM on Friday

September 7, 1984

Topics with new messages:

Lobby > That Night in Zyzzyzzywa > Derrida > dUMMY in WONDERLAND > Translator's Notes > 1–10 > Translator's Notes > Fuck Meaning > That Night in Zyzzyzzywa > Naïve Writing > Art A to Z > Ecriture from the Black Lagoon

Press ? for menu

You might press L for lobby and the cursor would spell out:

Lobby >

You'd type R and the cursor would leap to life and complete your command:

Lobby > Read new

And the contents of the topic, or "room" as we called it, would fill the screen. But instead of IN.S.OMNIA's Lobby containing, by BBS convention, welcome messages and organizational information from Sysop (system operator) about the board, the IN.S.OMNIA user might find:

"85mar09 from Lobby

All elevator "close door" buttons are only placebos to lend the illusion that you have some kind of control over your environment. The same thing goes for those crosswalk buttons and that key in the lower left-hand side of your keyboard. I don't know why I'm telling you this. I could get in big trouble by spilling the beans."[3]

The metaphorical term "rooms" was a term inherited from the dungeons of interactive fiction and encoded into the BBS software. I'm astonished that the clumsy, geographical "navigation" metaphor so readily adopted from early gameplay as users struggled to conceptualize the relationships among the flowing, glowing texts, is still with us.

Later in the 1980s, when faster modems became available, calling IN.S.OMNIA was accompanied by the song of the modem, which asked each answering serving computer what speed at which it would like to converse, by musically working its way up from 300 thru 1,200 to a blistering 2,400 bps.

The BBS's memory buffer was limited to 1,968 characters per message, and the total memory of the BBS was not large, containing approximately 100 messages at any one time. When the limit was reached, the system automatically deleted the oldest messages in favor of the newest. This meant that the original messages in any room might well be gone forever, especially if the room got popular. Each new message was the death sentence of an earlier one.

"Where's the first text? Scrolled into oblivion, friends. We are commenting on, and annotating, a quickly absent original. Truly no origins here. Beginning is always an interruption. ..."[4]

I, along with occasional others, made it my practice to periodically print current versions of each room directly from my Kaypro portable computer onto a dot matrix printer. It was from these archives that the print *IN.S.OMNIA* (IN.S.OMNIA), and later my full account of the period, *Invisible Rendezvous* (Wittig), published in 1994 by Wesleyan University Press,[5] was compiled.

Pseudonymity, Identity, and Virtual Community

On IN.S.OMNIA, 'nyms themselves varied in tone and style: from the reasonably recognizable Eugene Correct, Frank Function, and Clark Humphrey to Futures, cranky, actual size, "Big Phone" Bill, and contingent 'nyms summoned for a single appearance to support a scene in progress.

It was a game of skill to try to be able to hide the identity behind the 'nym and become truly invisible. This struck at the heart of the traditional writer's goal of cultivating a single, unique, identifiable voice in writing. Our goal, like that of the French writer and Oulipean Georges Perec, was to be able to write in as many styles as possible.

At an event commemorating Invisible Seattle at the MLA convention in 2012, I had the pleasure of reuniting with Invisibles Philip Wohlstetter, Paul Cabarga, and Tom Grothus for a raucous conversation about IN.S.OMNIA days, only then finding out for the first time which of them was the author behind some of the old 'nyms.

I moved away from Seattle in 1983, and IN.S.OMNIA became my lifeline to my creative friends. While the other Invisibles could gather at the coffeehouse to talk over their experiments, my connection to them was increasingly just through the BBS. I lived in a virtual community.

Born-digital Literary Styles and Genres

New technologies need new forms. That's what we believed. Although the goal of IN.S.OMNIA was the creation of new literary texts that could stand with the best literature of the past, we wanted at all costs not to merely transpose existing literary styles into a new medium. It was our explicit aim on IN.S.OMNIA to create rooms to explore beyond the suffocating limits of what Invisible Seattle godfather Philip Wohlstetter impertinently called the "holy trinity" of poem, short story, and novel.

Naïve Writing was a room that "continues a discussion started over a pleasant dinner but by no means finished. ... We began to wonder, over dessert, whether we can speak of a "Naïve Writing" and, if so, what is it?[6] Then Wohlstetter posted a message that set the course for IN.S.OMNIA away from being just a new venue for old forms.

I am trying to write differently ... I'm trying to lose my specific idea of what writing is ... I'm convinced that if you read supermarket lists, censored newspapers, the fine print on legal contracts and GREAT LITERATURE, your writing will be more interesting than if you just read GREAT LIT alone.[7]

The Derrida Room began as an attempt to do something now taken for granted but unprecedented for the Invisibles at the time: serious critical discourse in the marginalized, hacker/hobbyist technology of the BBS. The very idea was comedic.

After sparking a collective critical reading of *On Grammatology* and compiling exhaustive lists of writing surfaces and writing implements, one fine day the cathode rays slowly spilled out into the Derrida room:

85. feb08 from Desperado # One
Alright, nobody move!

85. feb08 from SOUND EFFECTS
(a crowd gasping. hub-bub. screams.)

85. feb08 from Desperado # 2
We mean it! We've got guns! Nobody move! ...

85. feb08 from Bystander # One
What's going on?

85. feb08 from Bystander # 2
A bunch of desperados are hijacking Invisible Seattle's BBS. They're holding the Derrida room hostage. ...

85. feb08 from Bystander # One
... What do they want?

85. feb08 from Bystander # 2
I don't know. Let's turn on the news and see.

85. feb08 from SOUND EFFECTS
(click)...

85. feb08 from [TV news anchor] Dan Rather
are holding a room on Invisible Seattle's IN.S.OMNIA BBS hostage, in an apparent attempt to hijack the board. They have yet to give any reason for their actions.

85. feb08 from Hapless User
Oh please, please won't you let my husband log out? He has a heart condition, I'm very worried for him, have MERCY!!

85. feb08 from Desperado # 3
Shut up![8]

This literal and figurative hijacking of the serious discourse, the co-existence and non-mutual-negation of these radically different modes of writing, was an example of an aesthetic the Invisibles admired and to which the uncharted neutrality of the CRT leant itself: the ability to code-switch, genre-switch, and, as we put it, "go anywhere from anywhere." James Joyce and Monty Python were our touchstones for this ability. This foreshadowed the remix culture of the 1990s and later.

While in Paris in 1985, Philip Wohlstetter saw an exhibition at the Pompidou Center, "Les immatériaux," curated by philosopher Jean-François Lyotard, which included a collaborative writing project on a BBS-like system, in which Jacques Derrida was a participant (CCI).[9] I wrote to Derrida, referencing our common experience of writing apart from ink and paper and our Derrida room, and we began a correspondence and series of conversations that resulted in the Fulbright project, which he and Lyotard kindly sponsored during which I drafted *Invisible Rendezvous*. I remember sitting with Derrida in a tea house in the Marais talking about the strange materiality of glowing, ephemeral words on a black screen. He, more than anyone else outside our group we had ever talked with, was aware of the profound dislocation of text that was taking place.

Real-Time Interdacts

With American Oulipean Harry Mathews and others, the Invisibles conducted live co-writing events we dubbed "interdacts" using early, direct connection software. Writers at the two machines shared a single writing space, shown identically at both ends of the connection. The screen accepted ASCII entries in the order received, often resulting in interleaved words. Harry and I agreed to the ceremonial opening word of interdacts being the word "Hoik!", an inadvertent concatenation of "Hi!" from one end and "OK" from the other. As with the BBS material, the text could only be saved by printing it immediately.

Collaboration and Crowdsourced Creativity

That Night in Zyzzyzzywa was a narrative room that developed out of a word game:

"85mar09 from Brent Olson
Define 'zyzziva.' Have fun.

85. mar09 from M. Valentine Smith
(E)Ava Gabor, dropped in a vat of 7-Up

85. mar09 from Brent Olsen (sic)
zyzzyva: noun, a tropical weevil. Six-legged variety.

85. mar10 from Sikuan
Zyzziva: In ancient Assyrian citadel, the antechamber where the Zizzi kept a close watch on the palace treasure."
(Anonymous, Sikuan [Jean-Paul Dumont], in *IN.S.OMNIA*)

and suddenly became a set of mutually contradictory accounts of a single place, beginning with:

"85mar13 from Multatuli
'Zyzzyzzywa,' the city of light sleepers. I've been there never ever having suspected that it was founded by Sikuan. ... So it was your equestrian statue I spied on the main plaza that late afternoon—only hours before the massacre." [10]

In our weekly meetings at a home or at the coffee shop of Elliot Bay Books, we had identified our dissatisfaction with the "add on" structure of literary collaboration. The writers of the first chapters have more leeway (and more fun). The writers who come later are increasingly hemmed in. Zyzzyzzywa taught us it is better for collaborators to disagree about events, thus allowing each full range to play.

This idea is defined nicely by Matt Besser, Ian Roberts, and Matt Walsh in *The Upright Citizen's Brigade Comedy Improvisation Manual*, where they make a distinction between game and plot. They write, "the Game is what is funny about your scene. It is a consistent pattern of behavior that breaks from the expected patterns of our everyday lives." They state, "The Who, What, and Where are the elements that make up the plot of the scene. The plot is not what is funny about the scene. ... What is funny is the game, a pattern that is played within the context of that Who, What and Where."[11] The game of Zyzzyzzzywa, in this sense, was, "We are all disagreeing with each other about what happened that night." This emphasis on game alongside story is a key characteristic of contemporary netprov projects.

Early Crowdsourcing: *Invisible Seattle: The Novel of Seattle, by Seattle*

The 1970s Seattle of the Invisibles was hopelessly remote and un-hip. Microsoft was a speculative venture only in its infancy; Starbucks (which sponsored the first year of Invisible Seattle performances) had only one store. Literary publishing happened in New York City. Period. Writers from the hinterlands mailed our manuscripts at the post office and waited patiently for our rejection slips. Seattle's identity was rugged and industrial: lumberjacks and jet planes. Seattle was so out of the way that ambitious young writers and artists like us felt both invited and compelled to invent an artistic scene from scratch. We had nothing to lose.

"Since 1979, Invisible Seattle has conspired 'to take over the city by hypnotic suggestion,'" begins an explanatory note found at the back of the bright orange, magazine-format printed sole issue of *IN.S.OMNIA, Invisible Seattle's Omnia*.[12]

Each of our acts has been an attempt to inject an element of "fiction" into the so-called "reality" of an unsuspecting average American metropolis. ... If the slogan of the modern movement is "Form Follows Function," Invisible Seattle's is "Function Follows Fiction."[13]

Wohlstetter's twin background in literature and political street theater led him to organize and inspire a series of Invisible Seattle events that took place in and around Seattle's annual Bumbershoot arts festival from 1979 through 1983. Featuring improv actors from the troupe Off The Wall Players, the events got increasingly ambitious, culminating with the construction of the participatory, crowdsourced novel *Invisible Seattle: The Novel of Seattle, by Seattle*,[14] which aimed to co-create the great novel that Seattle lacked. The novel project was the Invisibles' bridge to the digital world.

For a month, "Literary Workers" wearing hard hats and white overalls prowled the streets reading the prologue aloud and gathering texts from passersby ("Excuse me, we're constructing a novel, can we borrow some of your words?") and these texts, a million utterances searching for their authentic environment, were compiled into a novel on a gigantic literary computer called "Scheherezade II." This monolithic cyberno-ziggurat, designed by Clair Colquitt, kept hundreds of people actively engaged in the task of contaminating the real world with fiction during the four days of the Bumbershoot Arts Festival, reaching a total audience of 20,000 visitors.[15]

It was Colquitt who, after he had established a BBS to exchange information with hacker friends, offered it to Invisible Seattle for use overnight, and IN.S.OMNIA—conceived, in the words of Joyce in *Finnegans Wake* for "the ideal reader suffering from an ideal insomnia"—was born.[16]

The egalitarian, crowdsourcing spirit of the novel project found its way into the IN.S.OMNIA experiments as well. The prefatory "Notes for a New Medium (or) Is There Life After Literature?", keyboarded by Wohlstetter, talk about:

A true "Republic of Letters." Anyone can play, skilled or unskilled, pro or am. ... A sense that there is always someone else about to arrive and astonish us. You.[17]

The Invisibles were among the first to discover (and enjoy) the principle that was to surprise the first few generations of networked computing engineer-entrepreneurs (Minitel, AOL, etc.): people didn't want to be connected with information above all, they wanted to be connected, and play, with other people.

The netprov projects Mark C. Marino and I are currently concocting build directly on IN.S.OMNIA's people connections, this view of everyday creativity as a social activity from informal story swapping to parlor and role-playing games. As we migrate netprov projects to ever-newer social media platforms, we find ourselves continuing the same investigations IN.S.OMNIA started.

Center for Twitzease Control, a netprov, invited players to an awareness of the materiality of language—in the spirit of the abbreviations and acronyms rampant in Twitter—as they experimented with comic, Oulipean formulae for mangling language. "Twitzeases" ranged from the literal Dwidder Flu outbreak "Doze infegted wid the Dwidder Flu wride as tho they had a bonstrous head code sniff koff" (Marino, Wittig, et al.) to a dangerous vowel shortage in mnsvwls (minus vowels), to the mashup malady:

Symptoms: Two Tweets by two utterly unrelated Tweeters, often celebrities, are commingled or "mashed up" together to form a single, mutant Tweet.[18]

Marino led the exploration of psuedonymity and identity in the netprov *Temp-Spence*,[19] in which he used (with permission) reality TV star Spencer Pratt's Twitter account to enact a fictional, struggling British poet who had found (or perhaps stolen) Spencer's phone. First to the consternation, then to the delight, of Spencer's many fans, Marino's poet character initiated collaborative poetry writing games.

In the more serious, more controversial *Occupy MLA*,[20] Marino and I enacted a trio of fictional adjunct professors in the Modern Language Association's Twitter feed (Marino and I are adjuncts ourselves) and imagined not only a pro-adjunct protest movement but the poignant and chaotic personal lives of the three protesters. A firestorm of criticism matched by a groundswell of support occurred when Marino revealed the fiction at the MLA convention after an 18-month run.

The future of netprov as large collaborative (paid featured players) and crowdsourced (players) transmedia fictions has been successfully tested in projects such as the sprawling workplace comedy *Grace, Wit & Charm*,[21] which combined 24/7 real-time Twitter narrative and a website with live theater performance. Some form of netprov is sure to thrive at the intersection of social media meme creativity, roleplaying games, and traditional show business.

Notes

1. Rob Wittig, *The Fall of the Site of Marsha*, a novel in the form of three states of a vernacular home page, 1999. Available at http://robwit.net/?project=the-fall-of-the-site-of-marsha and http://robwit.net/MARSHA/.

2. Rob Wittig, *Blue Company*, a novel in email, 2002. Available at http://robwit.net/?project=blue-company/.

3. Anonymous, in *IN.S.OMNIA Invisible Seattle's Omnia*, 1985.

4. Ibid.

5. Rob Wittig and IN.S.OMNIA (Computer bulletin board), *Invisible Rendezvous: Connection and Collaboration in the New Landscape of Electronic Writing* (Middletown, CT; Hanover, NH: Wesleyan University Press; University Press of New England, 1994).

6. Multatuli [Philip Wohlstetter], in *IN.S.OMNIA*.

7. Ibid.

8. Paul Cabarga et al., *IN.S.OMNIA, Invisible Seattle's Omnia [IN.S.OMNIA]*. Number 1 (Seattle: Invisible Seattle Projects in association with Function Industries Press, 1985).

9. Centre de Création Industrielle (CCI), *Epreuves d'écriture* (Paris: Editions du Centre Georges Pompidou, 1985) (electronic collaborative writing).

10. Multatuli [Philip Wohlstetter], in *IN.S.OMNIA*, 1985.

11. Matt Besser, Ian Roberts, and Matt Walsh, *The Upright Citizens Brigade Comedy Improvisation Manual* (New York: Comedy Council of Nicea, 2013).

12. *IN.S.OMNIA*, 1985.

13. Philip Wohlstetter, et al. in *IN.S.OMNIA*.

14. *Invisible Seattle: The Novel of Seattle, by Seattle* (Seattle: Function Industries Press, 1987).

15. Rob Wittig et al., *IN.S.OMNIA*, 1985.

16. James Joyce, *Finnegans Wake* (London: Penguin, 1959), 120.

17. Philip Wohlstetter, in *IN.S.OMNIA*.

18. Mark C. Marino and Rob Wittig, *Center for Twitzease Control, a netprov*, 2013. Available at http://robwit.net/center/.

19. Mark C. Marino and Rob Wittig, *Tempspence, Reality: being @spencerpratt, a netprov*, 2013. Available at http://writerresponsetheory.org/wordpress/2013/01/26/reality-being-spencerpratt-a-netprov/.

20. Mark C. Marino and Rob Wittig, *Occupy MLA*, 2011–2012. Available at http://markcmarino.com/wordpress/?page_id=117/.

21. Rob Wittig, *Grace, Wit & Charm, a netprov*, 2012. Available at http://robwit.net/?project=grace-wit-charm/.

12 Art Com Electronic Network: A Conversation with Fred Truck and Anna Couey

Judy Malloy

Initiated by Carl Loeffler, the director of the San Francisco alternative artspace La Mamelle/Art Com, who had been working on artists' telecommunications projects since 1977, Art Com Electronic Network (ACEN) was implemented on The WELL by digital and information artist Fred Truck and began operation in the spring of 1986. Art Com's vision was to create a virtual environment for contemporary art that included the electronic publication of art journals, a conferencing system for art-centered discussion, and the interactive online publication of computer-mediated artworks and electronic literature.

How Carl Loeffler (1946–2001) and Fred Truck began a collaboration that resulted in Art Com Electronic Network; how ACEN brought artists online, incubated online conversation and early net art, published interactive art and electronic literature, hosted early "netprovs," and mounted a seminal traveling exhibition of artists' software, is detailed in this interview with Fred Truck—with the participation of Anna Couey, who was involved in the process at La Mamelle/Art Com and began editing the online version of *Art Com Magazine* in 1990.

The interview was created for *this* book and conducted by email.[1] The conversation with Fred Truck began in January 2014:

Date: Tue, 7 Jan 2014 15:27:52–0500
From: jmalloy@well.com
To: fredtruck@msn.com
Subject: Art Com Electronic Network: An Interview with Fred Truck

Anna Couey was invited to join us on February 5, 2014, and because her voice and vision were significant in shaping the system, she not only talked about her work bringing *Art Com Magazine* but also answered variations on the questions I asked Fred, such as, "What was important about ACEN in the field of artists' networking as a whole?"

Judy Malloy (JM)   Begin.

In 1984, artist Fred Truck, who lives in Des Moines, Iowa, was working on a computer-mediated performance art database, the *Electric Bank* (at that time *The Performance Bank*), when he went to San Francisco for the performance art festival *Inter Dada 84*. While he was in San Francisco, he stopped by the artspace La Mamelle/Art Com and demoed the *Electric Bank* to Founder and Director, Carl Loeffler.

Loeffler had been working on artists' telecommunications projects since the *Send/Receive Project* in 1977.[2] In 1980, La Mamelle/Art Com and the San Francisco Museum of Modern Art (MOMA) organized the *Artists' Use of Telecommunications* conference at MOMA. Inspired by Canadian telematics and with an online component carried by the pioneering Canadian telecommunications company I. P. Sharp (IPSA), *Artists' Use of Telecommunications* brought together international artists working in telecommunications under the direction of telematic artist Bill Bartlett. Participants were located in San Francisco, New York, Toronto, Vancouver, Vienna, and Tokyo.[3]

In 1983, La Mamelle/Art Com was a node for Roy Ascott's *La Plissure du Texte*, a collaborative fairytale told across multiple telecommunications nodes in the United States, Canada, Europe, and Australia.

Along the way, Carl had been thinking that online databases might be an important way to make performance art, video art, experimental literature, and information art accessible. He had already received a printout of the *Electric Bank* in the mail from Fred, and he was impressed.

Fred, to begin this interview, can you talk about the *Electric Bank*, what it was, how it worked, and what happened when you demoed it to Carl Loeffler in 1984?

Fred Truck (FJT):   The basis for the *Electric Bank* and the two earlier versions of this event, *The Des Moines Festival of the Avant-Garde Invites You to Show!* (1979) and *The Performance Bank* (1980), was an agreement between myself and the contributing artists: send me a performance proposal, and whether I perform it or not, I will write a review, publish documentation in a catalog, and send the catalog to you. Eventually, using the Osborne I computer, dBase II, and an additional software package that enabled graphics and animations, I migrated this information to *The Performance Bank*. The interface included a list of search terms and a list of artists' names; I worked hard on making the search experience enjoyable on the Osborne's very small (about 5.5-inch) screen.

This is the database I demoed for Carl at *Inter-Dada 84*. I actually showed it to him twice: once at a place I can't remember because there was a party going on, and then the next day at his office at 70 12th Street. I think Anna Couey was there, and Carl and Nancy Frank. Darlene Tong was at the first demo.

Art Com Electronic Network

Figure 12.1
Fred Truck, *The Performance Bank*. The printouts were created on a DEC mini-computer. This is the computer-generated catalog that Fred showed Carl Loeffler and the Art Com staff before ACEN began. Photo: Fred Truck.

Carl actually said very little. He and I had corresponded before and were somewhat aware of what each was doing, but this was different. He told me later I knocked his socks off, but I really got no indication of that at that time. I think we ran a few searches on selected artists and a couple on subjects such as Dada.

In December 1984, I bought a Fat Mac, a 512k Macintosh computer. With this machine, using a visual database that made it possible to draw simple pictures and attach information (accessible by clicking) to each part of the picture, I began migrating *The Performance Bank* to the *Electric Bank*.

I was working on this project when, after Christmas of 1984, I got a surprise call from Carl.

Essentially, Carl asked me whether I would program a database for him. He made no mention of what machine this database might run on or what software was to be used, but it quickly ran through my mind that Carl had an organization, contacts both here and abroad, none of which I had, and his success with La Mamelle made an impression on me. I told him I'd definitely be interested.

JM: So, sometime after Christmas in 1984, you were surprised to get a call from Carl Loeffler asking you to program a database for Art Com. And you decided to go for this, even though you had a few hesitations. Great start!

Meanwhile, I have retrieved an old Art Com *Media Distribution Catalog* and am reading Carl's paper, "Telecomputing and the Arts: A Case History."[4]

The vision for ACEN was electronic publishing, a conferencing system, an online exhibition space, and the fostering of online projects. The project, Carl emphasizes, began with solving two problems: distribution and storage. So in late 1984, the Art Com staff, along with Fred Truck, an artist and computer programmer residing in Des Moines, Iowa, began to resolve the problem of distribution and storage.

Can you tell me about the initial discussions, how the idea of an electronic network for contemporary art publication, networking, and exhibition began to take shape?

FJT: For about the first year—that is, 1985—Carl and I did a lot of socializing, both in real time and online. We had to find out who we were in relationship to each other. We explored a lot of things to find out whether we could work together. We discussed technology issues, of course, but it was always secondary to finding out where each of us fit in this collaboration.

Through this getting-to-know-you phase, I became aware of Carl's vision for ACEN. He wanted an electronic version of the La Mamelle/Art Com space and all the activities it engaged in. Art Com published *Art Com Magazine* in paper for a number of years before I came on the scene. I think the notion of publication in electronic form was present in ACEN practically from the beginning because of Carl's desire to see *Art Com Magazine* continue.

The large-scale vision for ACEN came, for the most part, from Carl. I materialized his vision through programming and tweaks here and there. Also involved in the process at La Mamelle/Art Com were Anna Couey, Nancy Frank, and Darlene Tong, and I worked with the support of my wife, Lorna, who is a librarian. We all had concerns about interface, making it easy and friendly to use, but the wording used was one of my contributions. I had used almost exactly the same wording in the interfaces I designed for the *Electric Bank* database. I also designed the iconic logo we used to head each section of the magazine:

/////////////////////////////

ART COM ELECTRONIC

/////// NETWORK ///////

JM: The collaboration between you and Carl and the time you took to create an interface for ACEN that was both user-friendly and elegant are important. There were artists' telecom projects and early BBS systems, but ACEN was seminal in creating a comprehensive pre-web online environment that encompassed the publication of online journals, the interactive distribution of computer-mediated artworks and electronic literature, and a conferencing system for discussion.

Where ACEN would be hosted was an important decision.

In "Telecomputing and the Arts: A Case History," Carl wrote, "If ACEN was to succeed, it needed wide distribution that would be inexpensive for our projected national and international user base." Pointing out the importance of interfacing distribution and storage, he also emphasized, "both of these concerns had to merge in a way that could be accessible to artists and artists' organizations. No small problem."[5]

The goal was to find a service that was not as large as CompuServe but had the potential to survive. So how did you and Carl choose The WELL, which was founded in the same time period, 1985, to host Art Com Electronic Network?

FJT: In late 1985 or early 1986, Carl asked me, as a computer consultant, to check out medium-sized computers and database software, make some recommendations on those, and find out how much such an outfit might cost. In retrospect, I am sure Carl was thinking about being as independent from outside sources for computing and telecom as possible. At the time, my understanding was straightforward. Art Com wanted to buy a powerful computer for the database I was going to program. Computers like my Macintosh were not powerful enough to handle multiple users, so the medium-sized computer made sense.

The mini-computer was the medium-sized computer of choice. The best of the mini-computers available in those days was the VAX, manufactured by the Digital Equipment Corporation. A VAX might cost $120,000 to $160,000, and the software about as much. Additionally, the VAX really needed a full-time programmer to run it.

In the early spring of 1986, I went to San Francisco armed with my report, which as I recall was somewhere between 20 and 40 pages long. I also had a couple of software catalogs for the VAX from different suppliers.

I gave the report to Carl, and he said he would look into it. A year or two later, Carl told me that he had indeed looked into it and just about died when he found out how much money we were talking about. That was the end of the idea of owning our own mini.

After trying other services, we found The WELL by an interesting, almost accidental combination of events. Nancy Frank, who seems to know everyone in San Francisco, had a friend who was running the Legal Conference on The WELL. She happened to tell Nancy about The WELL and mentioned they were looking for nonprofits or artists to manage a conference. Nancy told Carl, and the rest is history, as they say.

As I recall, Carl told me he had an interview with someone who was administering The WELL. It was a brief interview. He agreed to maintain our space and our files. We could use The WELL's conferencing system, but we had to do our own programming.

Nancy Frank remembers this a little differently. In correspondence among her, Darlene Tong, and myself, Nancy writes, "I was the one that told Carl that I was tired of selling print ads and I made the appointment with Stewart Brand who just launched the well and he shared his server with us and our ideas. Carl was blown away that I made the call."[6]

The WELL's conferencing software ran on a Berkeley UNIX operating system. At the time, I knew nothing about UNIX, but through the guidance of Matthew McClure and John Coate, WELL administrators, I was able to do enough UNIX to set up ACEN. Other people who helped me from time to time were Cliff Figallo and David Hawkins. After learning the basics, it was a matter of uploading the files to directories I set up, writing the menus, and installing anything else we needed to make it work smoothly.

JM: Great details about the process!

To augment the story, I'm going to fill in with a few details from "Keeping the Art Faith, Interview with Carl Loeffler," on Art Com Electronic Network, on The WELL, beginning on January 11, 1988.[7] I was the interviewer. Participants, in addition to Carl and you (logging on the first day from Iowa, wearing, you tell us, "a nice purple wool sweater"), included, among others, John Coate, Abbe Don, Janey Fritsche, David Gans, Freddy Hahne, Raul Gilbert MinaMora, and Howard Rheingold.

In the course of the interview, Carl notes that Nancy Frank's friend who suggested The WELL was Donna Hall, and he emphasizes the importance of The WELL's philosophy in the decision to go with The WELL.

"Why The WELL?" Carl observes. "The main reason was that we could agree with the philosophy of what was and is happening on the Net."

As you have done here, he also thanks WELL staff Matthew McClure, David Hawkins, John Coate, and Cliff Figallo for help in working on the platform, noting that "The WELL helped us out beyond belief."

Summarizing the major influences, including Canadian telematic art, Carl says in this interview:

"So ACEN is here because of Canada, Donna Hall, Truck, and The WELL, and we just love the opportunity. ... Thank you all!"[8]

Thus, after several years of planning and research, ACEN went live on The WELL in April 1986. You set up the initial menus (that you programmed with UNIX shell scripts). Designed to offer art publications and artworks to anyone who logged on to The WELL, these menus were a distinguishing feature of the project.

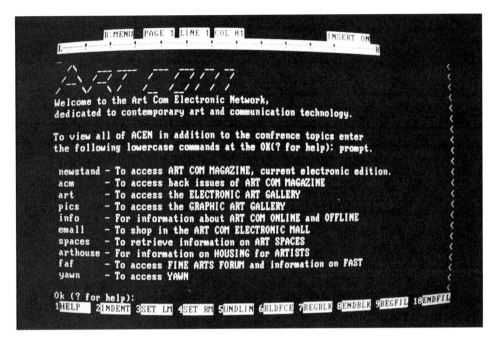

Figure 12.2
Art Com Electronic Network Start Menu by Fred Truck, 1990 version. Photo: La Mamelle, Inc.

Carl explains that "We launched by carrying *Art Com* and *Metier Magazine,* plus a bulletin board that offered feedback topics and an art project."[9]

April 1986 was also when I came on, and the art project was the collaborative *Bad Information,*[10] which I followed with my narrative database *Uncle Roger.*[11]

Let's talk about the conferencing (BBS) aspect of ACEN. The primarily online communication platform for ACEN and for all WELL users was the conferencing system. The software was PicoSpan, created by Marcus Watts.

Can you talk about your experiences with the conferencing system? What was it like using it?

FJT: PicoSpan was very easy to use. For the first online issue of *Art Com Magazine,* I wrote "Operating Instructions," an introduction to using The WELL. This text had a lot of information about accessing The WELL from outside San Francisco and so on; a lot of the text was quoted from *THE NEW USER'S GUIDE TO THE WELL* by Mick Winter.

Although the information in the magazine was accessible to anyone using The WELL at the command line Ok: prompt, we restated it in order to ease the way for artists, many of whom were computer novices. We thought it would be easier for beginners to access from a menu than the prompt.

Essentially, the instructions to using the conferencing system informed the user how to see the topics, how to access an individual topic, and how to respond to it. The list of commands also revealed how to start a topic in the ACEN BBS system and how to name the topic.

Of all the conferencing systems I had seen to that date, PicoSpan was the easiest to use and the most transparent. It did take a little confidence to face the blank Ok: prompt, but almost everyone who tried mastered it.

JM: Yes, I think that PicoSpan's ease of use (for a text-based conferencing system) was one of the many reasons for The WELL's success.

The participation of artists and art projects was another reason for ACEN's success. For instance, ACEN was a node for Planetary Network, a telecommunications project organized for the 1986 Venice Biennale by Roy Ascott, Don Foresta, Tomaso Trini, Maria Grazia Mattei, and Robert Adrian X.

Were you involved with *Planetary Network*? Do you remember the details?

Can you talk about early ACEN participants?

And how did the ACEN publication of John Cage's *First Meeting of the Satie Society* evolve?

FJT: Judy, you ask: "Were you involved with *Planetary Network*? Do you remember the details?"

Yes, I participated in *Planetary Network*. It used ARTEX (IPSA's Artists' Electronic Exchange System). Here is what we published in the second online issue of *Art Com Magazine*:

"PLANETARY NETWORK Venice Biennale, Venice, Italy

June 25, 26, 27, 29, July 1–30, September 1–29, 1986

As part of the Venice Biennale which takes place June—September, 1986, a 'Planetary Network' involving artists from twenty one locations around the world is established. 'Planetary Network,' one of six sections of the 'Ubiqua' laboratory/workshop of the Biennale, incorporates the use of various media: IP Sharp/ARTEX electronic mail network, Telefacsimile, slow scan TV—all of which are linked by phone. The theme of the activity is 'The News,' which can include montage, fantasy, critical analysis, distortion, correction, humor, etc. The Biennale is held under the direction of Maurizio Calvesi. Commissioners include Roy Ascott, Don Foresta, Tom Sherman, and Tomaso Trini. (c/o Roy Ascott, 64 Upper Cheltenham Place, Montpelier, Bristol, England, BS6 5HR or contact through ARTEX: ASCOT)."[12]

I happened to find a big printout from the *Planetary Network* project. A lot of it concerned one artist's hostile attack on everything connected with America and my response to it. Carl and Anna edited my response so we wouldn't spend all our time talking about politics. We wanted to focus on telecommunications.

You ask: "Can you talk about early ACEN participants?"

Four people come to mind immediately: Abbe Don, Joe Rosen, Robert Edgar, and Freddy Hahne, that is, besides yourself, Anna Couey, and Carl Loeffler.

All four of the people I listed had remarkably different talents, capabilities, and, as it turned out, futures. Freddy Hahne was and is a personality, the primeval jester. In his identity as <Are We Really?>, he was presence personified. He works in the printing industry. Abbe Don was a student at NYU in the Tisch School of the Arts. She was studying Interactive Telecommunications. Very intelligent as well as personable, she connected Carl and I to the Telecom program at NYU and helped make our appearances there possible. She is now Vice President, User Experience, ePocrates division at athenahealth. Joe Rosen was also an NYU student. He is an extremely talented programmer and, from what I saw, a builder of digital accessories, such as a virtual reality glove that worked very well. His current project is the GPS-A-Sketch App. On a long walk, this app turns your movement into a sketch that is reminiscent of an Etch-A-Sketch drawing. Joe has a wacky sense of humor based on pop culture. He is at Stanford University.

I became acquainted with Robert Edgar by writing an article about his *Memory Theatre One* application. He is also a programmer, video artist, musician, and Renaissance Man. Robert is difficult to describe because he engages in so many different art forms and enterprises. He is currently Senior Instructional Designer at Stanford University.

Of course, there were many other users of ACEN. A particularly interesting small group worked at SRI International and included Tom Mandel and Maria Syndicus. The WELL staff also frequented our space, along with David Gans and quite a few Grateful Dead followers.

You ask: "And how did the ACEN publication of John Cage's *First Meeting of the Satie Society* evolve?"

One night, I was working on the Macintosh and the phone rang. Lorna answered and said, "It's for you. It's John Cage."

And it was. Getting a phone call from John Cage right out of the blue wasn't as odd as it may seem. I knew a lot of Fluxus people in New York and environs. The Fluxus people knew I was working a lot with computers, so I'm pretty sure he got my phone number and a recommendation from them.

We had a nice chat. Essentially, Cage wanted to put *The First Meeting of the Satie Society* online. I thought it would be great and ran it by Carl and Anna.

I think Cage sent me the disks. I then had to format the text of his poems, which follows a very idiosyncratic pattern. For this particular work, Cage wanted to create a vast notational score in homage to composer Eric Satie that could either be read or musically performed. Textually, the piece is a mesostic, a form of writing that can be poetically read horizontally and vertically.

To create the notation, Cage had two custom computer programs written: IC, a program that selects excerpts from predetermined text sources, in a manner similar to the chance decision-making process of the I Ching; and MESOMAKE or MESOLIST, a program that establishes a mesostic from the select excerpts. Cage also considers the notation as an illuminated manuscript, produced by the glow of computer text on the screen. He hopes that telecomputing will become "a happy habit."[13] The two programs were written by Andrew Culver and Jim Rosenberg, who had also previously worked with Cage. I wrote the UNIX scripts that The WELL used, as well as some formatting code on the Mac to run SATIE through.

JM: Thank you for bringing back this important era so vividly!

To begin with art telecommunication politics, my recollection is that data from the *Bad Information* project was included in the News section of *Planetary Network* at the Venice Biennale, which Roy Ascott was working on, and from that (my) perspective, it was an

```
/////////////////////////////////
A R T   C O M   E L E C T R O N I C
///////// N E T W O R K /////////

Welcome to THE FIRST MEETING OF THE SATIE SOCIETY
by John Cage. Please enter your choice from the
selections below:

1. Information (3K)      10. Musikus (12K)
2. Introduction (10K)    11. Relakus (4K)
3. Cinekus (4K)          12. Sonnekus (7K)
4. Mesdamku (18K)        13. V_I_V (30K), Pt. 1
5. Messe (26K), Pt. 1    14. V_I_V (34K), Pt. 2
6. Messe (21K), Pt. 2    15. V_I_V (30K), Pt. 3
7. Messe (30K), Pt. 3    16. V_I_V (24K), Pt. 4
8. Messe (27K), Pt. 4    17. V_I_V (21K), Pt. 5
9. Messe (21K), Pt. 5    18. Satie Replies (4K)
c  19. Exit to the Datanet Artworks Main Menu

Enter your selection, M for previous menu
or control-d to exit.
```

Figure 12.3
John Cage, "First Meeting of the Satie Society," Art Com Electronic Network, 1986.

exciting project. But we were coming from different approaches. The Canadian ARTEX system, carried by I. P. Sharp, was a pioneering system of the art and telecommunications era. ACEN represented a different pioneering approach to arts networking: the influx of personal computers into artist-mediated telecommunications. It was no longer necessary to be at a designated node to participate in worldwide telecommunications projects. It was a watershed time.

To the artists whom you mention as core participants in the early days of Art Com Electronic Network, I would add Jim Rosenberg, whose spatial hypertext systems were of great interest to me and have also been influential in the electronic literature community. Some of the other artists, writers, and musicians who participated in conversations and collaborative projects on ACEN were writer and social media journalist/ and currently Stanford professor Howard Rheingold, who was instrumental in bringing different WELL communities together on his "The Mind Conference"; David Cannon Dashiell, later the painter of *Queer Mysteries*; John Coate, who later was important in bringing the *San Francisco Chronicle* online as sfgate.com; musician and writer David Gans (whom you also mention), host of *The Grateful Dead Hour*; and Dan Levy, Neal Margolis, and Raul Gilbert MinaMora. Pioneering interactive artist Sonya Rapoport wasn't involved in the online conversations, but her work on ACEN Datanet and for the traveling exhibition, *Art Com Software*, was important.

Thanks for this wonderful story of how John Cage called you about putting *The First Meeting of the Satie Society* online and your memorable account of formatting this work for publication on the ACEN system. The Art Com celebration party for the publication of *The First Meeting of the Satie Society* was a big and exuberant party. I remember talking with Cage about the reception of computer-mediated work—in that era, it was not prevalent—and he told me about some of the hostile reaction to his *4'33"*, and that there were people who didn't talk to him for years after the 1952 performance of *4'33"*.

Fred, I'd like to know more about the ACEN experience for you as an artist, as well as about the issue of software as art. Can you talk a little about how your own work progressed during the early years of ACEN?

Also, I recall the *Software as Art* topic as important in bringing artists' authoring systems to the forefront. We discussed both the software we used to create our works and how software itself can be art. Can you talk about the *Software as Art* topic?

FJT: To my knowledge, the *Software as Art* topic on the conferencing system is lost. I remember discussions with Bob Edgar, Joe Rosen, Anna Couey, Abbe Don, and yourself.

For my part, in 1986, I began *ArtEngine*, a piece of software for the Macintosh computer.

In *ArtEngine*, it is not the physical code that is art. The process the code sets in motion as it is interpreted or compiled is art. *ArtEngine* works on the premise that visual objects and words that describe them are the same. Write the words and a picture will appear. Given two visual objects, and words describing them, the Engine will compare the objects, determine what congruencies there are between them, and then make a new list that creates a new, third visual object. *ArtEngine* encapsulates my perceptions about art.

JM: Your discussions of your *ArtEngine* were core to the ACEN artists' workshop environment; your words about the *ArtEngine* in this interview serve as an example of how innovative projects initiated discussion on ACEN. Discussions about the authoring systems that you, I, Robert Edgar, Jim Rosenberg, Abbe Don, Joe Rosen, and others were developing permeated many other topics. Indeed, most of the discourse about my creation of the authoring system for *Uncle Roger* was actually in the "Feedback re: Uncle Roger" Topic that Howard Rheingold started when I first began to upload *Uncle Roger* in 1986.

I am also reminded that your talking about the *ArtEngine* online was a core initial component of ACEN as an online place for collaborative discussion of the making of computer-mediated art. Sharing/learning in online discussions was one of the most important aspects of ACEN. There was the excitement of discovering each other's work; there was the open sharing of our software and process.

To continue, I'd like to ask you as founding Sysop about your impressions of Art Com Electronic Network as a whole. What was important to you as a Sysop? What was important in the field of artists' networking as a whole? What has been missing from your life as an artist since the days of ACEN?

FJT: You ask: "What was important to you as a Sysop?"

I wanted to do programs and routines for the user that were easy to operate and not intimidating. Making ACEN as transparent as possible would encourage participation in my view.

I wanted all the scripts I wrote for ACEN to work. Crashes and dead ends are very intimidating for those new to computer technology, and in those early days, almost everyone was new to it.

I wanted to stimulate conversation and participation: That meant logging in frequently and responding to as much as I could. I tried to help users who were having problems logging on or getting the conferencing software to perform for them the way it should.

It was also important to me to be sociable. I learned to engage more with people, to find out who they were and what they were doing. This, in turn, attracted more users by way of positive comments to others by those I tried to help.

In other words, the items listed above were important to me as Sysop, but by learning to be important to ACEN by engaging others ensured, at least on my part, that the project would be successful. Of course, Anna and Carl carried their responsibilities well, and we were successful.

Judy, you ask: "What was important in the field of artists' networking as a whole?"

The element that ranked first in importance to me was to get artists to see that no less in the past than today, the world of art is dominated by a dozen or so major centers and to see why using microcomputers would benefit them.

We saw ACEN as a network that could link artists all over the world. We could really start the global village. Then we could do art, converse with friends, and even show our art in electronic space in a way that catapulted us beyond geographic limitations of having to live in New York or London, or the gallery and museum structures.

We met a lot of resistance. A lot of artists really feared computers and technology in general. Gradually, through articles we wrote, demos we did, and contacts we made with other artists, we made inroads, and artists and administrators saw you could actually have fun with these little machines and get good work done, too.

Judy, you ask: "What is missing from your life as an artist since the days of ACEN?"

Well, first off, I have missed the sense of being in on the ground floor of something exciting, innovative, with the potential to alter culture as part of a group. I've had that sense as an individual artist since then, several times, but there is something contagious about being in a group that is more easily communicated to others. It is much harder to do as an individual.

ACEN was inhabited by people from all different fields, not just art. One thing we all had in common was an interest in digital technology. Digging a little deeper, another thing we had in common was ACEN, a place we could do what we wanted in any way we wanted. It is a little idealistic to compare ACEN to Camelot, but I can tell you, it was a realized dream that will never happen again in my life. We took that moment, went with it, and never looked back.

JM: Winter weather today in Iowa and New Jersey! A good day for artists' networking.

Fred, you say: "One thing we all had in common was an interest in digital technology. Digging a little deeper, another thing we had in common was ACEN, a place we could do what we wanted in any way we wanted."

Yes! This is important in online community (as in many other communities). We look at the experience differently, we bring different skills to the community, we see our roles and the whole differently, but we are all together on the same platform.

And today is the day to introduce Anna Couey!

Hi, Anna. Welcome to this conversation about Art Com Electronic Network!

Can you introduce yourself and tell us how *you* became a core part of the ACEN team?

And because your voice was important in shaping the big picture, I'd also like to bring you in with variations on the questions I asked Fred.

What was important to you about Art Com Electronic Network?

What was important about ACEN in the field of artists' networking as a whole?

Anna Couey (AC): I got involved with the Art Com Electronic Network by virtue of being a staff member at La Mamelle/Art Com. I was just a few years out of college, and I'd never used a computer. I remember Carl telling me I was going to learn. I had no idea what I was getting into.

In school, I had been making fiber art and studying dada. Dada was my conceptual entry point for being interested in technology. I was hired to be Art Com's business manager and handled Art Com's book distribution project, Contemporary Arts Press Distribution. I also participated in the organization's publishing projects—proofreading *Correspondence Art: Source Book for the Network of International Postal Art Activity*, helping to produce *Art Com Magazine*—at that time in print—and writing articles for it.

Initially, I coordinated content production for ACEN, primarily *Art Com Magazine* and some of the electronic art works. I wasn't a regular participant in the conference. Over time, I became involved in shaping the development of ACEN, using it as an art-making space, and becoming part of the community. But I was hooked early on. I was convinced that we were making history, and it was a heady time.

Bad Information is one of my favorite examples for describing what was important about ACEN. Judy, your open invitation to contribute bad information for your humorous database project engaged artists and people who didn't consider themselves artists in making art. That collaborative and boundary-crossing approach to making art distinguished ACEN and has informed all of my work since. It was because of ACEN that I began to think about the politics of art making—who was an artist and who wasn't. Who got to speak and who was to listen. Those ideas have carried me from producing art that connected diverse cultural perspectives to participating in social justice initiatives that bring marginalized voices and ways of knowing into our collective understanding of reality. The common thread has been using communication to reshape how society functions.

JM: Ah, there are many threads here. But in this early online community, in any online community, the coming together of different visions was/is important. At times on ACEN, this differing was expressed in argument, but in my experience, the argument was usually diffused by a sense of this amazing online environment where we were participating—the possibilities, the idea that we were in different places but communicating on the same platform.

And here we are again in this forum, reliving these times! For myself, since the *Artwords and Bookwords* exhibition that Judith Hoffberg created at LAICA in 1978, I had been exchanging mail art with the artists book community. Coming from this background, my vision for collaborative works, such as *Bad Information*, was the idea of using the online platform for collaborative writing, with colleagues in an online cultural community, such as artist David Cannon Dashiell (*Queer Mysteries*) and poet Jim Rosenberg. On ACEN and on the larger WELL community, there were visual artists; there were writers; there were computer professionals; there were techno-creators, musicians, journalists, and people from all walks of the creative community who participated partially because the concept was important: we take information from the computer as true, but there is no guarantee that that is the case.

Contingently, I also saw ACEN and The WELL as a public square to tell my own stories, as I did in *Uncle Roger*. So for my vision, ACEN was an extraordinary place for creating electronic literature, both individual (*Uncle Roger*) and collaborative (*Bad Information*).

Carl's vision, as a whole, was also important in the (then mostly offline) Bay area art world, so it wasn't as if his vision arose with ACEN. The login <artcomtv> represented Carl's 10-year involvement with the video art archive, which started in 1977; it represented La Mamelle's PRODUCED FOR TELEVISION, which was artists video on real TV—Chris Burden, Lynn Hershman, and others—as well as the 1981 *Performing/Performance* series. However, Fred, the artists initially involved and then Anna, added different visions to the mix, and this was also very important.

Fred brought with him online his prior work with the *Electric Bank*, his work with the design and implementation of the central menu, which published interactive artworks online, and the sharing of the development of *ArtEngine*, and he was the one who learned to use UNIX shell scripts and showed many of us how to do this. The fact that, thanks to Carl and Fred, we actually had a menu where artworks could be interactively published—and this was in 1986—is amazing.

And then, Anna, you had been working at Art Com, but you weren't actually "visible" on the system until you began posting online. When you became visible (when was this?), you brought to ACEN an idealistic vision that encompassed community and diversity and looked to wider social networking. Then as editor of *Art Com Magazine* online, one of the first online arts publications, you asked artists to guest-create issues. Can you talk about *Art Com Magazine* as an early online arts publication? What were the issues of *Art Com Magazine* like? How were they distributed?

AC: Judy, your perspective on creative community made me reconsider. Some of the people on ACEN were active in the art world and situated their work in that context. Others did not, or did not at that time. I am reminded of the shifts that have taken place—not only in what constitutes art but how we define artist. The rigid definitions we pushed against have given way to more fluid understandings, hybrid identities, and functions.

To continue the <artcomtv> story—Art Com stopped presenting performance and video art in 1983, and Carl began to focus on licensing artists' video to broadcast television. This initiative was Art Com Television. As I understood Carl's vision, television art was a way for artists to shape the cultural mainstream and make money. When ACEN was getting started, Art Com Television was a cutting-edge approach to artists' use of video. The <artcomtv> login refers to this project.

I got my own WELL account in 1989, and that's probably when I began actively posting. My vision for ACEN, and for online art, focused on concepts of interactivity and collaborative artmaking and what those ideas meant in a social context. Although collaborative artmaking wasn't new, the dominant paradigm was that artists presented their vision to the public—a one-to-many communications model. Mass media was similar—newspapers, television, and radio all delivered master narratives to mass audiences. Online communication systems were much more conversational—many-to-many. Developing that participatory paradigm seemed essential to me. I believed artists had a tremendous opportunity and responsibility to shape online culture, and that what we did online would shape physical world cultures as well.

Art Com Magazine went online with the launch of ACEN in 1986 under Carl's editorship and was accessible through the menu system. These early editions reflected Carl's aspiration of building a digital archive of performance and video art events. We substantially expanded the number of event listings we had published in print.

Another goal of going digital was to reach a larger readership. ACEN could be accessed locally across the United States and internationally using long-range telecommunications carriers—in theory providing much greater distribution than we could reach in print as a small press experimental art periodical. But we found that ACEN was expensive for artists and arts organizations located outside the San Francisco Bay Area. Also, the culture of The WELL was oriented toward conferencing. I don't know how widely read *Art Com Magazine* was on ACEN.

Around 1989, we changed the format and began distributing *Art Com Magazine* for free by email. This method allowed us to deliver it to people on other systems. The email editions were very short by magazine standards—typically 2–3 pages of plain ASCII text—and focused on the intersection of contemporary art and new technology. Soon after we shifted to email distribution, we adopted a guest editor model, and I became the magazine's editor. In this form, each issue offered a unique view of software art, interactive fiction, robotics, cyborganic jewelry, low-tech art, intelligent art, mail art, pataphysics, and cyber-art activism. Some issues were artworks. Others were theoretical texts or manifestos. For example, Fred Truck's *ArtEngine* wrote an issue on "what it thinks Marcel Duchamp's *Boite en Valise* is, as determined by its analytic process, and the information contained in its memory."

Another issue featured *Thirty Minutes in the Late Afternoon*, the group narrative produced by Judy Malloy on ACEN in which three characters—Mary, John, and Rubber Duck—were developed collaboratively by 15 writers.

Stelarc, the Australian artist whose work explores the physiological structure of the human body, contributed a text that argued for redefining what is human: "It is no longer meaningful to see the body as a s i t e for the psyche or the social, but rather as a s t r u c t u r e to be monitored and modified."

The issue I guest edited was a call to online communities to actualize our vision of electronic space as a decentralized and collaboratively created environment, inspired by the Computers, Freedom, and Privacy conference formed in the wake of law enforcement crackdown on hackers.

Other guest editors were: the Normals; Tim Anderson, Simon Penny, and Wendy Plesniak; John Gaudreault; Vernon Reed; Kalynn Campbell; Abbe Don; Gil MinaMora; Benjamin Britton; Jeff Mann; Carl Eugene Loeffler; Dave Hughes; Coco Gordon; Chuck Welch; and Fred Truck.

JM: Anna, your observation about your vision for ACEN, which begins:

My own vision for ACEN, and for online art, focused on concepts of interactivity and collaborative artmaking, and what those ideas meant in a social context.

and concludes:

I believed artists had a tremendous opportunity and responsibility to shape online culture, and that what we did online would shape physical world cultures as well.

leads into the next question.

With your vision in mind, can you talk about what discussions (topics) and projects on ACEN were of particular interest to you, including those that you created yourself?

AC: In addition to *Bad Information*, Roy Ascott's *Planetary Network* was an important inspiration for me. *Planetary Network* was the first cross-system project on ACEN. It involved establishing a communications connection—and interactivity—between users of two discrete online systems, ACEN and ARTEX. On ACEN, we invited anyone who wanted to become an exhibiting artist in the Venice Biennale by contributing to the *Planetary Network* topic that we ported to ARTEX! The culture clash Fred described earlier revealed the more difficult aspects of interactivity relating to cultural politics and power. But I also remember being fascinated by the different kinds of stories that were contributed by users from each system.

Das Casino, an online betting game that Carl Loeffler and Fred Truck initiated as a topic on ACEN, became collaborative theater. Over time, the fantastical story that emerged became woven into the fabric of people's relationships and the shared history of The WELL community.

Status Report was a seminal topic in light of how social media has developed—Facebook built its whole system around the *Status Report* concept! Carl and Fred created the *Status Report* as a place for people to share personal updates. *Status Report* was highly successful in online community building—creating intimacy among people who didn't already know each other, people who didn't see each other every day, and even people who did. The *Status Report* concept spread to other WELL conferences and continued on ACEN for years.

Software as Art is in my mind the theoretical core of ACEN. We were geographically dispersed, and we were making art that didn't have an established aesthetic or theoretical context. In *Software as Art*, we were feeling our way together to develop a critical language and framework for what we were making. It was hugely important.

There were many other memorable topics and interactive works on ACEN. Judy's projects *Uncle Roger*, *30 Minutes in the Late Afternoon*, and *YOU!*, the collection of sentences you gathered from a group of us and compiled into a love story with the help of a program you wrote that analyzed whether our sentences reflected the beginning, middle, or end of a relationship. Gil MinaMora's text-based *Exquisite Corpse*. The Normals' *Anna Couey Museum of Descriptions of Art*, which exhibited viewers' descriptions of artworks.

My projects on ACEN attempted to use art to shape social systems. Each of them used The WELL's topic structure; most also involved alt.artcom, the Usenet group we created. I saw the projects as communication sculptures. They were about structure—who was communicating with whom, and conversation as action—bringing new social practices into being. The first online project I initiated, *Virtual Cultures*, was an art intervention into social space. It connected two online communities (ACEN and alt.artcom on Usenet) and Cyberthon, a 24-hour conference in San Francisco on the cultural and social impact of virtual reality. The project was an attempt to bring the participatory values of online culture to a physical world event. *Virtual Country* is a project I initiated to engage online participants in designing a cultural-social-political system in the new online territory we were inhabiting. My idea was to develop virtual countries with two different online communities: ACEN and alt.artcom.

JM: Thanks Anna!

Particularly important are how Carl and Fred developed the *"Status Report"* concept to create community around the sharing of information about our lives, how the "netprov" energy of *Das Casino* became a central part of ACEN, and how your *Virtual*

Cultures topic and the *Virtual Country* project imaginatively brought elements of the community networking movement into ACEN.

Porting your topics to alt.artcom brought in a wider audience and infused new voices and new energy. I also want to focus on how Carl and Fred created a virtual casino that segued into text-based Second-Life-like theater.

So before we move to the next question, Fred, can you talk about the *Das Casino* netprov?

FJT: I was in San Francisco, staying with Carl and Darlene. One night on the way home, Carl said he was interested in investigating gambling. He was interested in how gambling worked and in creating a virtual roulette game for ACEN that would function as interactive virtual theater. We decided to see what would happen if we created a virtual gambling situation on ACEN.

I chose the game, which was roulette. I chose roulette because the operations of the game could be represented by a simple number generator. There are a lot of random number generators out there because such routines are a standard part of UNIX. The command is: RANDOM.

In our game, we didn't have two spinning wheels. Carl, who was the croupier, only spun one wheel. We simplified it as much as we could so it was very low maintenance. Another simplification we adopted was we didn't allow betting on a range of numbers. Of course, I added conditionals in the script to cover even or odd and red or black.

The running of this virtually staged game was fairly direct. Carl called for bets. Players bet. Carl closed the bets and asked me for a number. I was sometimes called the Numbers Man. I "arrived" in true pimp-style. Always in a huge car with lots of girls. I ran RANDOM and announced the number.

Bettors either exulted or wept. Lots of drinks followed.

The next game began.

It was fun to do because people enjoyed it, and then it really took off, leading to a real Das Casino party at Art Com!

We were really surprised at how inventive people were with the identities they created and how they worked with their virtual finances! For instance, Freddy Hahne (<really>) was in heavy debt to the House. Carl and I decided to form the Art Com Collection Agency. We showed up at <really>'s house very early one morning, looking like a couple of mismatched hitmen. We both had violin cases and the requisite trench coats. We demanded full payment from the amused <really>. This may have been when his avatar sought employment with Das Casino.

JM: It is time now to move outside the virtual environment of ACEN and talk about the traveling exhibition that Carl curated—*Art Com Software: Digital Concepts and*

Expressions, which began at the Tisch School of the Arts at New York University, in 1988 and also traveled to San Jose State University, the University of Colorado, ARS Electronica, Linz, Austria, and Carnegie Melon University.

According to Judith Hoffberg's coverage in *Umbrella*,[14] *Art Com Software: Digital Concepts and Expressions* premiered at the Interactive Telecommunications Program at the Tisch School of the Arts on November 4–22, 1988, with the work of Robert Edgar, Judy Malloy, Stephen Moore, Sonya Rapoport, Joe Rosen, Fred Truck, and Paul Zelevansky.

In his catalog article, "Telecomputing und die digitale Kultur,"[15] for the 1989 *ARS Electronica Catalog*, Carl documents the following works (in addition to the Art Com Menus) that were included in the exhibition:

Artworks Online: John Cage: *The First Meeting of the Satie Society*; Judy Malloy: *Uncle Roger*; Ian Ferrier und Fortner Anderson: *The Heart of the Machine*; Gilbert MinaMora: *Exquisite Corpse*

Artists Software: Fortner Anderson and Henry See: *The Odyssey*; Robert Edgar, *Memory Theater One*; Judy Malloy: *Molasses*; Sonya Rapoport *Digital Mudra*; Joe Rosen *Mr Prezopinion and Vibrabones*; Fred Truck: *Illustrated ArtEngine*; Paul Zelevansky: *The Case for the Burial of Ancestors*

Fred, can you begin by talking about your role in Art Com Software, your work that was included, the impact of the exhibition, and any stories you would like to relate about the travels of *Art Com Software*?

FJT: With pleasure.

In 1987, I think—certainly very early on—I suggested to Carl that we have an artist's software show. I knew of a number of works that were reliable and easy to operate by artists like Robert Edgar, Paul Zelevansky, and Joe Rosen, among others. I told Carl it didn't matter where we did it. We just needed to do it right away and document it.

My role in *Art Com Software* was, first, to encourage others to begin to think about software as art, and then to encourage these people, usually artists, to make software as art.

With Robert Edgar, Paul Zelevansky, and Sonya Rapoport, no encouragement was needed as they had already made such work. Once Joe Rosen saw what everyone else was doing, and that we had shows lined up, he quickly produced some pieces. Judy, your *Uncle Roger* was online, so the basic show was ready.

I produced a small work called *Squared Circle*, which was a very reduced *ArtEngine*. In the little book that accompanied the disk, I explained that I had described a circle in words to the Engine. A circle is an arc with all points equidistant from the center. Then I described a square in words: A square is a rectangle with all sides equal. With the

Engine, any visual object that can be described with words can be changed. The *Engine* considered each description and combined them in the new form of a squared circle.

The effect of *Squared Circle* was minimalist and subtle. It demonstrated that an environment could be created in which there was no distinction between words and objects. In that environment, to type a word was to create the object.

All the participating artists in *Art Com Software* had their own approaches to the challenges software as art presented. There was something in the show for every user. There were narratives, game-like structures, narratives in game-like structures, and randomly organized work—basic forms that reflected the thinking in art at that moment.

People liked the work we showed. I was present for most of the showings, and there was rarely a computer that wasn't in use.

For me, the climactic show happened at ARS Electronica in Linz, Austria, in the fall of 1989. At that time, the Iron Curtain was falling. Linz was full of Russian artists and onlookers. Most had never seen computer art before, and none of them owned computers like the Apple SE30 or other Apple machines, which were slightly more powerful.

Carl and I couldn't import machines, so we relied on the staff of ARS Electronica to borrow machines for us. Amazingly, they were successful.

The major challenge for us in the setup was getting the telecommunications to work. At that time, Austria still had the Post Telegraph & Telephone (PTT). Every computer that had a modem also had a little meter that came from the PTT that showed how many bits and bytes you had transmitted and received. You were charged accordingly.

I had done a lot of reading and made a few phone calls about how to blast out of the firewall Austria had installed that prevented anyone from telecommunicating outside the country. One of the sponsors of ARS was Austrian National Radio. I talked to a guy there who told me how to get to their gate to the rest of the world.

We could get to the gateway, but we had no money to pay for the bits and bytes. Carl's solution was simple. When the government official came to read the meter, Carl disconnected the meter and handed it to him. He announced that we weren't paying for anything.

As far as I know, nothing came of this. We weren't billed. Being able to get out of Austria and telecommunicating on our own ensured the success of our software and telecommunications efforts. We could show ACEN and all the art we had there, as well as the disk-based pieces.

I think that the impact of the show was mainly in the area of individual initiative. People kept telling us what we did could never happen there. They had to ask permission, but we just did it.

Of course, things were changing. The wall was coming down, and in a very brief time, PTTs disappeared.

You could tell the quality of life was ecstatic, bordering on chaotic. There were no police in evidence. Suddenly, the passport authorities on the trains were gone. The Brucknerhaus, where ARS Electronica was held, had no security, except for a retired policeman. People were pretty much dancing in the streets.

The biggest change for Carl and I was in evidence the next morning, when we returned to the show. Overnight, someone had walked off with all the computers the ARS staff had borrowed for us. They also took all the cords. For a moment, we panicked. Then we sprang into action because we knew we had to inform the staff and try to find new computers and new cords. On with the show.

It took a few hours, but the excellent staff at ARS got us everything we needed. All we had to do was reassemble things.

JM: I had a treasured copy of your *Squared Circle*, but it is now with my papers at Duke. It was particularly difficult to part with my collection of artists' software, but I am happy that it is in a good place.

Anna, one thing I remember is the panel we did at San Jose State. I think it was in 1989. I have photos of you, Fred, Carl, Sonya, and Bob Edgar sitting on the steps and one that I asked someone to take that includes me.

AC: My introduction to artists' software began with a conversation I had with Robert Edgar about *Memory Theatre One*—sometime around 1985–1986. We were at Art Com. Coca Cola and Bank of America buildings dominated the view outside the window. Robert described the intangible architectural space of rooms and memories he'd programmed—and he blew my mind. The idea that a nonphysical space could be built and virtually moved through was absolutely new to me. Art Com didn't have computers that could run his software, so I didn't experience his work for a number of years—probably not until the San Jose exhibition!

Unlike the two of you, I did not become much of a programmer. But the concept of software as art—and of artists writing code—was important. It marked a shift from the role of the artist as a maker of meaning through content to being a creator of tools. This was not the first time in history for the fusion of these two roles. But in the context of software development, artists who worked with code established a space for emotive, meaningful computational tools—not solely pragmatic ones. This practice blurred distinctions between function and meaning, art and life that have become embedded in our lives today.

That panel at San Jose State University was my first public speaking experience! Carl organized the panel on "Telecomputing and the Arts," and you, Fred, and I were speakers—as were Robert Edgar, Sonya Rapoport, Jeanelle Hurst from the Australian Network for Art and Technology, Derek Dowden from ArtCulture Resource Centre, John Coate from The WELL, and Howard Besser from University of California, Berkeley. Jeanelle's presence was significant—we were just beginning to build ties with electronic artists in Australia. The panel and the *Art Com Software* exhibition were part of the 1989 National Computer Graphics Art Conference hosted by the Computers in Art and Design, Research and Education Institute (CADRE) at San Jose State University. Abbe Don's *We Make Memories* was included in that iteration of the *Art Com Software* exhibition.

JM: Thanks, Anna! This conversation has brought the formative experience of Art Com Electronic Network to the forefront—the ways in which ACEN not only created an online space that was a model for the future, but also was an incubator for the creation of art, literature, and art texts in an online environment. As we look at how social media has developed, there is an interest in exploring innovative early social media. With this in mind, this interview concludes with a two-part question:

What happened to Art Com Electronic Network?

And how would each of you envision a social media platform of the present or future that creates a nurturing environment for computer-mediated art and literature?

Onward!

FJT: Judy asks: "What happened to Art Com Electronic Network?"

By the mid-1990s, ACEN had played out. Like a marriage in which one partner or both are unfaithful but are the last to know that it's over, the final dissolution in 1999 still came as a jolt to me. Carl called one afternoon and said, Let's end ACEN before the new century starts.

A lot had happened to bring us to that point.

Someone has written: The classic development of an American artist follows a pattern. A young artist begins alone and struggles to find support. The first support comes from like-minded artists with whom he or she forms a group. An explosion of creativity happens and in a while matures. Over time rifts appear. Members of the group develop individual styles or interests. There are arguments. Dissension is rife. The group dissolves, implodes, or detonates. Sometimes lifelong hatreds develop among specific members. Other times the social front is remote, frosty. Jealousy can take hold. As old age approaches for members of the group, most of the emotional noise dissolves. Sometimes friendships are begun again. And so it goes.

I would say that our group, ACEN, small as it was, followed this pattern quite closely. The element that is left out of this pattern is further development in technology, an outside force. I think any group of artists working with technology is subject to this condition. What happened to us was the approach of the World Wide Web. Suddenly pictures could be transmitted easily in as many colors as you wanted. Because Carl, Anna, and I had invested so much time and energy in establishing a conceptual, text-based telecommunications environment, it was difficult to accept, although I started sending pictures every chance I got.

As of 1996 or thereabouts, the handwriting was on the wall. Carl was traveling widely, so he was not around to participate. We did Usenet versions of the magazine and started an online newsgroup, alt.artcom. Of course these formats were all text based. We never made the transition to pictures. Our energy for it was gone.

In the early 1990s, Carl saw virtual reality as a possible area of development for ACEN. For me, this was pushing the boundaries of my programming capabilities. I had already spent five years working on *ArtEngine*. I certainly continued listening to Carl, but programming was not on my horizon anymore. I was still smarting from Apple's lack of support for Scheme.

I wanted to work in graphics and animation instead. I was particularly interested in 3-D graphics. I heard about a program for the Macintosh that would do just that, as well as virtual reality. It was called Swivel 3-D, written by Jaron Lanier. I began an animation project based on Leonardo da Vinci's Flying Machine. Soon I had an ornithopter based on da Vinci's designs flapping across the monitor of my Macintosh.

Sometime in the summer of 1990, Carl rang me up. He wanted to know whether I was interested in splitting a fellowship with him at the Banff Centre for the Arts in Banff, Alberta, Canada. They were going to do art projects in virtual reality. The Bioapparatus Seminar was a signal event in my life in art.

Shortly after the Bioapparatus, Carl left San Francisco to take a position at Carnegie Mellon University in the Studio for Creative Inquiry. I finished my da Vinci Flying Machine there under his aegis.

As far as I can remember, we were together only once more, and that was in Anaheim at Siggraph in 1993, across the street from Disneyland. Carl and I were demonstrating virtual reality: I was demonstrating da Vinci's Flying Machine (called *The Labyrinth*), and Carl was demonstrating an ancient Egyptian environment.[16] Anna was conducting sessions on women in computer art and narrative.

By the mid-1990s, ACEN was over, but we soldiered on to 1999, unconscious of the fact. After it ended, I was very sad for a long time.

Shortly after the turn of the century, in 2001, Carl died. I grew more reflective about ACEN and began to collect what materials I had from our inspired time.

AC: I left my staff position at Art Com in 1991. Carl was working at Carnegie Mellon University. Although he maintained his position as Art Com's executive director, he was no longer regularly in the office. I continued to produce *Art Com Magazine* for Art Com Electronic Network for a while after I left the organization. But when I got hired as Network Coordinator for Arts Wire, my participation in ACEN dropped off.

Our interests were diverging. Carl saw the future in virtual reality. I saw it in online communications systems. Carl and I could see ourselves moving along different paths and talked about how we would reconnect our work in the future. I remember being at the Third International Symposium on Electronic Art in Sydney in 1992. Carl was there too, but our paths didn't cross. I had organized *Cultures in Cyberspace*, a multisite participatory telecommunications sculpture, and I was on a panel speaking about ACEN and Arts Wire. Carl was presenting on virtual reality.

SIGGRAPH 1993 must have been the last time I saw Carl. Lucia Grossberger Morales and I were presenting *Matrix: Women Networking* there, an exhibition of telecommunications works by Lisa Cooley, Judy Malloy, Aida Mancillas, Lorri Ann Two Bulls, and ourselves using readily available low-tech tools to "expand the concept of technological advances to include their social and cultural underpinnings and effects."

Fred's comments about the impact of the web on ACEN intrigue me because his assessment holds true for all of us, even though we were no longer working together. I began to move away from online communications as a focal point of my work not long after Mosaic, the first graphical web browser, emerged. The graphical web was not just a multimedia version of what had been a text-based space—it was a different medium, and it ushered in an entirely different online culture. The shift from text to multimedia was complicated socially. On the one hand, the graphical interface and embedded media capabilities of Mosaic brought online communications into the mainstream, and millions of people got online. But the early web was more a publishing model than a conversational one, and it privileged those who had the resources and expertise to create graphical communications. Of course the ASCII text-based systems privileged writers, particularly English speakers, and readers. The development of more inexpensive tools and social media has helped break down to some degree the multimedia communications hierarchy.

There was another substantive shift that accompanied the emergence of the web and the popularization of the Internet. That shift was about money and power. Experiments in interactive art and democratic communications were not the driving force.

I began moving toward social justice organizations and offline community organizing as a venue for social sculpture.

FJT: Judy asks: "And how would each of you envision a social media platform of the present or future that creates a nurturing environment for computer-mediated art and literature?"

This is an interesting question, and one I've thought about in the past. Now, having been asked, I suddenly find myself with new perceptions.

Originally, I considered a walk-in TV, which could be a room in a house in which the walls, floor, and ceiling are giant monitors. One enters this space and connects with anyone or any group anywhere. Everything takes place there in a space much like a holodeck.

Then I came to my senses and realized that Google Glass®, with really serious modifications, is a much better vehicle. You would use the Glass as a heads-up display, as it is now, or by touching a switch; you might shift into an alternate reality that combines broadcast and videophone, satellite technology, Bluetooth, aural input, immersive video, haptic responses, and technologies not yet devised. New art is presented in this environment. People you know and people you don't know drop by to look at the art or interact with it, converse with you and others. Satellites could be useful for establishing a global perspective and perhaps a universal point of view. The new, improved Glass® may connect you with the Hubbell telescope! You visit other people and chat about the latest news. There are special "rooms" in this alternate reality. In one, someone has built a three-dimensional graph that displays the emotional state of everyone you know. You can change it as you please, as you feel. In another room, you've created an environment for thinking. All the books you love are there, as well as thousands you've never heard of, but might be pertinent to your project. There are computer constructs there that can guide and help you. Some might be historical figures while others are completely imaginary.

I could go on and on … but it is clear that the social reality I envision is a mix of times past, present and future, dreams, objects, art, music, and conversation … and … Sport!

You'll have to excuse me now. I'm off to the racquetball court where I'm about to play cutthroat with two well-known players, Elvis and Abe Lincoln, a terror in his day at handball.

AC: What drew me to computer networks was the potential for new kinds of relationships with people and mutually beneficial cross-cultural communications. It was the possibility of engaging across many kinds of other-ness and collectively building creative and inclusive social structures.

There is a reciprocal relationship between existing social practices and systems and the technology we design and use. Contemporary social media platforms exist in an increasingly inequitable social system. Although they use a participatory communications paradigm, that paradigm is facilitating nearly ubiquitous corporate and government surveillance even as it enables new forms of art, community, and spectacularly powerful social movements. Our communications platforms and tools tend to privilege the individual rather than the group as the organizing structure. The rise of embedded screen culture is reorienting our relationship to physical space, exacerbating social disconnection. In this context, I no longer see online communications alone as the leading strategy for shaping pluralistic cultures and societies.

I would like to see more flexible social media platforms designed for open group communications—platforms that are community controlled and surveillance-free, that facilitate relationship building across cultural and social difference.

What made ACEN so exhilarating was that it created a new kind of space where people were able to set aside their usual selves and experiment with each other. I'm not sure another social media platform will take us there. The sense of wonder and possibility that the physical Occupy spaces created for a moment are closer to what I dream of. The seemingly improbable bridging that happens in my current work—the close relationships I have been able to develop with formerly incarcerated people across radically different life experiences and expectations that reveal the depth and complexity of what it means to be human beings in all of our trauma, hatred, fear, love, and joy—social spaces and processes and time that could facilitate that for more of us: that's what I would love to see.

JM: Contemporary social media is amazing. We live in a time when on contemporary social media, the power of many voices to raise issues is important and where the social media platforms for making, discussing, and publishing art that ACEN envisioned have begun to be realized. In the last years, the work of artists created in contemporary social media include, among many others, Joseph DeLappe's Second Life reenactment of Mahatma Gandhi's *Salt March to Dandi*, the artist collective LAinundacion's *LA Flood Project* (cross-platform), and on Twitter An Xiao Studio's *Morse Code Tweets*.

But ACEN holds a unique position as a platform that, beginning in 1986, empowered artists, fostered electronic literature (including the work of John Cage, Jim Rosenberg, and myself), and hosted collaborative netprovs and in-depth online conversations with artists and people of all walks of life. I am so sorry that my old friend Carl Loeffler (1946–2001) is not here with us today to add his voice to this discussion. In his concluding words (circa 1988) in "Telecomputing and the Arts: A Case History," he writes:

We hope that our excitement about telecomputing is communicated to you by this text. Sitting here in San Francisco we are highly motivated by the technological advances being made in Silicon Valley, and desire to put these advances in service of contemporary art. For anyone reading this, we offer an open channel for your interest and participation. See you online![17]

Notes

1. After the last question was responded to on May 17, 2014, I wrote the concluding words above. Then, with the participation of both Fred and Anna, as regards their own responses, in

the following months, the interview was somewhat shortened to conform to the length of the chapters in this book.

2. Carl Eugene Loeffler and Roy Ascott, "Chronology and Working Survey of Select Telecommunications Activity," in *Connectivity: Art and Interactive Telecommunications*, Roy Ascott and Carl Loeffler, eds., *Leonardo* 24:2 (1991): 236–240.

3. Ibid.

4. Carl Loeffler, "Telecomputing and the Arts: A Case History," *Art Com Media Distribution Catalog* 1 (n.d.): 66–71.

5. Ibid.

6. In personal email, Nancy Frank adds that when she called Steward Brand, who had just started The Well, and asked whether Art Com could be the first art magazine to be published online, he said "Sure." Nancy also relates that Brand showed them the server and said "fill up as much space as you can!!" Nancy Frank, correspondence with J. Malloy, January 15 and 16, 2015.

7. Judy Malloy, "Keeping the Art Faith, Interview with Carl Loeffler," *Art Com Electronic Network*, on The WELL, January 11–20, 1988. Portions of this interview are available at http://www.well.com/user/jmalloy/artcom.html.

8. Ibid.

9. Loeffler, n.d., 68.

10. Judy Malloy, "OK Research/OK Genetic Engineering/Bad Information, Information Art Defines Technology," in *META-LIFE*, eds. Annick Bureaud, Roger Malina and Louise Whiteley (Cambridge, MA: MIT Press, 2014). This paper was first published in *Leonardo*, 21:4 (1988): 371–375.

11. Judy Malloy, "Uncle Roger, an Online Narrabase," in *Connectivity: Art and Interactive Telecommunications*, eds. Roy Ascott and Carl Loeffler, *Leonardo* 24:2 (1991): 195–202.

12. *Art Com Magazine* 27:7(3) (1986).

13. Loeffler, n.d., 69.

14. Hoffberg, Judith, "Art Com Software, Digital Concepts and Expressions," *Umbrella* 12:1 (1989): 24.

15. Carl Loeffler, "Telecomputing und die digitale Kultur," *ARS Electronica Catalog* (1989). Available at http://90.146.8.18/de/archives/festival_archive/festival_catalogs/festival_artikel.asp?iProjectID=9010/.

16. Loeffler, Carl, "The Networked Virtual Art Museum," Carl Frank-Ratchye STUDIO for Creative Inquiry at Carnegie Mellon University, 1992. Available at http://studioforcreativeinquiry.org/projects/the-networked-virtual-art-museum/.

17. Loeffler, n.d., 71.

13 System X: Interview with Founding Sysop Scot McPhee

Amanda McDonald Crowley

Amanda McDonald Crowley (AMC): System X was an Australian-based dial-up BBS where, like many BBSs at the time, users created a community of interest with a variety of text-based conversations. It incorporated a virtual gallery of images and sound that invited visual and sound artists and musicians to share work and collaborate via the system. At that time, it was, of course, uploading and downloading files. System X also sought to originate critical thought about the nature of information storage and control, data networks, and how art might be practiced in this media. Importantly, it provided a context for members of the community to upload their own content and share that content not only with a primarily local, Sydney-based community, but also with the growing national and international community interested in this data space. That's probably the key thing that got me involved, at least as an observer and sometimes contributor to the discourse.

Scot, why don't you start off by telling me when it was that you decided to start up System X and what influenced you to do that?

Scot McPhee (SM): My original motivation was music. I was a musician; I was composing/playing electronic music in Sydney in the mid-1980s. At this time, home computers had started to be used in the electronic music world, mostly in the form of the Commodore 64 and, later, the Atari ST, Amiga, and Macintosh computers. I was using electronic bulletin board systems—BBSs—which you would dial up directly with your modem attached to your computer, and exchange messages and files for the other users. It was crude, slow, and entirely text-based. After a while, I decided to start my own BBS, which was focused on the electronic music community in Sydney. It was called Sys-Ex ("System Exclusive"), which is the protocol of MIDI (Musical Instrument Digital Interface), so it would have been obvious to people working in that world. We started on FidoNet, which was a worldwide network of BBSs, and later moved to APANA, an Australian public Internetwork association in the days when only institutions like universities could get a "proper" Internet connection.

Because my contacts in the electronic music world extended to people working in electronic art, I was slowly learning about video art, machine art, mail art, performance art, and other types of art like that. I came from the early 1980s post-punk electronic music scene; bands like SPK, Severed Heads, Throbbing Gristle, and Cabaret Voltaire were huge influences on me, and all those bands had certain aspects of art performance to them. When the Internet arrived, or was more prevalent in the home, in the early 1990s, I was involved in an electronic music collective called "Clan Analogue," which had aspirations to extend to other electronic art forms. I'd been friends with Jason Gee, who was a video artist, for a long time, and he was my co-administrator on Sys-X, as it was known by then. Jason was a video artist with a fine arts degree, so he brought a lot of solid art world knowledge and skills to what we were doing. My background, my "day job" if you like, was computer engineering. The "X" now implied the unknown and mysterious; like Planet-X, we were System-X. We did an installation called TELE-MAT for the TISEA conference in Sydney, which I think was 1992.

AMC: Could you elaborate on what you were using BBBs for and why you felt you needed to start your own? (And do you recall when, precisely, you started Sys-X?)

SM: I can't recall the exact date; it was in the late 1980s, about 1988 or 1989. After I finished my film school short course on film sound, which was 1987. In 1989, I started a computer science degree, and it was in this same time period. At university, we had a big Unix mainframe-type computer that we used for coursework, email, and what have you. It had a modem so you could dial into it from home, and I already had the modems I was using in BBS land at that point.

The main reason I started it was because there were no available Sydney BBSs that specialized in electronic music. There were music forums on various big popular boards, of course, but I felt a board that specialized in music was lacking.

AMC: My understanding is that most BBSs were about sharing information and conversation. At what stage did it occur to you that System X was an art project, not just a site for information exchange?

SM: Probably it was the contact with works by artists like Barbara Kruger, Jenny Holzer, and so on that I realized communication could be a key part of art. I was always into word play and dumb puns. There was always an element in the BBS culture that relied on things like deliberate miscommunication and playful détournement of accepted BBS paradigms. These were often found in many Usenet newsgroups, particularly those in the "alt" name space. Later on, this devolved into what is now known as "trolling" (and particularly sites like 4chan), but at the same time, I could go to an art gallery and see a video work like Bruce Nauman's *Double No* and make the connection between that and experimenting with a text-only computer interface and (for example) programming it to do something childish like always run counter to the user's expectations.

Another part of this was Sys-X's audience, which was in the main people from Sydney's underground experimental electronic scene. These types of milieux always have contact with people from the art world and are sometimes the same people. Certainly in Sydney, a lot of the music was performed in spaces like Art Unit or Performance Space.

AMC: One of the things that was intriguing and groundbreaking for me about System X is that it seemed to be one of the first BBSs where artists and musicians were uploading, downloading, manipulating, and re-uploading music and images, making it in effect one of the early sites practicing share/share-alike copyright and indeed open source principles. Was that intentional? What effect did it have on the artists involved in the System X community?

SM: File sharing was an integral part of BBS culture, even of non-copyrighted self-developed lo-fi folk-art (not just, say, copyrighted material like music and films, which were anyway, in those early days, far too large in file size to be able to transmit via a slow 2,400-bps modem and store on a computer with only 300 Mb of disk storage). So in some ways, that was the natural expressive result of bringing together a lot of artists and musicians onto a BBS. The problem of copyright in this type of environment tends to lend itself to non-copyright outcomes.

AMC: Who were the key members of the community (aside from you and Jason Gee, the site administrators)? Where were most contributors?

SM: Tom Ellard from Severed Heads was an early contributor. There were also a lot of guys from the rave scene that was starting at the time and, also, some people like Michael Dagn and Rosie Cross, who I knew from public radio. I was also involved with an electronic music collective, Clan Analogue, so those guys were also big contributors, especially Brendan Palmer and others involved. Artists included people like Francesca da Rimini, Julianne Pierce, Josephine Starrs, and Leon Cmielewski.

AMC: Who was the key audience for the site? How do you think you reached them and why?

SM: See above. We got to them mainly through word of mouth at events and via other electronic means.

AMC: System X exhibited TELEMAT at TISEA (The Third International Symposium for Electronic Art) in Sydney in 1992[1]—alongside a "telecommunications" conference stream that included American Indian Telecommunications/ Dakota BBS, ArtsNet, Arts Wire, Usenet, and The WELL, exploring culture in cyberspace.[2] I believe it was the first time that Internet-based art was "exhibited" in an ISEA. What was the impetus behind realizing a real-time real-space iteration of an online environment? What was the installation? What did it look like and how did it operate?

SM: Well, it was a very simple text-based interface. The idea, I recall, was that users could leave simple messages that were combined into a big single text file. When you logged in, you'd get a random message extracted from the System X database and were invited to respond. I think the responses were posted into a news group.

AMC: You also "exhibited" System X at The Performance Space. As I recall, you'd recently lost a hard drive and had no backup of a lot of the content, so the hard drive became part of the installation. How was that installation received? What else did you display in a gallery space that helped an audience to understand what participating in an online community/environment meant creatively?

SM: Well, I recall the platter was exposed to the air. It was a big, 5.5" full height hard disk, so quite an object—not like today's micro-miniatures. The idea was that people were invited to admire the shiny, highly engineered precision object and think about the loss of the years of archives that were exposed to the air so the metal oxides that encoded them would rust.

AMC: By 1994, with the popularization of the Mosaic browser, you moved a lot of the activity over to the web. How did this affect the kind of work that artists were making, and how did it affect the community?

SM: The main effect was first to allow a lot more visual work, I think, and, second, to give an explicit form for the exploration of hypertext (i.e., the linking of documents). Before the web, a lot of work online tended to be about text-based communication, so while the idea of hypertext art was not unknown (e.g., through things like HyperCard), online the art (in my recollection) was more concretely focused on ideas of the poetics of text communication, human interaction, and disruption. Those currents still existed with the coming of the web, but I think the visual nature and self-linked multipath visual narratives started to become more dominant.

Also Usenet newsgroups started to fade at that point.

The move to the web as the medium had another effect: the need for even more bandwidth!

AMC: At what point did you decide to wind System X down, and why? What do you think its legacy has been in the Australian digital arts community?

SM: I think by the end of the 1990s, it was clear to me that its necessary purpose had passed. By this time, anyone who was living above subsistence levels in a reasonably sized town could afford a basic Internet connection from a commercial provider and get web space and a domain of their own for not a lot of money. Institutions saw the potential, and so many could offer servers and bandwidth to the artists and educators affiliated with them. At that point, System X had less of a coherent reason to exist than in the early 1990s, when home broadband connections were not so common and many institutions thought their few servers to be precious technical resources reserved for things such as computer science, physics, or medical research.

Its legacy? I don't really know. I guess we gave initial impetus to many early Internet art pioneers in Australia. I think that we also brought to people the idea that it could be a viable platform for the construction of artistic meaning. In the late 1980s and early 1990s, when I used to talk to people about this stuff, the most common response was, "Why would I want to talk to a bunch of computer nerds?" But within 10 years, nearly everyone in the developed world under the age of 40 had an email address and a website, and artists' web projects, art resource sites, mailing lists, and so on proliferated everywhere. I guess we forged the start of a path that others could follow.

Notes

1. *TISEA: Cultural Diversity in the Global Village* (Sydney, AU: ANAT, 1992*)*.

2. Anna Couey, "Cultures in Cyberspace: Communications System Design as Social Sculpture." [*Social Media Archeology and Poetics*].

IV Networking the Humanities

14 In Search of Identities in the Digital Humanities: The Early History of Humanist

Julianne Nyhan

It is widely agreed that Digital Humanities (DH) is in the ascendant. Reaction to this is to be found in traditional scholarly literature as well as on social media and in blogs, gray literature, press articles, and even Internet memes.[1] While some literature triumphantly asserts that DH will revolutionize the Humanities,[2] others portray it as an agent of the creeping, and seemingly inexorable, computerization of all aspects of modern life and admonish that "literature is not data."[3] Groups like 4Humanities argue that DH can help the Humanities to "communicate with, and adapt to, contemporary society,"[4] whereas others predict a dystopian future where Humanists "wake up one morning to find that they have sold their birth right for a mess of apps."[5]

The popularity (or, depending on your perspective, infamy) of DH that underlines such arguments may be new, but the discipline itself not. Its origins can be traced back to 1949 at least, when Fr Roberto Busa, with funding from IBM, began work on an index variorum of some 11 million words of medieval Latin in the works of St. Thomas Aquinas and related authors.[6] However, the history of DH has, with a few notable exceptions, been mostly neglected by the DH community itself as well as by the mainstream Humanities.[7] Of the many research questions that wait to be addressed, one set pertains to the history of the disciplinary formation of Digital Humanities. What processes, attitudes, and circumstances (not to mention knowledge and expertise) conspired, and in what ways, to make it possible for DH to become disciplined in the ways that it has (and not in other ways)? What might answers to such questions contribute to new conversations about the forms that DH might take in the future? Here I will make a first and brief contribution to answering such far-reaching questions by identifying and analyzing references to disciplinary identity that occur in conversations conducted via the Humanist Listserv in its inaugural year.

Humanist was set up in 1987 and is "an international online seminar devoted to all aspects of the digital humanities … [a forum where] the technology, informed by the concerns of humane learning, can be viewed from an interdisciplinary common ground."[8] It was set up by Willard McCarty, who, at the time of writing, remains its editor. He is also Professor of Digital Humanities in the Department of Digital

Humanities, King's College London and Professor in the Digital Humanities Research Group, School of Humanities and Communication Arts, University of Western Sydney, Australia. As well as having numerous highly cited publications, he has won various international awards for his scholarship, including in 2006 the Richard W. Lyman Award (from the National Humanities Center and the Rockefeller Foundation) and in 2014 the Roberto Busa Award (from the Alliance of Digital Humanities Organizations).

Within the context of DH, Humanist can arguably be categorized as a protosocial media platform due to the ways it has enabled information, knowledge, and social connections to be made (and perhaps unmade) and transferred. More to the point, however, is that newer and slicker social media have come (and, in some cases, gone), but Humanist has endured. Indeed, it arguably remains DH's most vital locus of long-form questioning, imagining, and reflecting on and about itself and its many interdisciplinary intersections. Therefore, this chapter takes as its starting point that Humanist is likely to contain a wealth of evidence about the disciplinary formation of DH. It makes two further assumptions that should be briefly considered. The first is that it is here assumed that DH is either a discipline[9] or is on the verge of becoming one. Should this not prove true, the issues explored here will still be useful because understanding why a field does not successfully transition to a discipline is also valuable.[10] The second is the assumption that developments that took place at a time when the field was mostly known as Humanities Computing are, to some extent, relevant to the disciplining of DH. Space does not allow me to explore this presumed relationship in detail. Yet we must be careful not to assume that the two are points, merely separated by time, on a linear trajectory. Mahoney has made especially clear the inherent flaws of such an approach:

> When scientists do history, they often use their modern tools to determine what past work was "really about"; e.g. the Babylonian mathematicians were "really" writing algorithms. But that's precisely what was not really happening. What was really happening was what was possible, indeed imaginable, in the intellectual environment of the time; what was really happening was what the linguistic and conceptual framework then would allow.[11]

Also, such an assumption would attribute to both Humanities Computing and Digital Humanities a sense of internal cohesion and unity of purpose that neither term is likely to have in any practical way, a point that I will pick up on in the conclusion below. I will now give an overview of Humanist itself.

Humanist: An Overview

Humanist was established in 1987 on the BITNET/NetNorth/EARN node in Toronto, Canada,[12] and run on Listserv software. Following test messages that were sent on May

12 and "13 May to approximately two dozen people in three countries" (Ibid.), on May 14, a longer message about Humanist was sent:

From: MCCARTY@UTOREPAS
Subject:
Date: 14 May 1987, 20:17:18 EDT
X-Humanist: Vol. 1 Num. 5 (5)

Welcome to HUMANIST

HUMANIST is a Bitnet/NetNorth electronic mail network for people who support computing in the humanities. Those who teach, review software, answer questions, give advice, program, write documentation, or otherwise support research and teaching in this area are included. Although HUMANIST is intended to help these people exchange all kinds of information, it is primarily meant for discussion rather than publication or advertisement. ... [13]

At that time, the Listserv technology was fairly new,[14] but neither the form that Humanist took nor the need it met seem to have been. Humanist was described as an "electronic seminar," a label that was chosen deliberately to "evoke the academic metaphor of a large table around which everyone sits for the purpose of argumentation," in the etymological sense of "making clear or bright."[15] One also suspects, and I will build a wider case for this below, that the choice of the seminar form signaled a deep commitment to Humanistic epistemology and disciplinarity, notwithstanding any expectations there may have been of how Humanities Computing research would, in time, augment these.

Humanist was initially founded for those who worked in computing support and who encountered, among other things, a "lack of proper academic recognition."[16] The idea for a forum like Humanist came about during an impromptu meeting held after the International Conference on Computers and the Humanities in Columbia, South Carolina, in April 1987 (Ibid.). It was created under the auspices of the newly formed Association for Computers and the Humanities (ACH) "Special Interest Group for Humanities Computing Resources."[17] The "executive committee ... consisted of George Brett (North Carolina), Michael Sperberg-McQueen (Illinois at Chicago)" and McCarty.[18]

It is interesting to contrast this reported lack of recognition with developments in the wider field, as well as the image of Humanities Computing that was now being portrayed in publications like the ACH Newsletter.

By 1987, Humanities Computing was becoming reasonably well established. Regular conferences were being held and some important centers had already been set up,[19] as had some teaching programs.[20] It had its own scholarly societies (The Association for

Literary and Linguistic Computing [ALLC] was formed by Joan Smith and Roy Wisbey in 1973, and the ACH was founded in 1978). The field's first journal, *Computers and the Humanities*, would soon celebrate its 25-year anniversary (in 1991), and ALLC founded the journal *Literary and Linguistic Computing* in 1986. In the ACH newsletter, the tone is upbeat. For example, Harris writes,

> In 1970 I was accused of trying to "destroy literature" and now the 1984 MLA Convention Program lists 10 meetings devoted to Computer Assisted Instruction and Research. ... It is an indication of the changing times ... the profession has largely accepted computers as a part of the discipline ... even though not everyone is involved in using computers, they are no longer viewed as threatening.[21]

External evidence seems to largely corroborate this. For example, in addition to the points noted above, a National Endowment for the Humanities (NEH) report on Computer Uses in Research carried out just two years later was described as "illustrating the changing nature of humanistic research in the age of the computer."[22] Such was the interest in the field that in 1986, the ACH advertised "Speakers Bureau," which aimed to

> facilitate contact between its members and people who can benefit from their expertise. Approximately fifty people from all geographical areas of the United States as well as Canada and Europe have listed their names with the Bureau, all of who are available to give papers, seminars, workshops, or presentations on computer use in the humanities.[23]

Although there is much talk about the changing role of the computer in Humanities research, there is not much talk about the changing role of the Humanities Computing specialist or the institutional contexts in which they worked. In this way, we may view Humanist as representing a parallel role and giving a voice to those who might not otherwise have been heard. Nevertheless (and unsurprisingly when seen in the context of the statement of Harris above), within a few months, the membership of Humanist had expanded well beyond those working in computing support roles:

> By September of that year, however, tenured faculty, directors of computing centres, and other well established sorts began to join HUMANIST in significant numbers. This was a crisis of identity for the new group, ... I decided not to constrain HUMANIST to its original purpose but to let it find its own identity. Had I kept it "on track" it would, I think, have died of exhaustion against the thick, hard, and very cold walls of the institution.[24]

Why the name "Humanist" was chosen has not, to the best of my knowledge, been recorded. To this author's European ears, the term "Humanist" sounds somewhat arcane; indeed, OED indicates that "Humanist" is often used in a specialized way: "A person who pursues or is expert in the study of the humanities, *esp.* a classical scholar." The more salient point, however, is that the name does not evoke the groups' connection to computing; indeed, this contrasts with how a number of the associations and journals mentioned above were named. A discussion about the naming of the SIG from which Humanist emanated seems relevant:

> The name should ... distinguish our work from older forms of support for computing, which has established administrative niches inappropriate for our circumstances and personnel. These older support structures are not specific to the humanities. ... Since they were not designed for people with backgrounds and interests in the humanities, for example, they did not tend to offer such people the opportunities they require for developing as humanists.[25]

Considering this, it seems reasonable to speculate that when McCarty described Humanist as "a voice representing a minority to those in power,"[26] it was not only the established Humanities that was being addressed but also the Humanities Computing status quo.

It was in the broad context sketched above that Humanist came about; we will now look at its early content in greater detail.

An Overview of Humanist Posts

Summaries of Humanist from 1987 to 1990 can be found in the ACH Newsletters for those years. Rockwell and Sinclair have analyzed the Humanist corpus from 1987 to 2008 using text analysis and distant-reading methods like correspondence analysis.[27] Their findings include observations about the nomenclature of the field and hypotheses about the impact of the web on its development. They also observed three phases in the corpus sample. The first, from 1987 to 1995, is of a Humanities Computing that was interested in computers, software, hardware, and texts, and that took place in English departments and/or in English. There followed a transitional period from 1996 to 2000, when words related to the web occurred more often. This continued in the third phase, which ran from 2001 to 2008. In light of this, they argue that the increasing use of the web had a transformational effect on Humanities Computing, and that it helped to create the conditions in which Digital Humanities came about. The latter is, they argue, "not only an administrative term but one that signals a detectable change in the way electronic texts were used" (Ibid.). They conclude that this may explain why hardware and software are now much less discussed, whereas the discussion of web services has taken on a new prominence.

Notwithstanding this shifting focus (so convincingly argued by Rockwell and Sinclair), in the transactions of Humanist we can trace the field's enduring concerns as much as their fault lines. For example, a through line of Rockwell and Sinclair's phases is, of course, programming because it is the foundation of software as much as the web. Numerous debates about programming (ranging from the virtues of specific languages like Prolog and Snobol (for example, Humanist 1:117), over the issue of whether programming should be taught to Arts and Humanities students (for example, 1:97) and on to the uses of thinking programmatically (for example, 1:121) take place on Humanist alrcady in its inaugural year. Space will not allow me to trace the development of such debates throughout Humanist. Instead I will point out that programming remains a contentious issue in DH, as evidenced by, among others, the ongoing "Hack versus Yack" debate.[28]

It is impossible here to give an indication of the mixture of topics discussed during Humanist's 27 years. Instead I will mention that, during its first year, issues like professional recognition, electronic publishing, desktop publishing and markup languages attracted much attention in addition to conference calls, project announcements, requests for information and an advertisement of a job.[29]

Terms Related to Disciplinary Identity

In this section I will return to the main aim of this paper, namely to identify and analyse references to disciplinary identity that occur in Humanist during its inaugural year. The terms listed below were used in posts that discussed, directly or indirectly, an individual's understanding of Humanities Computing, his or her role in it, or Humanist itself (which I read as an extension of the same thing). They were identified and tagged during a close reading that I performed of the Humanist corpus and were subsequently extracted, analyzed, and categorized according to the loose groupings listed below. For reasons of space, I here focus on the first year of Humanist; ultimately, I hope that this will form part of a larger study. In many cases, such terms occurred in posts almost as "off the cuff" comments, analogies, or allusions. I tagged terms as I encountered them and did not intentionally exclude any even if they were used in a post that seemed to contain a logical error or other defect. I have not taken the identity of correspondents into account, and this is probably a limitation of the approach used here given what is known about patterns of contributing and lurking on social media. Where comments were made in a tongue- in-cheek or ironic fashion, I have tried to note this, even though it can be difficult to detect in email exchanges.

Terms used to describe the group seem to signal how idealistic and personally involved a number of the early practitioners were. These included: "Suppor[t]ers of computing in the Humanities" (Humanist 1:44), "free people" (1:80), "true believer[s]" (1:1035), and the lament, "I thought we were all in this together" (1:661).

The activities of the group sometimes seem typical of the Humanities; for example, the terms "skeptic," "interpreter," and "Socratic" are used. Others activities are less so; these include: "the break[ing] down of artificial barriers" (1:187), determining whether the computer lets us do "something new" (1:214), "consciousness raising" (1:347), the opportunity to explore a "new opportunity" instead of the "humanistic tradition of isolated individualism"(1:222), "[pursuing] the notion of real computing in the humanities" (1:344), making "computing an accepted part ... of humanistic scholarship, teaching and research" (1:347), "overcoming ... compartmentalization of knowledge" (1:782), "we don't so much go where no man has gone before but continually return to basic questions" (1:198), "contributing to an emerging discipline" (1:182, 1:344), and "defining our work academically and raising its standards" (1:144).

Notwithstanding such aims, the group is not necessarily a cohesive one. Some are more technical than others and are referred to as "Digital cognoscenti" (1:809), and a reference to "Computing Hegemony" seems to include some members of Humanist too (1:871). Discussions that are predominately technical sometimes cause unease, leading one member to ask "where the 'humanities' in 'humanities computing' had gone?" (1:697). The nature of the group is occasionally questioned: are we "enough of a community"? (1:532) The dominance of some disciplines on the list prompted discussion of whether Humanist would become a (literary) ghetto (1:776, 782, 787, 856). The word "types" occurs, as a collective noun, with some frequency and refers to the group or subsets of it. Sometimes it seems to be used for want of a more specific or widely agreed collective noun, for example, "What is needed is some sort of alliance between the computing types and the professional librarians." Sometimes it indicates technical preferences of members, for example, "IBM types" (1:483). While the context can sometimes be ambiguous, it is also used in a disparaging way, for example, "ever more self-assured ACH types" (1:152) or "... think of themselves as humanities types" (1:557). It is interesting to note—and my interpretation of the wider issue it references will be picked up in the conclusion—that it is the questioning of participants' Humanities credentials, rather than their technical ones, that rankle: "I will note in closing that Joe Raben deserves better than to be accused implicitly of not being a 'real humanist'" (1:144).

The result of being in the group is sometimes described as a kind of marginalization. For example, Humanists describe themselves and are described by others as being "revolutionary outcast[s]," "academics without a proper job" (i.e., those in support roles) (1:98), "people from somewhere else" (1:227), and [working in an] "atmosphere of poverty" (1:508) (a characterization that is disputed by others). Their knowledge also separates them from others "something that many of us in our pride of knowledge tend to forget: that most such users have no real interest—and will never have any real interest—in computers." Numerous references to the lack of recognition their work received from the mainstream Humanities occur (1:44, 46, 47, 49, 349, 351). The dangers and limitations of using computing in Humanities research are also broached: "I recognize the dangers of computing—of being 'seduced' by the apparatus" (1:1080; see also 156, 198); "Machines should work. People should think" (1:715, 991); and "we all know of cases in which well-known and respected humanists have been forced out of the profession" (1:47).

Yet marginalization is not a wholly negative process. The possibilities it opens for comparing and contrasting their work with contemporary and historical groups allow issues about identity to be reflected on and articulated in new ways. Indeed, according to Crozier, "in general a discipline defines itself through a process of differentiation."[30]

For example, it is argued that the word "Unprofessional" can be used in a positive sense because it can result in solutions that may not occur to computer scientists (1:845). Humanist is compared to members of the Republic of Letters, "who were the equivalent of modern functionaries and could, therefore, take advantage of the mail systems developed for kings and princes" (1:744). Some of their rank may also have time for research that those in the academy will not have due to their teaching commitments.

It must be commented that, in the period covered, a feeling of entitlement or superiority is difficult to detect.[31] For example, one correspondent may have wanted to do Humanities Computing but reflects, "I was not hired to do Humanities Computing" (1:47). Indeed, the tone that comes through is more often one of anxiety and self-reflection, as will be discussed below.

The references above emphasize what some saw as the newness of their endeavor, but two references to historically important research occur in the period under discussion. One is (the somewhat tongue in cheek) call to "Let us pause for a moment of silence and all give thanks for the networks, which have given us facilities that only sixteen years ago were only a visionary dream" (1:703). It is interesting that religious metaphors and language arise with some frequency. There are also various references to belief, "if they did not believe in the value of this forum" (1:233 [Digest no. 6]), and of concepts explained using religious metaphors, "The true believer has a natural tendency to convert the infidel; I would rather think that in my father's house there are many mansions" (1:1035 [Digest no. 2]). "As Zacour said, the believer is driven to convert the infidel, and few believers see any reason why they should understand the infidel's scripture. (It's the devil's work anyhow and therefore dangerous to mess with.)" (1:1036 [Digest 1]). "What one must do to keep the faith, and keep it intelligently, in a time of little recognition or outright rejection" (1:98). One mentions "people of the computer faith" (1:874), whereas another talks (again tongue in cheek) of "infidelity towards my IBM PC clone" (1:1051).

Conclusion

From the labels and phrases above, we begin to get a sense of how disciplinary identity was perceived, expressed, and performed in the Humanities Computing community of the day. It is interesting to note that, in the period surveyed, direct questions about disciplinary identity were not put to the group. The questions that were asked addressed the kinds of knowledge that the group was creating, for example, their scholarly contribution to the Humanities (1:98) (a question that went essentially unanswered [cf.

1:185]). Questions about how the characteristics of the group may have shaped, and been shaped by, their activities were not directly explored, yet a wealth of observations, both direct and indirect, can be found in their communications.

In some ways, the picture that emerges is not an unexpected one: identity is a complex and shifting concept that is still in the process of emerging (and here I don't wish to imply that this process has since reached an end point). Indeed, this issue is central to present-day Digital Humanities, where publications, formal and informal, that attempt to define what Digital Humanities is (or, less often, is not) abound.[32] Much remains to be done in order to identify and analyze the dynamics of Digital Humanities' disciplinary formation and to contextualize this with reference to a broader comparative context.[33] Nevertheless, literature from the domain of the sociology of science emphasizes that neither disciplines nor disciplinary identities are monolithic structures or concepts. This is emphasized by Powell et al. in a discussion of the naming of systems biology that draws on much wider understandings of the nature and role of disciplinarity: "Systems biology ... exemplifies how a name unites and gives special strength to a broad array of technologies and intertwined methodological and epistemic practices, even when the array lacks an established paradigmatic core."[34]

Notwithstanding such diversity, when the labels and phrases cited above are considered in the aggregate, one notices that certain traits, which seem to be characteristic of the group, emerge. Curiosity and a degree of idealism arguably characterize a number of the remarks that I categorized above as "terms used to describe the group." A certain degree of resilience can be noticed in the way the group persists in its activities despite what it perceives as a lack of recognition from the academy proper; the ways that some perceived their work almost as a matter of faith is perhaps connected to this experience. Self-reflection seems to be another characteristic trait. Indeed, in due course, it could prove instructive to investigate whether such traits are still characteristic of Digital Humanists and whether they played a meaningful role in the discourse that surrounded the formation of the discipline. However, I will now focus on one characteristic in particular: anxiety.

In the period studied, many comments can be identified that reveal anxieties about the quality of their work and its relationship to the wider Academy and the Humanities in particular. For example, references are to be found to what is perceived as "the poor quality of [software] reviews" (1:344), "the often poor quality of the writing (and sometimes thinking) associated with computing" (1:49) and the field is described as "disorganized" (1:381). The question of whether it is at all legitimate to offer a PhD in the area is discussed (see 1:662, 1:667, 1:725) and conversations about whether their work constitutes research can be found. As I will now argue, perhaps the most notable aspect of these conversations is that they seem to be conducted and judged with reference to established standards of the Humanities, a fact that seems to sit somewhat uneasily with some of the more radical aims of the group cited above.

For example, an early conversation that attracted numerous contributions concerned the nature and status of Humanist and whether it should or could function as a formal publication of some sort (e.g., an academic journal). It is interesting that the group did not, during the period in question, give much consideration to attempting to reform the publication practices of the Humanities or to trying something completely new (in fact, I counted some 10 posts before the idea of doing something completely new was considered: "The new medium makes possible new forms of intellectual work, forms of collective research and collaborative writing that have not yet been defined, professionally or institutionally," 1:51). What might count as research also seems very much to be defined with reference to the Humanities proper; for example, "I have not really used the computer as a tool in my own research; I have worked on the improvement of the tool itself. ... But programming, running computer centers, etc., is not, and probably should not be, valued as research" (1:47). The forms that research outputs might take seem to reference accepted Humanities standards; for example, "I don't expect my marked up text to be of interest to anybody either. Nevertheless, if I'm successful, the final result (an essay or book) will say something valuable to others" (1:542). Discussions about proper use of Humanist also seem to reference unarticulated assumptions about Humanities norms, as revealed in one powerful post:

> Some recent comments, referring to "shallow" conversation and "inane chit-chat" are quite threatening ... it also seems to be far from the values inherent in the literature that many of us have devoted our lives to—in which metaphor represents a way of freely expressing one's view of the human condition. We may use tools, such as computers, to tear apart the imaginative worlds created by our language(s), but we are not trying to destroy them. Mutability, chance, the protean nature of sensible things—unconstrained discourse—these are what keep us from just building monuments to ourselves. (233: 6 [digest format])

A superficial interpretation of this evidence might indicate that in the early record of Humanist, a lack of ambition and imagination on the part of Humanities Computing people is to be found, along with a deep discord between their aims and the ways they sought to implement them. Instead, I wish to argue that Humanist shows us the opposite. In essence, it demonstrates that the community combined its idealism with a deep sense of pragmatism. This seems to be encapsulated in a comment made by McCarty: "the trick is to exploit rather than be thwarted by the characteristics of the medium. A change in how research in the humanities is done could result."[35]

Again, looking to the wider literature from the sociology of science and the history of computing, this is not especially surprising. For example, Mahoney's research on the formation of the fields of theoretical computer science and software engineering argues that "people engaged in new enterprises bring their histories to the task, often different histories reflecting their different backgrounds and training."[36] In the example under discussion, we may say that it is participants' backgrounds and training in the Humanities that they bring to the task of forming Humanities Computing: true to the name of Humanist, they often approached the creation of this new field as Humanists. Indeed, this hypothesis should be investigated further when a more sustained study of the disciplinary formation of DH is undertaken. Another important point to make, even though it is an obvious one, is that institutionalization is a necessary part of disciplinary formation. Indeed, in the Powell et al. examination of how "twentieth- and early twenty-first-century disciplines were created and institutionalized in relation to disciplinary naming stories," they relate such practices (while emphasizing that they are not definitive characteristics) to "other elements of disciplinary formation such as paradigmatic achievements, defining technologies, and institutional recognition."[37] From this we may construct another hypothesis with which to move forward. Namely, the institutional recognition that Digital Humanities has won came about via a process that seems to have already been initiated by Humanities Computing and that was dependent, initially at least, on conforming to some institutional norms of the Humanities.

However such hypotheses may, in due course, be modified, two issues are clear. First, Humanist is a largely untapped historical resource for exploring, among other things, issues relevant to the dynamics of disciplinary formation. Second, now that Digital Humanities has achieved a degree of institutionalization and disciplinary recognition, is it time for the field to confront the question of how its research, social, and intellectual achievements relate to the more idealistic aims discussed above? It should be noted that such "radical" aims have remained part of the discourse of the field, where, for example, talk of its "revolutionary" status can still be found.[38] Yet as the field is arguably moving more toward the mainstream and away from the margins, how should current notions of disciplinary identity be reassessed? Might this lead to a more ambitious or radical research agenda?

Some four years after it had been set up, Raben wrote:

A conference called Humanist operates through Bitnet at Brown University. In their brief lives, these two operations seem to be still struggling to identify their role in research and even to define their audience. Whether either or both of these services will survive the period of their novelty and mature with the evolving technology, or be replaced by new ones based on fax or another technology, is part of the larger question of whether computer communications has any meaningful role in humanities research and instruction.[39]

This chapter aimed to cast some light on a hitherto neglected facet of the history of Digital Humanities. Further to Raben's comments, which I believe to have been at that time most reasonable, it also demonstrated that Humanist has, in fact, played a unique and multifaceted role in the development of Digital Humanities.

Notes

1. See, for example, UCLDH, "Memes," *UCL Centre for Digital Humanities*, n.d. Available at http://blogs.ucl.ac.uk/definingdh/digital-humanities-and-design/memes/.

2. Todd Presner et al., "The Digital Humanities Manifesto 2.0" (UCLA Mellon Seminar in Digital Humanities, 2009). Available at http://www.humanitiesblast.com/manifesto/Manifesto_V2.pdf.

3. Stephen Marche, "Literature Is Not Data: Against Digital Humanities," *Los Angeles Review of Books*, October 28, 2012. Available at http://lareviewofbooks.org/essay/literature-is-not-data-against-digital-humanities/.

4. Humanities, "Mission," *4Humanities*, 2014. Available at http://4humanities.org/mission/.

5. Adam Kirsch, "Technology Is Taking Over English Departments," *The New Republic*, May 2, 2014. Available at http://www.newrepublic.com/article/117428/limits-digital-humanities-adam-kirsch/.

6. R. Busa, "The Annals of Humanities Computing: The Index Thomisticus," *Computers and the Humanities* 14:2 (October 1980): 83–90.

7. Julianne Nyhan, Andrew Flinn, and Anne Welsh, "Oral History and the Hidden Histories Project: Towards Histories of Computing in the Humanities." *Digital Scholarship in the Humanities* 30(1) (July, 2015): 71–85. Available at http://dsh.oxfordjournals.org.libproxy.ucl.ac.uk/content/30/1/71/. First published online in *Literary and Linguistic Computing*, July 30, 2013.

8. Willard McCarty, "Humanist: An Online Seminar for the Digital Humanities," 2009. Available at http://dhhumanist.org/announcement.html.

9. See, for example, Geoffrey Rockwell, "Is Humanities Computing an Academic Discipline?" (presented at *Is Humanities Computing an Academic Discipline?*, University of Virginia, 1999). Available at http://www.iath.virginia.edu/hcs/rockwell.html; Willard McCarty, "Humanities Computing as Interdiscipline" (presented at *Is Humanities Computing an academic discipline?* University of Virginia, October 22, 1999). Available at http://www.iath.virginia.edu/hcs/mccarty.html; Susan Hockey, "Is There a Computer in This Class?" (presented at *Is Humanities Computing an academic discipline?*, University of Virginia, September 1999). Available at http://www.iath.virginia.edu/hcs/hockey.html.

10. Alexander Powell et al., "Disciplinary Baptisms: A Comparison of the Naming Stories of Genetics, Molecular Biology, Genomics, and Systems Biology," *History and Philosophy of the Life Sciences* 29:1 (January 1, 2007): 27.

11. Michael S. Mahoney, "What Makes History?", in *History of Programming Languages II*, ed. Thomas J. Bergin and Rick G. Gibson (New York: ACM Press, 1996), 831–832.

12. Willard McCarty, "Humanist So Far: A Review of the First Two Months," *ACH Newsletter* 3:3 (1987): 1.

13. Humanist, "Humanist 5/87-5/88," ed. W. McCarty, 88 1987, vol. 1:5. Available at http://dhhumanist.org/Archives/Virginia/v01/.

14. D. A. Grier and M. Campbell, "A Social History of Bitnet and Listserv, 1985–1991," *IEEE Annals of the History of Computing* 22:2 (April 2000): 32–41.

15. Willard McCarty, "HUMANIST: Lessons from a Global Electronic Seminar," *Computers and the Humanities* 26:3 (June 1992): 207.

16. Ibid., p. 209.

17. See George Brett and McCarty, Willard, "A New Special Interest Group in the Support of Computing in the Humanities" *ACH Newsletter* 9:2 (1987): 1–2.

18. McCarty, "HUMANIST," n. 20, p. 281.

19. Susan Hockey, "The History of Humanities Computing," in *A Companion to Digital Humanities*, ed. Susan Schreibman, Raymond George Siemens, and John Unsworth, Blackwell Companions to Literature and Culture 26 (Malden, MA: Blackwell Pub, 2004).

20. n.a., "Computing in the Humanities Course," *ACH Newsletter* 5:3 (1983).

21. Mary Dee Harris, "Observations on Computers and Poetry," *ACH Newletter* 6:4 (1984).

22. Anne Jamieson Price, "NEH Grantees Report on Computer Uses in Research," *ACH Newletter* 8:2 (1986): 1.

23. Donald Ross, "Do You Need Information about Computing in the Humanities?", *ACH Newletter* 8: 2 (1986): 5.

24. McCarty, "HUMANIST," 209.

25. Brett and McCarty, Willard, "A New Special Interest Group in the Support of Computing in the Humanities," 1.

26. McCarty, "HUMANIST," 209.

27. Geoffrey Rockwell and Stéfan Sinclair, "The Swallow Flies Swiftly Through: An Analysis of Humanist," in *Hermeneutica.ca*, forthcoming.

28. See, for example, Bethany Nowviskie, "On the Origin of 'hack' and 'yack,'" *Nowviskie.org*, 2014. Available at http://nowviskie.org/2014/on-the-origin-of-hack-and-yack/.

29. McCarty, "Humanist So Far: A Review of the First Two Months," 2.

30. M Crozier, "A Problematic Discipline: The Identity of Australian Political Studies," *Australian Journal of Political Science* 36:1 (2001): 22.

31. For an interpretation of the present-day situation, see Adam Crymble, "Digital Hubris, Digital Humility," July 15, 2014. Available at https://www.insidehighered.com/views/2014/07/15/essay-backlash-against-digital-humanities-movement/.

32. See, for example, Melissa M. Terras, Julianne Nyhan, and Edward Vanhoutte, eds., *Defining Digital Humanities: A Reader* (Farnham, Surrey, England: Ashgate Publishing Limited; Burlington,

VT: Ashgate Publishing Company, 2013); note also that the emphasis on outputs rather than individual or collective disciplinary identity remains.

33. See, among others, Powell et al., "Disciplinary Baptisms," in Tony Becher and Paul Trowler, eds., *Academic Tribes and Territories: Intellectual Enquiry and the Cultures of Disciplines*, 2nd edition (Philadelphia, PA: Open University Press, 2001); Kate Bulpin and Susan Molyneux-Hodgson, "The Disciplining of Scientific Communities," *Interdisciplinary Science Reviews* 38:2 (June 1, 2013): 91–105; Neil McLaughlin, "Origin Myths in the Social Sciences: Fromm, the Frankfurt School and the Emergence of Critical Theory," *The Canadian Journal of Sociology /Cahiers Canadiens de Sociologie* 24:1 (January 1, 1999): 109–139.

34. Powell et al., "Disciplinary Baptisms," 26.

35. Michael S. Mahoney, "The Histories of Computing(s)," *Interdisciplinary Science Reviews* 30:2 (June 1, 2005): 120.

36. Ibid.

37. Powell et al., "Disciplinary Baptisms," 8.

38. Nyhan, Flinn, and Welsh, "Oral History and the Hidden Histories Project."

39. Joseph Raben, "Humanities Computing 25 Years Later," *Computers and the Humanities* 25:6 (1991): 348.

Bibliography

4Humanities. "Mission." 4Humanities, 2014. Available at http://4humanities.org/mission/.

Becher, Tony, and Paul Trowler. *Academic Tribes and Territories: Intellectual Enquiry and the Cultures of Disciplines. 2nd edition*. Philadelphia, PA: Open University Press, 2001.

Brett, George, and McCarty, Willard. "A New Special Interest Group in the Support of Computing in the Humanities." *ACH Newsletter*, 9:2 (1987): 1–2.

Bulpin, Kate, and Susan Molyneux-Hodgson. "The Disciplining of Scientific Communities." *Interdisciplinary Science Reviews* 38:2(June 1, 2013): 91–105. doi:.10.1179/0308018813Z.00000000038

Busa, R. "The Annals of Humanities Computing: The Index Thomisticus." *Computers and the Humanities* 14:2 (October 1980): 83–90. doi:.10.1007/BF02403798

"Computing in the Humanities Course." *ACH Newsletter* 5:3 (1983).

Crozier, M. "A Problematic Discipline: The Identity of Australian Political Studies." *Australian Journal of Political Science* 36:1 (2001): 7–26.

Crymble, Adam. "Digital Hubris, Digital Humility." July 15, 2014. Available at https://www.insidehighered.com/views/2014/07/15/essay-backlash-against-digital-humanities-movement/.

Grier, D. A., and M. Campbell. "A Social History of Bitnet and Listserv, 1985–1991." *IEEE Annals of the History of Computing* 22:2 (April 2000): 32–41. doi:.10.1109/85.841135

Harris, Mary Dee. "Observations on Computers and Poetry." *ACH Newletter* 6:4 (1984).

Hockey, Susan. "Is There a Computer in This Class?" presented at the Is Humanities Computing an academic discipline? University of Virginia, September 1999. Available at http://www.iath.virginia.edu/hcs/hockey.html.

Hockey, Susan. "The History of Humanities Computing." In *A Companion to Digital Humanities*, ed. Susan Schreibman, Raymond George Siemens, and John Unsworth. Blackwell Companions to Literature and Culture 26. Malden, MA: Blackwell Pub, 2004.

Humanist. "Humanist 5/87-5/88." Edited by W. McCarty, 88 1987. Available at http://dhhumanist.org/Archives/Virginia/v01/.

Jamieson Price, Anne. "NEH Grantees Report on Computer Uses in Research." *ACH Newletter* 8:2 (1986): 1, 3, 4.

Kirsch, Adam. "Technology Is Taking Over English Departments." *The New Republic* (May 2, 2014). Available at http://www.newrepublic.com/article/117428/limits-digital-humanities-adam-kirsch/.

Mahoney, Michael S. "The Histories of Computing(s)." *Interdisciplinary Science Reviews* 30:2 (June 1, 2005): 119–135. doi: .10.1179/030801805X25927

Mahoney, Michael S. "What Makes History?" In *History of Programming Languages II*, ed. Thomas J. Bergin and Rick G. Gibson. 831–832. New York: ACM Press, 1996.

Marche, Stephen. "Literature Is Not Data: Against Digital Humanities." *Los Angeles Review of Books* (October 28, 2012). Available at http://lareviewofbooks.org/essay/literature-is-not-data-against-digital-humanities/.

McCarty, Willard. "Humanist: An Online Seminar for the Digital Humanities," 2009. Available at http://dhhumanist.org/announcement.html.

McCarty, Willard. "HUMANIST: Lessons from a Global Electronic Seminar." *Computers and the Humanities* 26:3 (June 1992): 205–222. doi:.10.1007/BF00058618

McCarty, Willard. "Humanist So Far: A Review of the First Two Months." *ACH Newletter* 3:3 (1987).

McCarty, Willard. "Humanities Computing as Interdiscipline," presented at the Is Humanities Computing an academic discipline? University of Virginia, October 22, 1999. Available at http://www.iath.virginia.edu/hcs/mccarty.html.

McLaughlin, Neil. "Origin Myths in the Social Sciences: Fromm, the Frankfurt School and the Emergence of Critical Theory." *The Canadian Journal of Sociology /Cahiers Canadiens de Sociologie* 24:1 (January 1, 1999): 109–139. doi:10.2307/3341480.

Nowviskie, Bethany. "On the Origin of 'hack' and 'yack.'" Nowviskie.org, 2014. Available at http://nowviskie.org/2014/on-the-origin-of-hack-and-yack/.

Nyhan, Julianne, Andrew Flinn, and Anne Welsh. "Oral History and the Hidden Histories Project: Towards Histories of Computing in the Humanities." *Digital Scholarship in the Humanities* 30(1) (2015): 71–85. Available at http://dsh.oxfordjournals.org.libproxy.ucl.ac.uk/content/30/1/71/. First published online in *Literary and Linguistic Computing*, July 30, 2013.

Powell, Alexander, Maureen A. O'Malley, Staffan Müller-Wille, Jane Calvert, and John Dupré. "Disciplinary Baptisms: A Comparison of the Naming Stories of Genetics, Molecular Biology, Genomics, and Systems Biology." *History and Philosophy of the Life Sciences* 29:1 (January 1, 2007): 5–32.

Presner, Todd, et al. "The Digital Humanities Manifesto 2.0." UCLA Mellon Seminar in Digital Humanities, 2009. Available at http://www.humanitiesblast.com/manifesto/Manifesto_V2.pdf.

Raben, Joseph. "Humanities Computing 25 Years Later." *Computers and the Humanities* 25:6 (1991): 341–350.

Rockwell, Geoffrey. "Is Humanities Computing an Academic Discipline?", presented at Is Humanities Computing an academic discipline? University of Virginia, 1999. Available at http://www.iath.virginia.edu/hcs/rockwell.html.

Rockwell, Geoffrey, and Stéfan Sinclair. "The Swallow Flies Swiftly Through: An Analysis of Humanist." In Hermeneutica.ca, forthcoming.

Ross, Donald. "Do You Need Information about Computing in the Humanities?" *ACH Newletter* 8:2 (1986): 5.

Terras, Melissa M., Julianne Nyhan, and Edward Vanhoutte, eds. *Defining Digital Humanities: A Reader*. Farnham, Surrey, England: Ashgate Publishing Limited; Burlington, VT: Ashgate Publishing Company, 2013.

UCLDH. "Memes." UCL Centre for Digital Humanities, n.d. Available at http://blogs.ucl.ac.uk/definingdh/digital-humanities-and-design/memes/.

15 Echo

Stacy Horn

It was 1982. I was working as a telecommunications analyst when a coworker pointed to what was then called an electronic bulletin board system (BBS) and said, "Check this out." It was filled with teenage boys, and while it was charming in a "Big Bang Theory" sort of way, it wasn't a place I'd want to visit again. When I told the boys I was female, they practically imploded. Most didn't believe me. They had good reason to doubt my word. There were few women online in 1982. Still, there was an allure to that BBS which I never forgot.

A few years later, in 1986, I entered a graduate program at NYU called the Interactive Telecommunications Program (ITP). There I was introduced to a whole other kind of BBS—an online community in California called The WELL. For people under a certain age, the extent to which The WELL blew my mind is going to be hard to comprehend. Today, everyone has 24/7 access to people all over the world. Then, only a small percentage of people owned a computer, and almost no one knew what a modem was. Unlike the BBS I'd visited years before, The WELL had a diverse group of smart, funny people who talked about every imaginable topic. It was a text-only version of what is perceived as the Internet today. In the 1980s.

By the time I got to my last year at ITP, I still hadn't decided what to do with the rest of my life, when someone on The WELL said, "I heard you were going to start a WELL-like service in New York." That had never occurred to me. "Yes," I immediately typed back, "I am." I spent my last semester writing a business plan, and by the fall, the new online service I'd started, Echo, was up in running. In early 1990, it was officially opened to the public.

I couldn't get funding to launch Echo initially. In 1989, I was unable convince a single person that the Internet would catch on. I was laughed out of every office I entered. "Who wants to talk to and meet other people on a computer?" men behind desks asked, with pity and disdain, like I was a loser for even thinking anyone would. I must have been the worst salesperson in the world, or the poorest presentation-giver, because you didn't need to be a visionary to see the Internet explosion coming. It was obvious to anyone who took the time to log into The WELL and start interacting.

So Echo began with every penny I had in the bank: $20,000. It was a struggle. The attitude I found when trying to raise capital was there whenever I tried to entice anyone online. If people had even heard of the Internet, they thought it was of interest only to socially challenged geeks. The only thing I had going for me was that I was not a socially challenged geek (or, rather, I didn't look like one). I cornered people at parties, art openings, lectures, conferences, and everywhere else I went and got them online one-by-one. As quickly as I could, I gauged what kind of people and conversations might be attractive to them and then convinced them they'd find these people and discussions on Echo. I'd explain what a modem was, talk them into buying one, teach them how to install it, and finally show them how to use the commands to navigate what was then a completely text-based world. In 1990, there were no graphics, no websites as we know them today, no pointing and clicking. It was a lot of work. I didn't sleep for years.

Getting people online one at a time was not an effective growth strategy, however. Without enough money for any kind of real marketing campaign, by the early 1990s, Echo was in danger of going under. I was terrified. Starting a business is kind of like starting a family. You put your entire heart and soul into it, and you don't want to lose it. You want to see it thrive. But I was running out of on-the-cheap ideas.

Then Bill Clinton was elected. He and Al Gore immediately started talking about the "information superhighway." This was a term for the Internet that had already been in use, but Al Gore popularized and made it mainstream. The Internet and Echo and places like it had already gotten some attention in the press, but now articles started appearing all over the place almost daily. It wasn't long before everyone was talking about the information superhighway. People began to feel a vague sense of unease. If they didn't get online, they'd be left behind. "How do I get on this highway?" they all started asking, anxious for directions. I was able to say, "Here! Echo is one of the stops along the information superhighway." When I met the director of the Whitney Museum and told him the Whitney should explore cyberspace, it wasn't such a hard sell. Geek became chic.

People still smirk from time to time about Al Gore claiming to have invented the Internet, something he never actually did. What he was taking credit for, and deservedly so, was his support, both economically and legislatively, for the Internet, support that began in his senate days. He introduced the High Performance Computing Act, which passed in 1991, officially making Al Gore the best presentation-giver ever with respect to the Internet. Unless you were an academic or a government or defense worker, chances are you weren't online until Al Gore got everyone talking about the information superhighway. The Internet would have prevailed eventually, but growth would have been delayed, and in that time, a lot of startups like Echo would have failed. The people who had room and encouragement to innovate would have found themselves in an entirely different world. Maybe they wouldn't have even begun. Who knows what that delay would have ultimately cost. As it was, when Marc Andreessen and friends developed Mosaic, the market was ready for them.

The phone companies weren't. I remember endless battles with NYNEX (now Verizon) about the quality of the lines. These were the days when you got online by dialing up using your phone line, and they were very noisy back then. It didn't matter for a voice phone call, but it was hell for data transmission. The phone company would insist the levels were just fine, and for almost a hundred years, they had been. Fortunately, I had a background in telecommunications, and I knew what I was talking about. I kept escalating my issues, and our lines started getting cleaner and clearer.

Echo didn't go under, thank you very much, Al Gore (and Bill Clinton). Twenty-four years later, we're still here, just barely. We have a small group of people who will probably stay here as long as the machine and software hold out. I like to say we're the dive bar that has hung in there even though the neighborhood all around it has changed. You want to know what the Internet looked like in the early days? Log onto Echo. It's all still there. Our real-time reactions to news and the evolving world around us have been saved, preserved in silicon.

A quick overview: Echo is a social network, although we didn't call it that at the time; we called them virtual communities (I was never able to come up with a term I liked better). It is composed of conferences, and each conference has a general heading like Culture, New York, or Movies & TV. Within each conference are topics of conversation that fit within that general heading. For instance, in 1993, someone started an item in the New York conference called "Terrorism in Our Town." This was on the day of the first World Trade Center bombing. It begins, "I am getting no TV reception except channel 2." Remember TV reception? Something was going on in the Towers, but we didn't know what. "I doubt that it was a bomb," another person said. "We'll see." Sadly, years later, on September 11, 2001, at precisely 8:47 a.m., I typed in, "A PLANE JUST CRASHED INTO THE WORLD TRADE CENTER."

In the Politics conference, there's a record of our stunned reactions as we watched the Clarence Thomas confirmation hearings and listened to the testimony of Professor Anita Hill. There are entire conferences devoted to things that no longer or barely exist, such as mail art. Or zines. In the Music conference, you'll find posts about who is playing at CBGB's. We started the "The NEA Death Watch Item" in the Performance conference in 1995, but fortunately the NEA has managed to survive. In a private conference for women only, there are initial reactions to the first transgender member who was about to be admitted. What was so earth-shattering then is so commonplace now. Response was mixed, and I think most people would be mortified now about what they said. I know I am. As an amateur historian, though, I can only dream about what it would be like to have this wealth of data from previous centuries. Think of what we'd gain in understanding. We wouldn't have to guess at how people thought or lose our eyesight deciphering handwriting in aging diaries. It's all right there.

When I started Echo, there were a few things I planned to do differently. Because most online communities were predominately male, I was going to do everything I could to get as many women on Echo as men. The estimate for the percentage of women online at the time was 10%. Echo quickly grew to 40% female. My success was due in part to the fact that I was the only one trying. Around the time I was launching Echo, I went to a conference about computer games. All the big developers were there. I stood up and asked, "What games are you developing for women?" They all looked at each other inquiringly and then back at me. Blank faces. Not a single one of them was exploring games for females. The commercial online services at the time were not targeting women either.

It helped that I was female. When I went up to women or addressed women's groups about getting online, it was easier for me to be heard. And believed. But there was one more thing I did that I believe was crucial. I'd gone into this whole thing with this image burned into my brain: I'm sitting at the end of a long conference table ringed with men in suits. I'm the only woman on the data side of Mobil's corporate telecommunications department (where I worked before leaving to start Echo), and as usual, I'm struggling to be heard. I knew the key to gender diversity was power and I made sure that women on Echo had it. Every conference on Echo has a host who oversees the place and is responsible for keeping things lively and settling disputes. They set the tone. Half the hosts on Echo were women. When women got online, they saw women in charge everywhere.

This really wasn't about politics to me. I wanted Echo to have a more diverse voice. Having more women would simply make Echo better. I also wanted fewer tech-oriented discussions and more talk about the things I was interested in. To make this happen, we opened conferences like Art, Movies & Television, Books, and Culture, and invited organizations like PEN and newspapers and magazines like *Ms.*, the *Village Voice*, and *High Times* to host conferences as well.

Looking back, what stands out for me is how utterly unprepared I was for the role I had to assume. At the time, people were still talking about the Internet in utopian terms, as the digital equivalent of a perfect world over the rainbow, forgetting that there were wicked witches and scary flying monkeys in the wonderful world of Oz. Almost immediately I was faced with the same problems that continue to plague the Internet today: trolls, harassment, love affairs gone bad, meanness, pettiness— basically all the problems and evils of society. I didn't think to put *that* in my business plan.

It wasn't fun. Much of my time was spent mediating arguments that could get quite ugly. Someone on Echo who'd made racist comments learned that one of her online friends was a member of that race. When she failed to have an awakening due to that discovery, I lost whatever innocence I had left about this new medium. That was when "we learned why cyberspace wasn't going to stop wars," I later wrote in *Cyberville*,[1] a book about my experiences creating Echo. It "wasn't going to bring peace and understanding throughout the world, tra-la-la-la. Cyberspace does not have the power to make us anything other than what we already are. Information doesn't necessarily lead to understanding or change. It is a revealing, not a transforming, medium." And you can't pry people's personalities from their furiously typing fingers.

"I was almost going to call my book," I joked, "We Will Never Learn and I Have Proof." While everyone else was talking about how great it was going to be when everyone was online, I pointed out that "they forgot how fucked up most people are. Think: Grover's Corners—the Dark Side."

As the de facto mayor of this new town, I had to come up with ways of managing every problem that arose. On the fly. And without possessing any particular skills or training to do so. I worried all the time about making the right decisions, ones that would minimize harm and hopefully not inflict any either. I did my best to develop guidelines for online behavior to reduce conflict without censorship. The most effective rule we established was no anonymity. If you were going to make ugly comments, you couldn't hide. You had to put your name to everything you said. If you failed to provide your real name when you opened your Echo account, you'd get kicked off. Another rule we instituted was "attack the idea and not the person." It created just enough distance that people could argue and disagree strongly and yet still maintain a small degree of civility. Not that you always want civility. We also decided later that attacks against groups (i.e., racist, sexist, or homophobic comments) were not welcome either.

We were pretty strict about harassment. If an Echo user asked another Echo user to stop trying to contact them, via email or something we call "Yo's" (real-time messages that would pop up on the person's computer, like a private Twitter post) or through constant references in posts, then that user had to stop. Continued attempts at contact would get the offender kicked off. This also helped contribute to the growing percentage of women on Echo. Think about what goes on in comments sections today. Now imagine what it was like 25 years ago, when even reasonably intelligent people were a great deal less enlightened about what constituted sexism and harassment.

We also created a number of private conferences to give different groups safe spaces to talk, such as the conference for women only mentioned earlier and one for men. I got people to meet in person, which helped defuse tense situations. We were constantly struggling to establish a balance between the right and need to express yourself and how to behave in a virtual world of mostly text.

The last thing I wanted to do differently exposed my biggest blind spot about the Internet. When Echo began, most people were talking about how great it was that the Internet was breaking down geographic and cultural boundaries. I wanted to celebrate and feature them. Logging into The WELL was so much fun for me because of its West Coast flavor and the ability to talk to people I wouldn't get to know otherwise. When people visited Echo, I wanted them to get a NYC vibe the minute they walked through the virtual door. What good is it, I thought, to be able to go all over the world if you don't feel like you've arrived somewhere? I wanted to retain our cultural identity.

For better or worse (mostly better), the Internet and all the applications that go with it have been slowly, ever so slowly, flattening out those cultural differences. A global voice is emerging, and you can't always tell if you are talking to someone from the United States, India, or Saudi Arabia. This has the potential to increase understanding, so it's hard to regret it too much.

A professor of mine at NYU once said, "Users get very attached to their particular system. It's like being citizens of different countries. They develop loyalties." But that is less true these days. Now we have more virtual middle grounds, where people of all sorts congregate, such as Facebook.

The Internet has worn down more than cultural boundaries. The boundaries of power are changing as well. The United States will become less and less important, but in my opinion not in a bad way. It's just that the rest of the world has risen and will continue to rise in importance, and in power and influence. Many things have contributed to this growing shift, of course, but the ever-present global interaction and sharing of information has played a big part in empowering individuals, groups, and nations. The Internet is the ultimate power-distribution medium.

No one gives up power without a battle, however, or accepts change without resistance. Some groups are more entrenched and recalcitrant than others. Down South, some people still fly the Confederate flag. The divide between liberal and conservative in this country is at its most acrimonious right now, just as America is slowly becoming more fair and more just (we have a black president, increased civil rights for gays, etc.). For that reason, as tolerance grows around the world, groups such as ISIS and others will arise. There are going to be problems everywhere, as we all continue to adopt from one another's culture, and to blend into each other a tiny bit, and then more and more, some rising a little, some falling (or perceiving that they are falling). There will always be those who fight change. But it is happening, and it is inexorable.

So I was wrong. It is a transforming medium. It's just that it was happening so slowly, I didn't see it. There will always be differences, of course. I don't want to overstate it. Some people will go to their graves flying the Confederate flag, and they will give birth to people who continue to fly that flag. But in time, more will see why that isn't such a great idea. Someday a softer version of the most entrenched and recalcitrant groups will be sharing the future Facebooks with the future Occupy Wall Streeters and Black Lives Matter activists, and someone else will have to struggle, as many of us did and continue to do so now, to establish guidelines to make *that* work. Perhaps one thing won't change. I suspect that reading the comments section—reading or whatever form communication takes in the future—will always be a bad idea. There will always be a dark side.

Note

1. Stacy Horn, *Cyberville: Clicks, Culture, and the Creation of an Online Town* (New York: Warner Books, 1998).

16 MOOs and Participatory Media

Dene Grigar

Twenty years ago, I defended my dissertation online with more than 50 people in attendance from the United States, the United Kingdom, Australia, and Norway in a virtual environment called a MOO.[1] MOOs circa 1995 were a net-based, multi-user technology that, before the ubiquity of the browser, offered textual interfaces and spaces for people to communicate and connect socially with one another. The idea that I was broadcasting a live defense, certainly a serious event in one's academic career, over the net to an audience of textual avatars was unique, but even more interesting in light of today's interest in participatory media was the fact that the event was interactive, and the audience engaged with me in a way uncommon in traditional defenses. In a sense, this event—the first documented online dissertation defense—represents a kind of social practice seen readily in social media networks today.

This chapter revisits aspects of the event, tying them to theories and current practice of social media. In doing so, it situates MOOs in a historical, cultural context as an early form of participatory media related to social media environments prevalent today. In doing so, it establishes that popular social media platforms, such as Facebook and Twitter, are not new but, rather, are part of an evolution of technologies that foster the human impetus to connect with one another across any mode of communication.

Situating Ourselves Then

In 1995 when this event took place, graphical user interfaces (GUIs) were not yet prevalent. Mosaic, the first commercial web browser, had been on the market for only a year and a half, and Netscape Navigator, the browser that popularized the web, only seven months. MOOs, during this early period of the web, were still accessed through the "net" and all objects, people, and conversation found in them, instantiated textually. Later, some MOOs would convert to web-based environments, but at the time of my defense, this technology was not yet possible.

MOOspace was often structured and articulated as "rooms" described textually in great detail as a way to humanize the online experience, and these descriptions were

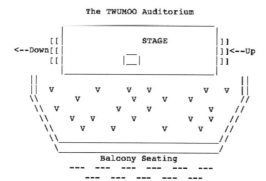

Figure 16.1
The Auditorium located in TWUMOO, circa 1997.

available to visitors who "teleported" into the space. Communities emerged around particular MOOs. LinguaMOO, the MOO created and maintained by Cynthia Haynes and Jan Rune Holmevik where I defended my dissertation, was a gathering place for scholars interested in digital culture and writing. The room in which I defended my dissertation was called the Auditorium. MOOs could contain as many rooms as the server space on which the MOOs were managed would allow. Considering that each member of a MOO community could potentially create and manage a series of rooms, it is easy to see that MOOs like LinguaMOO with hundreds of members could easily become large complexes of virtual offices, personal residences, and public spaces.

The graphics used to visually represent rooms and objects were created as ASCII art and, so, made of letters and other keyboard elements. Communication occurred by typing in a buffer. Our comments were represented in a convention that began with the phrase "Dene says," thus emphasizing the human interaction taking place. With no access to a spellchecking or grammar tool, typos were common and commonly ignored by necessity. Shortcuts for words and phrases, such as ROFL or "rolling on the floor laughing"—arose as a way to make communication more immediate and easy. Many of these shortcuts, such as LOL ("laughing out loud"), are still used today. Additionally, text-based MOOs were silent spaces with sound limited to the pounding on the keyboard when we typed. Full-color graphics, now ubiquitous in our online lives, were not possible in these virtual worlds until later, and even then not all MOOs embraced the GUI that web-based MOOs offered. While people today living in a world of Facebook may regard text-based MOOs as unappealing, we denizens of MOOs coming from "the late age of print," as Jay David Bolter had coined this period of our cultural history,[2] were comfortable in this black and white, "bookish" environment. What made them special to us were the social interactions that took place in them and the fact that they were indeed a cutting-edge technology that we all knew was a step in a digital revolution. Thus, despite the lack of color, movement, and sound, the experience we had in and with the MOO environment imbued us with a sense of the unlimited possibilities of human expression.

A Social Defense

Anyone who has ever defended a dissertation in the United States knows that it is not the time to ignore errors of any sort or "roll on the floor" doing anything, especially "laughing." Conversation is generally limited to a formal presentation by the candidate, followed by a series of questions and responses between the committee and candidate. Visitors, if present, ask questions following the committee's own. True, while defenses may often provide a friendly, relaxed atmosphere, the kind of raucous give and take found in MOO discourse is incompatible with the notion of polite participation found in a traditional dissertation defense.

Also, defenses are not generally viewed as social events. They may be advertised through the department or college with a general invitation to attend. But in the 20 years I have been on faculty at various universities, I can attest that attendance by others outside of the committee and candidate is sparse, and certainly audience participation is limited to questions posed to the candidate and does not include speaking aloud when the candidate is talking or visitors mumbling among themselves.

From the start, my defense in LinguaMOO constituted a social, participatory event that did not follow traditional protocols. Notices about the event, for example, were disseminated across the net to listservs like *Humanist* and *MBU* (Megabyte University). Scholars began writing me directly to inquire about the event and reposting the invitation on sites like the Center for Programs in Contemporary Writing listserv. Close friends promised to attend the event, but many others whom I did not know RSVP'd from places that I had no idea my defense had been announced.

Fifty people from across the globe came to see the first online dissertation defense. They assembled in the Auditorium, the room that Jan had created specifically for this event, and talked among themselves before and during defense. They appeared to me as avatars in the space. I could see them "finding" seats, "tiptoeing" in and out of the room, and "clapping" for me. Conversely, they could see me "showing" slides, speaking my words, and "smiling"—or I should say that we all saw textual representations of us taking part in these activities. However, once the event began, visitors' comments were moderated so that no one could send me answers to questions or interrupt me while I spoke. In a sense, "the panel mode" that Jan programmed for the space

Dissertation Defense in MOO

```
From janruneh@utdallas.edu Fri Jul 14 18:34:17 1995
Date: Fri Jul 14 17:32:09 1995 CDT
Subject: Message from LinguaMOO

Dear friends,

It's a great pleasure for us to tell you of Dene Grigar's on-line
doctoral defense to be held here in LinguaMOO on *JULY 25TH*. The
defense will start in the Auditorium at 4pm CDT (5pm EDT, and 10pm GMT)
so be sure to be there at least 10 minutes before it starts. This event
will coincide with Lingua's 6 month anniversary, and there will be a short
ceremony in the Courtyard shortly after the defense to commemorate this.

Dene Grigar's study, "Penelopeia: The Making of Penelope in Homer's Story
and Beyond" traces the artistic response to Penelope in literature, the
visual arts, and music, from the Middle Ages onward. To the best of our
knowledge, this will be the first defense held on-line.

Participating in the panel discussion at the defense will be Grigar's
committee, comprised of Dennis Kratz, Tim Redman, Rainer Schulte, and
Deborah Stott, as well as a few scholars who are currently involved in
research on Penelope. Thus far, Patricia Moyer (UNC) and Katie Gilcrest
(Oxford) have both agreed to participate.

Cynthia Haynes-Burton and Jan Rune Holmevik
```

Figure 16.2
Cynthia Haynes-Burton and Jan Rune Holmevik, "Dissertation Defense in MOO." Announcement dated 7/14/1995 for Dene Grigar's defense of her thesis. Available at http://writing.upenn.edu/~afilreis/88/diss-defense.html.

provided some sense of structure and ensured that I was, indeed, responsible for my own responses.

The physical space where my defense took place at the University of Texas at Dallas was the computer classroom, a lab filled with Macintosh computers where others and I taught writing and literature courses. In the physical space, I stood at the instructor's podium and gave a formal presentation using MOOslides—that is, textual slides that provided information about my research. Participants in the computer classroom could read them on the room's projection system or, like the online audience, on their computer monitors as slides presented in the virtual Auditorium. The questions and responses that followed the formal presentation were also rendered as text my examiners and I typed into the buffer. Because members of my committee and others who wanted to join us from online locations were not all familiar with the MOO environment, Cynthia, Jan, and I recruited assistance from colleagues acquainted with virtual environments. In the room with me at the defense were Cynthia Haynes and John Barber, who had just completed his dissertation on online communication; online were Jan Rune Holmevik and Jeff Galin, who was a member of the MediaMOO community. During the event when some audience members could not see the MOOslides, Jan went to work to fix the bug in the software, and Jeff jumped in to explain to the online audience the reason for the "silence." Having four active participants involved in the proceedings of my defense was untraditional and added to the collaborative and social nature of the event.

The defense was recorded as a MOOlog and then archived and disseminated later for others to study. John and I published an essay about it for Cynthia and Jan's book, *High Wired*,[3] and we gave papers about it at various conferences on topics ranging from MOO defenses to research methods.

Defending the Social

Although certain traditions like the formal presentation and the interrogation by a committee were adhered to, my defense was essentially a social one, taking place in an early form of a social network.

It was attended by invited, immediate friends. My collaborator on the textual analysis of the *Odyssey* that led to my first published essay was present by invitation to the face-to-face event. Invited to the online portion were colleagues with whom I had met and collaborated in LinguaMOO, like Jan. Friends of friends and people who had heard about the event from the online postings to listservs also attended. Some of these scholars I knew or had produced work I had studied; others were as unknown to me as people who see my comments on other people's Walls on Facebook. Ironically, as Katie Gilchrist, one of two Classicists who acted as outside examiners, pointed out in an email to me following the event, the inclusion of so many people in the "room" reminded her and others of an "old-fashioned sort of open viva" that had, indeed, been replaced in contemporary times by a small committee of examiners.[4] So, the future met the past in this event.

As strangely social that this event was to colleagues in the late 20th century, such an experience is not so odd today. While, to my knowledge, no dissertation defense has been held on Facebook, academic events are commonly advertised, and debates take place on social media sites. But a social event reflected precisely the kind of experience those of us who frequented MOOs and spent so much time online envisioned and believed to be appropriate for our work.

I had conducted my research, for example, from the beginning, as a socially constructed experience. I spent hours online visiting library catalogs as they were being posted online. At the time, university and other research libraries were racing to make catalogs of their holdings available on the Internet. Using "gophers," systems now rendered obscure today by the web, I tunneled through the net finding sources for my dissertation on Homer's Penelope. Because my research focused on finding as many interpretations of her in the arts from the middle ages onward, my exploration was broad and access to premier research library holdings imperative. I had no budget for travel to Harvard, Princeton, or the Library of Congress just to see whether any of them housed images, recordings, or videos of opera, ballet, or theater performances I needed, but I could "see" their online catalogs containing holdings that referenced works. My dissertation chair had told me that if I could discover 30 to 40 resources, then he would be happy. Within six months of online research, I presented him with 346.

Once I had developed this robust list of primary sources, I wanted to share it with others. This ethic is one that stemmed from my experience with MOOs where one shares the objects of our creation freely with others. The MOO community embraced collaboration and sharing; key to success was the production of useful creations, and fame was predicated on the recognition that what one made was deemed useful by others. The fact that Cynthia and Jan built a room for my defense in their MOO exemplifies the *esprit de corps* that built up around virtual environments and is present even now in open source communities. So, I created a webpage. To do this in 1994, I had to build it with a Unix account on my university server. The articulation of my site was essentially an ASCII graphic of a woman representing Penelope and the long list of works I had discovered that related to my topic. This webpage was considered the first research webpage at my university and in the city of Dallas. I disseminated the URL around the net so that others could access my research. I also posted specific information about my work on listservs and in email messages to colleagues, adding the URL to the website for others to access.

In the days before web analytics, I had no idea how many people actually visited my webpage. But people who did visit sent me suggestions for additional references. My website generated an additional five resources, bringing my total number of representations of Penelope in the arts to 351; later it would grow to 356. My dissertation committee, while delighted with my research findings, expressed concern that I was giving away my work and making it possible for others to publish it before I had a chance to defend or publish it myself. But I felt that publishing on the net was no different than publishing in print. Besides, I told my Chair, I had already read, viewed, listened, and experienced most of the sources on my list. It would have taken others an immense amount of time to catch up.

Today, on Facebook, we can find pages related to all kinds of research projects. I maintain, for example, a Facebook page for my digital preservation *Pathfinders* project, listing and posting up-to-the-moment findings about my research for others to access and use. Email messages from colleagues who requested access to *Pathfinders* videos resulted in my collaborator, Stuart Moulthrop, and I uploading the rough cuts of videos from our interviews on YouTube and placing a link to them from our website, which I also maintain for us. When the videos were finally edited and polished, we shared them immediately on Vimeo, updating our website to reflect this new addition to our research project. To date, 10,234 people have frequented our *Pathfinders* Word Press site. Additionally, we've logged over 10,000 visitors to the project's eBook, published in June 2015. The research has been cited in numerous essays and has found its way into classrooms of scholars interested in digital preservation of electronic literature.

Thus, because I had been so active online with my research, it made perfect sense to my committee and me that my defense should be conducted online. I will say that most of my committee members were unaware of how social MOO environments were and how vocal participants could be in a space where sound did not exist. I do not think any of them were familiar with the concept of "holding up signs" and other antics common in MOO discourse. My committee also did not expect the waving and chatting in the midst of such a serious event, but MOO denizens were neither surprised nor rattled by the activity. In fact, I was comforted by the fact that I had so much support and that there was such wide interest in my work. We see this kind of positive reinforcement in social media sites where people actively "Like" our comments and "Share" our postings. On the flip side, negative comments, bullying, and harassment existed even then. Flaming was common, and rules about online etiquette, or "netiquette," developed over time to address personal conduct. Today, my students recognize that using all caps in Twitter is akin to yelling at someone. They also know that "troll" is a negative term used for people who make a habit out of attacking others on Twitter and Facebook. What they may not know is that the term was used for people who behaved badly in MOOs. Julian Dibbell's famous essay, "A Rape in Cyberspace,"[5] chronicles one such notorious occasion. In the MOO I administered at a woman's university, it was not uncommon for me to find people teleporting in specifically to harass female students. In those cases, I had options for dealing with such conduct, like @eject, @newt, @toad, and @blacklist, as Twitter does today in its handling of trolls. Social media does not always mean socially polite media but rather socially human ones. But the way Jan had constructed "the panel mode" for the defense meant that negative comments would have been moderated before they reached me. I did later read in the MOOlog archiving the event that colleagues had wondered about a comment I made or teased me when I lagged in responding to a question. By that time, I had passed my defense and earned my PhD.

Finally, I wanted to make my work public, and 20 years ago, posting on listservs, emailing colleagues, publishing a text-based webpage, and hosting an event in a MOO constituted what it meant to be a "public" digital scholar working in the Humanities. I did, indeed, follow traditional scholarship and gave presentations about my work at conferences, but the reach of this early social media traveled farther and wider than any paper I gave at the Modern Language Association conference.[6] When I began collecting references to my defense for the article in *High Wired*, I remember being surprised by the number. I had a collection of emails from people I had never met saying they were interested in the event—some because of the subject matter, and others because of it being an online defense. But I had no way to track the total number of references the defense sparked. Today, when I post about my work on Facebook, the information goes out to my 1026 Friends and then is viewed by their own Friends. Facebook Insights provides a detailed look into the effectiveness of my communication. My Twitter tweets go out to 1900 Followers, many of whom are not Friends on Facebook. These Followers have hundreds, even thousands, of Followers of their own, so my research quickly becomes *very* public. Social media sites are now deeply supported by universities and are not the oddities they were 20 years ago. My own university maintains numerous Facebook and Twitter accounts, and those hired to manage these accounts are also responsible for reposting and retweeting information that faculty and students make available. Individual departments and faculty own and manage social media sites such as Facebook, Twitter, YouTube, Linked In, and many others. In truth, the experiment we undertook with my online defense in LinguaMOO was, indeed, a precursor to the kind of investment universities are making today with their promotion of education and research.

Demise of MOOs, Birth of Public Spaces

LinguaMOO disappeared when the university shut down the server on which it was hosted. I pulled the plug on my own MOO, Nouspace, in 2006 when I moved from Texas to take a position in Washington State. At the time, I had only a handful of people left in my community, down from 200 at its height in the early 2000s. It was expensive to host my MOO, costing me $300 a month to maintain a business Internet connection that allowed for the speed and port access needed. Because I never charged for access, I bore the cost of maintenance myself. Expense may have been a reason that MOOs managed by individuals shuttered, but universities lost interest when hacking became a concern. The open port needed to administer a MOO made them potential targets. However, the kind of social, participatory activities they encouraged continue today in Facebook, Twitter, and other social networks.

As I type this chapter, I am watching the activity on my Facebook page, which reminds me of managing my MOO. The green dots besides my Friends' names indicate when they are coming and going online. Friends message me with questions. One Friend has "Liked" another Friend's post about her anniversary. Different, of course, are the photos and videos my Friends have posted. But MOOs also allowed for this media when they morphed into web-based environments. Nouspace MOO was originally built on enCore technology created by Jan and used in LinguaMOO in the late 1990s, rendering it a web-based environment that could take present sound, images, and animations. Visitors to Nouspace's *Fountain of the Avatars* could view an animation that had water spouting up and on to the stones below and hear music playing in the background. When *Second Life* emerged, many of us viewed it as a natural development of virtual environments. Blogs and, later, social media seemed only to be different expressions of what we had already developed, theorized about, and played in. In 2006, when I gave up my MOO, I joined Facebook and "Friended" colleagues and students; colleagues working in traditional disciplines with no experience of virtual environments were shocked by my "informality." But to me, Facebook offered a way to build a community of scholars, share information quickly, and provide support much like MOOs did. It does not surprise me that many of these critics now have social media sites of their own and are also now "Friends" with students.

Influencing the decision to undertake my dissertation defense in a MOO was the understanding that we were in the midst of a digital revolution affecting every part of our lives, from communication to social relationships. We denizens of MOOs were interested in learning whether they and other spaces like them would hold their own against real environments in educational, academic experiences and what the future of such social spaces held. Sites such as Google+'s Hang Out are now used by scholars for communicating at a distance with one another. Remote options for defenses are, indeed, occurring for online degrees. MOOs represent an early iteration of social, participatory communities that current social media reflect. As I write these concluding words, I wonder how I would utilize Facebook for a dissertation defense. Certainly, it is possible, and academe is ripe for such an endeavor. The digital revolution is far from over.

Notes

1. The event was documented as "Penelopeia: Penelope in Homer's Story and Beyond," Lingua-MOO, July 25, 1995. It is also referenced in Jan Rune Holmevik, *Inter/Venion: Free Play in the Age of Electracy* (Cambridge, MA: MIT Press, 2012); and in Dene Grigar and John Barber, "Defending Your Life in MOOspace," in *High Wired: On the Design, Use, and Theory of Educational MOOs*, ed. Cynthia Haynes and Jan Rune Holmevik (Ann Arbor: University of Michigan Press, 2001), 192–231.

2. Jay David Bolter, *Writing Space: The Computer, Hypertext, and the History of Writing* (Hillsdale, NJ: Lawrence Erlbaum Associates, 1991).

3. Grigar and Barber, 2001.

4. Ibid, 196.

5. Julian Dibbell, "A Rape in Cyberspace," *Julian Dibbell.com*. 1993. Available at http://www.juliandibbell.com/articles/a-rape-in-cyberspace/. Accessed October 21, 2014.

6. I gave a paper on the research methodology I used in a presentation titled "On Site Research and On-Byte Investigation: Three Views of Conducting Thorough Academic Research." 1995 Modern Language Association Convention. December 1995. Chicago, IL.

17 Hacking the *Voice of the Shuttle*: The Growth and Death of a Boundary Object

Alan Liu

This is the tale of two hacks. The first was building my *Voice of the Shuttle (VoS)* starting in 1994—a "Web page for humanities research" (as I subtitled it) that collected online resources in the humanities alongside selected scientific, social-science, and other resources. Each item was glossed with a brief description, and the whole was organized in a classificatory framework that both stayed faithful to academic categories (field, period, authors, etc.) and blurred them into new configurations (e.g., on my pages for "Cyberculture" and "Technology of Writing," which were de facto placeholders for the yet to emerge "new media studies" and "digital humanities" fields).[1] One of the earliest attempts to create a portal of Internet resources for the academic humanities, *VoS* in its prime in the 1990s and 2000s had mirrors in Italy, Japan, and the United Kingdom and was widely used among not only scholars but a broader public. Among its other awards, it was named to *Forbes Magazine*'s "Best of the Web Directory" in 2002 as a "premier online destination for the humanities and social sciences, for casual surfers and die-hard researchers alike" because of its "deep research links" and balanced attention to "classics" and "contemporary topics like Cyberculture, Technology of Writing or Postindustrial Business Theory."[2]

The second was the malicious hack (actually a series of hacks) a decade later that destroyed—or, put another way, clarified—my dream of what *VoS* could and could not be. *VoS*, I now understand, was a transitional version of what Susan Leigh Star and her colleagues theorized as a "boundary object."[3] It was a sunrise, and sunset, boundary object spanning disciplinary and geographical scholarly communities in hopes of creating a global collaborative in which humanists (and the public) could freely take and add knowledge across their divisions.

As defined in Star's groundbreaking article of 1989, "The Structure of Ill-Structured Solutions: Boundary Objects and Heterogeneous Distributed Problem Solving,"

Boundary objects are objects that are both plastic enough to adapt to local needs and constraints of the several parties employing them, yet robust enough to maintain a common identity across sites. They are weakly structured in common use, and become strongly structured in individual-site use. (Star 1989, 46)

That is, boundary objects are artifacts invested with concepts and practices that facilitate epistemological transaction between communities (e.g., research biologists and amateur naturalists) because, like mental magnets, they have the property of being able to attract local interpretive paradigms from different knowledge enclaves and stick them together with just enough, if partial or "ill-structured," overlap to allow everyone to "cooperate without having good models of each other's work" (Star 1989, 46). Specifically, *VoS* fit the bill of two kinds of boundary objects identified by Star: "repositories" and "forms and labels" (Star 1989, 48, 50).

In general, Star observes, boundary objects are key to "open systems" in which "different locales have different knowledge sources, viewpoints, and means of accomplishing tasks based on local contingencies and constraints" (Star 1989, 45) and where there is thus "no global temporal or spatial closure" or "central authority" (Star 1989, 40). Translated in terms of the metaphor of the "Web" (which I spun out in the title of my site by alluding to Aristotle's, and later Geoffrey Hartman's, mention of Sophocles' lost play about Philomela's "voice of the shuttle"), boundary objects like *VoS* are creatures of an open rather than Arachnean weave.[4]

Over the last 20 years, *VoS* lived and died this dream of a boundary-spanning open system. Today, it survives as an eroded monument to an early Internet era that dreamed of free, "ill-structured" systems but was "so rudely forc'd"—as T. S. Eliot's *The Waste Land* describes Philomela's violation—to wake to something else.[5] The following are the two hacks of *VoS* that taught me this lesson.

Hacking *VoS*, 1994

I hacked *VoS* into existence in late 1994—where "hack" in this context has the *maker* and *bricoleur* senses idiomatic among early computer academics and hobbyists (and more recently among the digital-humanities community in expressions such as "more hack, less yack").[6] The site began as a local resource at the University of California, Santa Barbara (UCSB), where I had been researching the Internet while working on what became my *Laws of Cool* book. This research spilled over into creating courses that included the Internet as a topic, starting a Web-authoring hobbyist group called the Many Wolves, and selling in the campus bookstore my lengthy, self-published *Ultrabasic Guide to the Internet for Humanities Users at UCSB*.[7]

All this spillover activity was inspired by the fact that, like the self-aware and newly unfettered artificial intelligence at the end of William Gibson's *Neuromancer*, I felt rather lonely online.[8] There was a big, new wired universe out there, but my local humanities community at UCSB (like most such communities at the time) had barely discovered email. To seduce my community onto the Internet, therefore, I hawked my *Guide*, sent out proselytizing emails disguised as how-to tips, and otherwise tried to pull back the veil of technical arcana that then obscured the Internet. I typically began with housekeeping skills (Unix commands, downloading and uploading, etc.) and then proceeded through various facets of the Internet from FTP and telnet through Gopher and the Web.

Soon, however, it became clear that I would do best to concentrate on just one facet of the net. That was the Web, of course, which had begun to prove its general appeal and was quickly subsuming other facets. My agenda thus simplified: I standardized on recommending the Web as the path to follow beyond email, and I tried to make visible enough compelling academic Web content to entice my community to take the bait.

It was to surface this content that I started *VoS* in December 1994 as a simple file of HTML links circulated locally on campus—a restriction necessitated by the fact that my university's humanities server computer at the time (humanitas.ucsb.edu, now extinct) was not yet a Web server. My links file became an accretion disk growing ever larger and more densely organized as I continued collecting (using mainly the Lynx text-only browser for sheer speed in the days of dial-up modems). I also solicited and HTML'd the writings of UCSB people for a "Featured Works" section, rotating among academic departments so that I could lure fresh newcomers onto the Web to read the work of colleagues or mentors.

When on March 21, 1995, the campus Humanitas machine became a web server, *VoS* made its global debut as its home page. It remained there until July 22, 1999, when it migrated to its present campus top-level domain (vos.ucsb.edu). As *VoS* went global, I adjusted my descriptions of pages, categories, and topics on the site to make them more useful for a larger audience (i.e., to make the site a better boundary object).[9] *VoS* now addressed the UCSB humanities community *plus* a widening archipelago of other user communities, each discrete in disciplinary field or geographical location yet all cross-linked through my evolving ontology of categories and levels. Due to my background in literary studies, the disciplinary communities that gathered around my literature pages (sending me correspondence, making suggestions, etc.) were fullest and deepest. But my Art and Art History, Religious Studies, and several other pages also early on acquired robust, discrete user communities. A similar archipelago pattern emerged geographically. Most repeat visitors in the early years hailed from the United States, Canada, and the United Kingdom, but increasingly I saw in my log files sizable user communities from Italy, Australia, New Zealand, Japan, Mexico, and several other countries (plus a constant stream of first-time visitors from all over).[10] The fact that such globalism was a matter of boundary spanning rather than homogenization was suggested by events after a *VoS* mirror site went up in Italy in 1998 (one of several mirrors around the world).[11] Suddenly, I began receiving emails from Italian users accompanied by suggestions for resources authored in Italy. It was clear that a discrete community had been added to my user base. In terms of sector and age groups, too, communities were boundary-spanning in the sense of being at once differential and intersecting. As might be expected, most visitors and contributors came from higher-education institutions. But many also came from the elementary and high school educational communities, adding a distinct flavor of their own. Even the surprisingly large number of general public users (from the .com, .gov, .org, and .mil domains as well as America Online, Prodigy, Compuserve, Netcom, etc.) asserted a discrete personality while also reaching hands across the aisle to academe. I often received feedback from such users in a persona self-conscious about its niche outside higher education but determined to bridge the boundary (e.g., the user fooled by the "shuttle" in my site title who wrote to say that "as a working engineer I wouldn't normally have stumbled onto your site, but I just had to see what NASA had to say about the humanities" or, again, the correspondent who began a letter, "I'm just an AOL'er, but I have a literature site that may interest your readers").

Only occasionally did usage spikes sparked by the mention of *VoS* in a mass-circulation newspaper, magazine, or website (as occurred at various times when *CNN.com*, *Forbes.com*, *Los Angeles Times*, *New York Times*, or *USA Today* cited the page) create an apparently undifferentiated global response. For instance, when the CNN website in 1996 linked to an essay I had posted on *VoS* the previous year ("Should We Link to the Unabomber? An Essay on Practical Web Ethics"), more than 3,000 visitors stopped by each day, which at the time was a large number for an individual academic site.[12]

By June 1996, *VoS* had grown to some 70 web pages covering 23 to 30 disciplines (depending on how one counts disciplines), and the time I spent working on the site averaged two hours each day. The total bulk of *VoS* at that time was about 3 Mb of links plus another 1 Mb of images and other matter. But it was the range of *VoS*'s user and contributor communities—widening and intersecting, enacting the "world wide web" as an open weave suited to the "interdisciplinary" enthusiasm of the time—that most excited me. I remember in those early days, when I still had shell access to the server (something unthinkable today), that I watched my log files in real time to see the hits coming in from user communities around the world. People also sent me scores, then hundreds, and ultimately thousands of emails each year to contribute links, request subject coverage, make corrections, and so on—to the point that suggestions I once fielded by email had to be redirected through a web form to an automatically aggregated "New Links Suggested By Users" page from which I harvested material.[13] I felt that I had hacked the humanities for the better and that *VoS* could become the crossroads site for all like-minded, if distinct, communities interested in the humanities.

On October 16, 2001—with the programming assistance of Robert Adlington and Jeremy Douglass (at the time graduate students in my department)—*VoS* took the fateful step of migrating its static HTML pages to a dynamic database-to-HTML platform (built on a Microsoft SQL Server backend with a custom-programmed content management system created by Adlington and Douglass). The goal was not just efficiency (allowing me to edit via web forms to propagate changes) or more customizable searches and views for users. It was also to broaden and deepen community. The plan—still stated on the *VoS* home page as a relic—was to let users sign up for accounts that would permit them to submit links for specific locations on the site and then, once these had been edited and approved, maintain links and even in future create custom versions of *VoS* for local purposes. I quote from the full plan that appeared in December 2003 when one pressed the "Helping Edit VoS" link on the *VoS* front page:

Contributing to VoS: Once users sign up for a free account on *VoS*, they will be able to use Web forms to suggest links for particular areas of the site. ... Suggested links appear right away on *VoS* under Unvetted Submissions. They later [appear] on the regular pages after being reviewed by *VoS* editors. ...

Editing VoS (for those with accounts): In a later implementation phase of the new *VoS*, users who have signed up for accounts will have additional editorial privileges allowing them to maintain/revise links that they have contributed. There will also be group accounts that enable classes, organizations, conferences, etc. to build subsets of *VoS* resources that will appear both on the regular *VoS* pages and on a special page set aside for the group. ... *VoS* will thus be an open platform serving the needs of both general and specific communities of users. ... (Internet Archive copy of "VoS Help Page," December 20, 2003)

Hacking *VoS*, 2004–2005 ("Bad Hack")

That was the open dream. But the dream died—or, as I put it, clarified—when the dark side of the force of "hacking" emerged in two hacks by outsiders in 2004–2005. On January 31, 2004, I woke to find all content and the category structure of *VoS* erased, leaving a single category populated by the stark phrase, "Bad Hack." It was a SQL injection attack—a mode of exploiting online SQL databases that came into fashion beginning circa 1998.[14] We restored *VoS* from tape backup and then put in place a number of protective measures. However, on December 11, 2005, a further attack resulted in another catastrophic hack, deleting all content and retitling the top category in the *VoS* hierarchy, tauntingly, "Site Hacked." After detailed forensics by Jeremy Douglass (working with our longtime department sysadmin, Brian Reynolds), we traced the attack's pathology and probable origin. In all likelihood, it was initiated from a part of the world well known to target Western and other websites. The modus operandi—a kind of dark-arts digital forensics known as error-based SQL injection—was to inject bad SQL statements (via http requests appended as queries to a target site's URL) into a website's database, provoking the database into returning error reports that revealed enough about internal structure (columns and tables) and ultimately also user logins that the hacker could take control.[15] Over several months (during which further SQL injection attacks of a "script-kiddie," rote sort were attempted), Douglass worked to harden the *VoS* site—one of the most unfortunate diversions of intellectual talent for purely nonproductive, reactive goals I have ever witnessed (although the chapter on "The Aesthetics of Error: IF Expectation and Frustration" in Douglass's dissertation on interactive fiction owes its inspiration to his forensics on the *VoS* error-report exploit).[16]

The result of these "bad hacks" was a fatal hardening of the arteries of *VoS*. We hardened the site in innumerable ways by revising table relationships, constraining permissions, turning off error reporting, and restricting access routes for editors, with the overall result that both the quantity of editor-users and the quality of their editing experience were severely curtailed. This stopped the hacks, but it also killed plans for a relatively open system. Of course, if *VoS* had been a .dot com firm with vast engineering staffs, it could have created a secure custom platform. If its homemade content management system had not preceded later ones supported by open-source communities (WordPress, Drupal, etc.), it could have adopted a platform whose security is constantly tightened by others. Yet again, if *VoS* had been a Wikipedia with a large group of administrators, "bureaucrats," and others able to evolve an increasingly elaborate system of governance (policies, criteria, talk pages, administrative and enforcement roles, arbitration processes, sanctions, protections, and bots to assist with the above), things also could have been fine. But *VoS* was just an academic site I had personally started and maintained with graduate students and a departmental sysadmin. We couldn't be the secure open system we wanted to be at the time. Today, the cost in work hours that would be needed to reimplement *VoS* on a new platform designed for better compromise between openness and security is simply too great to justify when the site's original function (helping humanists discover links) has in great part been superseded by algorithmically driven search engines (although, of course, these sacrifice the edited organization and context provided by *VoS*). Plus, of course, the Web has just gotten too big, making my original vision of "editing" it naive. When *VoS* started, there was Yahoo with its (at the time) hierarchically collected and arranged resources. I thought I could not only match Yahoo in the area of humanities resources but outdo it in providing an organizational framework better suited to scholarship. Now there is Google, Bing, and so on, and I have neither the ability nor ambition to impersonate them, which is also to say that as a scholar, I actually do not wish to be a corporation. For corporations today, after all, the great boundary object is the stock market and, more basically, capital.

VoS was a scholar's dream of the Internet born in an era when no one imagined how big the web would become or how divergent, hostile, or capitalized some of its digital population could be. Boundary crossing, after all, also invites invasion. In Star's terms, *VoS* was an "ill-structured" open system that woke to the reality that such systems must eventually be guarded, thus revealing themselves to be what they perhaps always were: temporary displacements of modernity's closed organizations of knowledge for which, today, the dream is again to find boundary objects able to nudge "closed" toward "open."

In a section of her later "This Is Not a Boundary Object: Reflections on the Origin of a Concept" (2010), Star reflects on "the growth and death of boundary objects"—"their origin, development, and, sometimes, death and failure" (Star 2010, 613).[17] She hypothesizes a cycle:

> Over time, people (often administrators or regulatory agencies) try to control ... and especially, to standardize and make equivalent the ill-structured and well-structured aspects of the particular boundary object. ... [But] Over time, all standardized systems throw off or generate residual categories. These are categories that include "not elsewhere categorized," "none of the above," or "not otherwise specified." As these categories become inhabited by outsiders or others, those within may begin to start other boundary objects ... and a cycle is born [last ellipsis in original]. (Star 2010, 613-614)

As Star also says, "We live in a world where the battles and dramas between the formal and informal, the ill structured and the well structured, the standardized and the wild, are being continuously fought" (Star 2010, 614).

VoS was *wild*, and now it is tamed. A larger thesis that could be started at this point (something I am working on in a book on digital humanities and cultural criticism) is that in modernity, new media are experienced as "new" precisely when they serve as transitional boundary objects initiating such a cycle. Today, for example, open-system new media attract professional communities of "knowledge workers" (or students training to be such workers) laboring in standardized organizational institutions where earlier conventions of professional autonomy are increasingly subjected to "managerialism." The boundary-object new media spaces that such professionals develop on the side (their blogs, open-source projects, etc.) compensate for eroded autonomy by regenerating residual professional patterns of cross-organizational association, collegiality, shared governance, and so on (and related modes of intellectual, artistic, and—more basically—social association). But fast forward the time line, and such boundary objects evolve their own standardized protocols and governance structures uncannily mimetic of managerialism (e.g., "templates" for blogs). Then the cycle starts over as "not otherwise specified" exceptions incubate fresh boundary objects (e.g., today's social media) tantalizing institutionalized knowledge workers with visions of refuge from modernity's standardizations and, more uncanny, contemporary postindustrialism's *distributed* standardizations (e.g., password control points, template forms, and surveillance propagated at all levels of distributed systems).

To rewrite the myth: *VoS* had a tale to tell of interdisciplinary freedom. But the bad hacker violated it, cut its tongue out, and shut it up in a guarded place.[18] So *VoS* remade itself as a hardened site. Arachne took over—a spider at the center of a managed web. Seeking new open spaces, Philomela thought she might next become a "digital humanist," trying out fresh ways of making boundary objects such as topic models, semantic networks, social network graphs, and spatial visualizations able to reveal among disparate bodies of knowledge surprising correlations and connections.

Acknowledgment

Parts of my narrative of the early days of *VoS* in the "Hacking *VoS*, 1994" section of this chapter are adapted and expanded from my article, "Globalizing the Humanities: 'The Voice of the Shuttle: Web Page for Humanities Research,'" *Humanities Collections* 1:1 (1998): 41–56. An open-access version of the earlier article (the submitted manuscript version) is available at http://liu.english.ucsb.edu/wp-includes/docs/vosessay.pdf.

Notes

1. Significant milestones in *VoS*'s development include: *December 1994*: began as local resource at University of California, Santa Barbara (UCSB); *March 21, 1995*: became world-accessible on UCSB's Humanitas server (humanitas.ucsb.edu); *July 22, 1999*: moved to new server and domain name (vos.ucsb.edu); *October 16, 2001*: changed from static HTML pages to database-to-HTML site in custom-made content management system; *January 31, 2004:* "Bad Hack" SQL-injection attack; *December 11, 2005:* "Site Hacked" SQL-injection attack; *2004–5:* hardening and lockdown to restrict editing; *2006 and after:* gradual slowing of active development and maintenance. *2015:* Matthew Burton and Melissa Chalmers begin working on archiving *VoS*, preserving its taxonomy and re-referencing dead links to Internet Archive versions of external pages, as part of their research on advancing web archiving methods. (The earliest archived copy of *VoS* itself in the Internet Archive is dated November 21, 1996: https://web.archive.org/web/19961121003828/http://humanitas.ucsb.edu/. See this copy for the site's early taxonomy and content, including the mentioned pages on "Cyberculture" and "Technology of Writing.")

   Over most of its history, *VoS* was edited and maintained as a solo endeavor by myself, although from 1999 to 2002, I sponsored about 20 graduate students as editorial assistants with the aid of campus funds. (See http://vos.ucsb.edu/credits.asp for acknowledgments to these assistants and others.) While I did the early technical work on the site, later technical work relied on contributions from graduate students in my department working in concert with my department's longtime sysadmin, Brian Reynolds. Carl Stahmer developed such technical resources as a search engine, user suggestion form, and automated link management system to facilitate harvesting from my "New Links Suggested by Users" page. Robert Adlington and Jeremy Douglass performed the programming and database design work that changed *VoS* into a database-to-HTML dynamic site. Douglass (in league with Reynolds) then again crucially assisted in conducting technical forensics and security upgrades in response to the hacker attacks of 2004–2005.

2. For an archived copy of the *Forbes Magazine* review of June 2002, see The Internet Archive: https://web.archive.org/web/20021221085454/http://www.forbes.com/bow/b2c/review.jhtml?id=6392. Some other awards *VoS* garnered at the time—when the novelty of the medium and the lack of algorithmic ranking (e.g., Google's PageRank) led to many attempts to sift quality through awards and ratings—may be seen on this archived copy of the *VoS* awards page from 2001: https://web.archive.org/web/20010606163655/http://vos.ucsb.edu/shuttle/awards.html.

3. See Susan Leigh Star, "The Structure of Ill-Structured Solutions: Boundary Objects and Heterogeneous Distributed Problem Solving," in *Distributed Artificial Intelligence*, Vol. 2 Michael N. Huhns and Leslie George Gasser, eds. (Los Altos, CA: Morgan Kaufman, 1989), 37–54; Susan Leigh Star and James R. Griesemer, "Institutional Ecology, 'Translations' and Boundary Objects: Amateurs and Professionals in Berkeley's Museum of Vertebrate Zoology 1907–39," *Social Studies of Science*, 19:3(1989): 387–420; Susan Leigh Star and Karen Ruhleder, "Steps Toward an Ecology of Infrastructure: Design and Access For Large Information Spaces," *Information Systems Research* 7:1(1996): 111–134; Susan Leigh Star, "This Is Not a Boundary Object: Reflections on the Origin of a Concept," *Science, Technology, & Human Values* 35:5(2010): 601–617; and chapter 9 in Geoffrey C. Bowker and Susan Leigh Star, *Sorting Things Out: Classification and Its Consequences* (Cambridge, MA: MIT Press, 1999).

4. As Geoffrey H. Hartman wrote in "The Voice of Shuttle: Language from the Point of View of Literature," in his *Beyond Formalism: Literary Essays, 1958–1970* (New Haven, CT: Yale University Press, 1970), 337:

> Aristotle, in the *Poetics* (16.4), records a striking phrase from a play by Sophocles, since lost, on the theme of Tereus and Philomela. As you know, Tereus, having raped Philomela, cut out her tongue to prevent discovery. But she weaves a tell-tale account of her violation into a tapestry (or robe), which Sophocles calls "the voice of the shuttle." If metaphors as well as plots or myths could be archetypal, I would nominate Sophocles's voice of the shuttle for that distinction.

For *VoS*, I created a hypertext-fiction-like page that explained the myth behind the site's title through a hypertextual interweaving of passages from poets, critics, and theorists all playing on the theme of Philomela (http://vos.ucsb.edu/myth.asp). A significant early contribution to *VoS* was made by Patricia Klindienst when she allowed me to post as a "featured work" her "The Voice of the Shuttle Is Ours," first published in *The Stanford Literature Review* 1(1984): 25–53. The original post of Klindienst''s essay on *VoS* is now available at http://oldsite.english.ucsb.edu/faculty/ayliu/research/klindienst.html.

5. From T. S. Eliot's *The Waste Land* (New York: Horace Liveright, 1922; Bartleby.com, 2011, available at http://www.bartleby.com/201/1.html), ll. 203–6:

Twit twit twit
Jug jug jug jug jug jug
So rudely forc'd.
Tereu

6. On the early semantics of the word "hack" in computing communities, see the appendix on "'Ethical Hacking and Art'" in my *The Laws of Cool: Knowledge Work and the Culture of Information* (Chicago, IL: University of Chicago Press, 2004). On the "more hack, less yack" expression in the

digital humanities, see for example Natalia Cecire, "When Digital Humanities Was in Vogue," *Journal of Digital Humanities* 1:1(2011), web; and Adeline Koh, "More Hack, Less Yack? Modularity, Theory and Habitus in the Digital Humanities," *Adeline Koh*, May 21, 2012, web.

7. A relic of the UCSB Many Wolves web-authoring collective survives in the Internet Archive: https://web.archive.org/web/19961121021644/http://humanitas.ucsb.edu/projects/pack/webshow.htm. A PDF of my *Ultrabasic Guide to the Internet for Humanities Users at UCSB*, self-published in hard copy in 1994, is available at http://liu.english.ucsb.edu/wp-includes/docs/retro/ultrabasic-guide.pdf.

8. I refer to the desire of the freed artificial intelligence at the end of Gibson's novel to communicate with other machine intelligences in the galaxy. See William Gibson, *Neuromancer* (New York: Ace Books, 1984), 270.

9. An example is the description on the *VoS* "Science, Technology, & Culture" page, which explained the rationale for including these topics to a larger humanities audience: "This sub-page includes a selection of resources on science, medicine, technology, and cultural-studies/historical approaches to science designed for humanists interested in the relation between sci-tech and society. The emphasis is on materials that reflect upon, historicize, critique, collect, exhibit, or otherwise mediate (and mediatize) sci-tech rather than on scientific research per se."

10. An example of the international character of site visits can be seen in the "Domain Report" section of an archived *VoS* "Visitor Statistics" page of January 10, 2001. Available at https://web.archive.org/web/20010110182000/http://vos.ucsb.edu/access.html.

11. The site was kept at http://www.vol.it/mirror/humanitas/humanitas_home.html.

12. CNN.com included a link to my Unabomber essay (original of 1995 now available at http://oldsite.english.ucsb.edu/faculty/ayliu/research/whyuna.htm) on April 9, 1996, in its story, "Exhaustive Search of Kaczynski's Cabin Continues." Available at http://edition.cnn.com/US/9604/09/unabom_pm/.

13. For a sense of the robustness and density of user suggestions, see the archived version of my "New Links Suggested by Users" page of February 5, 2001. Available at https://web.archive.org/web/20010205000000/http://vos.ucsb.edu/shuttle/unvetted/unvetted.html. I enrolled names of correspondents publicly on my "Contributors" page before their number made it impractical. See, for example, the archived "Contributors" page of April 1, 2001. Available at https://web.archive.org/web/20010401043259/http://vos.ucsb.edu/shuttle/acknow.html#contributors.

14. For an explanation of SQL injection attacks, see Wikipedia: http://en.wikipedia.org/wiki/SQL_injection. On error-based SQL injection in particular, see "Anatomy of Error-Based SQL Injection" (by "nullvyte"), *Cybrary*, 21 September 2015. Available at https://www.cybrary.it/0p3n/anatomy-of-error-based-sql-injection/. The basic idea is that a user can call on a database via incorrectly filtered characters or other means in a web http request, thus injecting SQL statements asking the database to take actions otherwise disallowed to outsiders (e.g., reveal table properties or user information, delete or corrupt tables, etc.).

15. My gratitude to Douglass for his reverse engineering of the hack. I summarize key facts here from his long email of December 15, 2005, to myself and Brian Reynolds detailing the probable attack vectors and origins.

16. Chapter 3 in Jeremy Douglass, "Command Lines: Aesthetics and Technique in Interactive Fiction and New Media," Diss. University of California, Santa Barbara, 2007. Available at http://jeremydouglass.com/dissertation.html.

17. Cited above in n. 3.

18. Compare Edith Hamilton's telling of the Philomela myth: "He [Tereus] seized her and cut out her tongue. Then he left her in a strongly guarded place. ..." (*Mythology* [1940; New York: Signet, 1969], 270).

V Community Networking

18 Community Networking: The Native American Telecommunications Continuum

Randy Ross (Ponca Tribe of Nebraska and Otoe Missouria)

Computer-mediated communications has evolved exponentially each decade since the mid-1980s. Pre-Internet exploration in the era of FidoNet and supported by dial-up modem equipment running over x.25 exchange switching does not seem possible to have existed at all. With three decades of change to reflect on, questions remain today about whether the impact of technology and telecommunications has advanced tribal nationhood. Early dialogue about telecommunications and information technology leaned toward the metaphor of a double-edge sword. Change has never been easy for Native communities; there has always been a trust barrier given the traumatic histories experienced by nearly all Indian Nations. The litany and trail of broken treaties furthered by mistreatment and dishonor has further served to create hesitance. On the other hand, Native ways of adaption reflect great ingenuity and achievement—the ability to make attitude a driving force of social and cultural change. This might be as simple as taking a contemporary item, and with attitude and art it becomes distinctly "native."

From 1984 to 2014, this three-decade span has impacted Indian Country for good and for bad. It is time to explore that pathway and attempt to reconcile early thought patterns about shifts in tribal life-way and new federalism under a new economy driven by emergent broadband technologies and informatics. Innovation in Silicon Valley brought about a new wave of hope for an information society that would shift not only economic influence, but the great hope of a great electronic democracy ended in an illusion for many. The media technology and commercial ownership changes that accompanied network advancements and innovations seemed to have reduced the vision of an information super highway to that of profit-making and ownership, ownership of the switches themselves very much unlike the value of the Interstate highway system as was the misguided vision of the technology future in this country.

For three decades, we were witness to court battles, and corporate transitions and transactions that diluted competition and the price war of the cost of broadband along with impacts of convergence brought about an even more limited vision of a possible future of societal growth based on information freedoms.

For Indian Nations, their economies continue to be the worst in the United States. Out of the top 10 poorest counties in the entire United States, six of them are in South Dakota. On Shannon County or what is most of the Pine Ridge Indian reservation, the annual income of a family of four is less than $6,000 per annum. The unemployment rate has remained steady at 80% to 90% since 1984 and running steady through to today. Access to broadband and information services has remained unattainable by many households on these reservation areas.

It was not until cell phone technology changed to include Internet-based services that usage climbed, and today that usage is around social networking primarily. Broadband subscription services never were able to form an aggregate market share for Tribes to form their own re-seller entities for broadband services. Since the 1990s, most Tribes have not looked at public switched networks or formation of rural telephone coops for any number of barriers encountered, to include ETC status attainment all the way to financial access through RUS funding.

The email culture that has emerged across America has in all likelihood been among the most pervasive impacts to change in tribal communities, via cell phone technology. Commercial broadband services have not existed on Indian reservations due to infrastructure costs, lack of customer base, and cost per mile for deployment. The subsidy programs for universal service have not fairly reached Indian country in ways that have triggered tribal priority for development but in most cases have remained elusive.

Today, the Tribal economy centers around big box leaders such as WalMart and the commercial cell phone industry has introduced "pre-paid" and "pay-per-phone" products such as Straight Talk, unlimited text, voice, and data for $45 per month no contract, no credit, and with the purchase of inexpensive phone devices. We note that cell phone tower citing and placement on Indian reservations is still a void, and thus service is still spotty on many Indian lands across the country.

A conversation about what can be done in the next three decades based on what was learned in the past three decades would be a necessary conversation by tribal leaders to gain optimum ground for ownership, sovereignty, jurisdiction, and sustainability. This statement seeks to address some of these issues in deeper and greater context so that more thought can be provoked about the potential gains and what risks are to be noted given the recent patterns, trends, and history of the National Information Infrastructure paradigm that took center stage in the mid-1990s with the claim that Vice President Al Gore invented the Internet.

A greater level of convergence is occurring to include energy, environment, climate change, health care, and a new reservation economy that suggests it's time to move toward "off-grid" life and living. We seek to examine these challenges and opportunities as they have evolved in the turbulent financial underpinnings of this country as a whole and how Indian nations might better survive culturally, economically, politically, and socially as they always have survived since pre-reservation, on through the Industrial Age, through post–World War II, and now through the era of self-determination.

19 The Art of Tele-Community Development: The Telluride InfoZone

Richard Lowenberg

The InfoZone[1] was a program of the Telluride Institute, a not-for-profit research, education, and cultural organization ("world's highest altitude think-and-do tank"). The InfoZone was a site-specific response to the regional community's needs and desires, and an intelligently creative model for broad-spectrum rural community development and education, using information and telecommunications technologies. The InfoZone program was a living laboratory for the technical, cultural, political, and economic impacts and implications of the tele-mediation of our emergent local-global communities, intending by example to promote an ecological understanding of the Information Society.

Located high (8,745'+) in the southwest Colorado Rocky Mountains, Telluride is a former gold and silver mining town and site of the world's first (1891) commercial long-distance transmission of AC power. Following booms and busts, since the 1970s, Telluride has transformed to become a winter sports, summer festivals, ecocultural tourism and real estate development-based hometown to about 2,000 year-round residents, many part-timers, and visitors. The InfoZone uniquely coincided, integrated with, and openly informed all other community-building efforts.

The InfoZone began as a first-class BBS[2] network before connecting to the Internet via the National Science Foundation (NSF)-supported Colorado Supernet in 1992, to become the world's first rural (non-university) Internet PoP, adding the first rural community spread-spectrum wireless WAN in 1995. The InfoZone garnered international attention when it Internetworked the 1993 Telluride Institute Ideas Festival: *Tele-Community*," for about 200 onsite presenters, invited guests, and attendees, with live interactive forums conducted globally via The WELL.

Between 1992 and 1997, the InfoZone provided free Internet access to Telluride's populace and visitors, and it placed numerous designated public access sites around town. A small, dedicated cadre of contractors and volunteers helped build applications and content, provided training and user support, maintained technical systems, and promoted and stewarded the project. Before shutting down in the late 1990s, the InfoZone, with up to 1,200 subscribers, provided free email and hosted online local

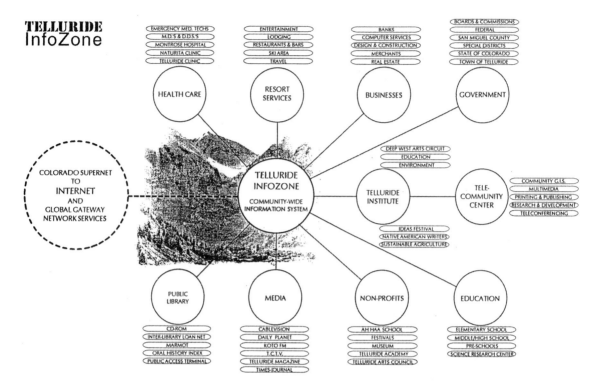

Figure 19.1
InfoZone Community Network Diagram, Telluride Institute.

Figure 19.2
Main Street, Telluride. Photo: Bill Ellzey.

government, healthcare, library, schools, arts, businesses, tourism, and other topical exchanges.

During its short life, the InfoZone generated lots of media, corporate, and academic interest for its uniquely creative efforts to integrate virtual community with a real community of place.

Background

Community networks demo'd emergent "social media" long before the term was commoditized.

The convergence in the 1970s of small, affordable personal computers, with bulletin board systems (BBSs) and eventually the Internet, also brought forth considerations and creations of networked localism amid a globally networked society. There was a seemingly coherent trajectory from de Chardin's *Noosphere* and McLuhan's *Global Village*[3] to ARPANET, SRI, and Xerox PARC, to the *Whole Earth Review/WELL*.[4] Artists and cultural activists globally were concurrently experimenting with analog/digital media, slow-scan, satellite telecommunications, software programming and hacking. *Community Memory*, in 1970s Berkeley, had direct personal connections with the Home Brew Club and the birth of "personal computing." Some of the BBSs of the late 1980s and early 1990s were designed as virtual versions of real communities, thereby facilitating the emergence of place-based community networks, such as the Cleveland Freenet (OH), the Blacksburg Electronic Village (VA), La Plaza (Taos, NM), Charlotte's Web (NC), Seattle and Vancouver Community Networks (WA + BC), Rete Civica (Milan, IT), and the InfoZone. Almost all were about more than simply having an online presence; they provided training and education, applications development, computer reuse, and bridging of digital divides. Few of the original community networks have survived into the present as the economies, technologies, and dynamics of our networked local-global society have evolved. They may be more needed than ever. The few that have endured, including Davis Community Network, Mountain Area Information Network, Vancouver Community Network, along with many new community open wireless networks around the world, primarily with local support, are important to take note of as they continue to be valuable living laboratories.

While the InfoZone was one of many community networking initiatives being realized in the early 1990s, its roots and influences were in large part personal, growing from involvements of myself and colleagues beginning in the late 1960s and 1970s, exploring convergent understandings of cybernetics, information theory, and computing; environment and ecology, community design, and planning; and new media arts, cultural motivations, creative collaborations, and networking.

In the fall of 1979, I was invited by relocated UK friends, John Lifton and Pamela Zoline, to move from the Bay Area to Telluride so as to join them in master planning of the new Telluride Mountain Village, along with other regional planning, and to continue our own creative works. John was featured in the exhibition and book, *Cybernetic Serendipity*,[5] and in 1976, we collaborated on bio-music/video sequences for the feature film, *The Secret Life of Plants*.[6]

From the outset, along with all other regional community master planning considerations, John, Pamela, and I discussed this as a unique early opportunity to deploy open conduit and fiber optic infrastructure, concurrent with deployment of all other earth-moving, utilities trenching, right-of-way permitting and financing. I had phone conversations about this with Admiral Bobby Inman and others involved with early fiber optic telecommunications test-bedding at Microelectronics and Computer Consortium (MCC) in Austin, Texas, and discussed satellite up/down capabilities to a community fiber head-end, with Joseph Pelton and others at UC Boulder and in the DC area. Although our plans were too early and technically unrealizable in the early 1980s because there was no public Internet yet, let alone cable TV in the Telluride area, the concept of a tele-connected community stayed actively on our minds.

The Telluride area had set an important precedent as the world's first site of long-distance AC power transmission in June 1891. We aspired to do the same with the Internet 100 years later. Pamela Zoline coined the name InfoZone around the time of the founding of the Telluride Institute in 1984. Having computers, modems, and personal BBS accounts in the late 1980s, a few of us participated in online community networking exchanges and began to take active steps forward, with the limited but rapidly developing technical capabilities then available, to make the InfoZone a program of the newly established Telluride Institute.

The InfoZone was well placed and timed to become a supported rural test-bed network, as the newly elected Clinton–Gore administration took steps to realize the National Information Infrastructure (NII) and to privatize the Internet, along with emergence of the worldwide web in the early 1990s.

Aside from our personal motivations and abilities, Telluride presented a natural confluence of needs and opportunities. The place was physically attractive but remote. Young, well-educated, adventurous people were discovering, settling, raising families, and creating economic lives in the Telluride region. Master planning and defined development of the area required bottom-up consideration of all physical and social infrastructure, including telecommunications.

These processes involved residents in an ongoing conversation about the nature of community.

The InfoZone was a bright but short-lived star. Following my departure in late 1996, lacking leadership and continued support, the InfoZone came to a slow end, like many of the other early community networks around the country. By then commercial ISP services were becoming available as an easy alternative, although Telluride and most small, rural western towns continue to lack the infrastructure, services, competition, and pricing available in more urban areas today.

The InfoZone Program: Telluride Institute Programs Support

The InfoZone Program had four primary components: Telluride Institute Programs Support, Regional Network Systems and Services, Special Projects, Support and Outreach, and Extended Community Networking.

The InfoZone supported all other Telluride Institute programs through basic Internet and web service offerings, through creation of dedicated public/private online work areas and forums, and by partnering on grant funding opportunities. The Institute, a nonprofit research and education organization, was not going to compete with or be an ISP. From the outset, the InfoZone, as a community R&D project, provided free accounts and 800# dialup access to partnering western states, environmental groups, and Native American schools. Local program support came in part from the Zoline Foundation, Town of Telluride, San Miguel County, and user donations.

From the outset, InfoZone technical systems were housed in the basement of the Stronghouse, a remodeled 1890s stone warehouse, which served as home to the Telluride Institute. As the InfoZone was put to the service of all other Institute programs, staff, and volunteers, the Native American Writers Program had an early website and participated on NativeNet. The annual Ideas Festivals, which began in 1985 ("*Re-Inventing Work*"), utilized the InfoZone BBS and later Internet services to enhance programming and to extend global outreach communications.

Regional Network Systems and Services

InfoZone's First Class client-server, BBS, was an interim approach to our community networking objectives. It offered cross-platform, client, and graphical user interface access to an increasing range of regional content and applications, including full Internet email, telnet access, and Usenet newsgroup feeds. It was operated as a form of public access media, much like community radio or TV, with user donations and fundraising events offering support.

The Colorado Advanced Technology Institute (CATI) initiated a Rural Telecommunications Program and in 1992 provided its first grant to support the InfoZone's connection to Colorado Supernet, with full Internet service via a point-of-presence (PoP) in the Telluride local calling area. This required preparation of a local demand-aggregation proposal, co-signed and submitted by the Bank of Telluride, to have then regional Bell phone company, US West, upgrade old analog party lines to a local digital switch. US West soon further upgraded to T-1 Frame Relay connections. The InfoZone, with the

Town of Telluride, registered the geographic domain, telluride.co.us, and created Telluride's first web presence.

As an early and attractive project, the InfoZone was able to garner additional technical support from high-tech companies. The president of IBM personally approved donation of an RISC-6000 server, US Robotics and Global Village Communications donated modems, SuperMac/Radius donated a large-screen monitor, Hewlett Packard donated printers, and InFocus donated a projector.

Community public access tele-computing sites were to be an important part of the system's overall concept, design, and intent. Soon after connecting to Colorado Supernet, we heard from Steve Cisler, with the Apple Library of Tomorrow (ALOT) program, who encouraged submission of an InfoZone proposal. Upon doing so, the Telluride Institute was quickly approved for an ALOT Award, providing eight public access computers, printers, peripherals, other systems, and the beginning of an inspiring, productive relationship with Steve, Apple, and a growing activist community.

In 1994, upon recommendation from Steve Cisler, discussions began with Dewayne Hendricks and company representatives from Tetherless Access, Inc., ultimately resulting in the InfoZone and Telluride becoming a pioneering test-bed for a new spread-spectrum wireless system, connecting all (8) public access sites at 160 Kbps.

Public access sites included Telluride High School, Medical Center, Library, KOTO Radio, Historic Museum, Bank of Telluride, Arts Council, Telluride Institute, and the Agricultural Extension Office in Norwood. Soon other public access systems were independently placed at the Steaming Bean Café, other local sites, and at appropriate locations in the neighboring communities of Montrose, Ouray, Rico, and Ridgway, where classes and events were also held.

Education was a key part of InfoZone's mission. With town, county, state, and federal support, InfoZone staff and volunteers offered free weekly lessons and start-up accounts, along with applications tutorials and telecommunications planning assistance throughout the region. Special projects included specialized applications and user classes.

Residents and InfoZone staff initiated many short- and long-term community networking projects and applications, utilizing the InfoZone and Internetworking capabilities as selectively listed. (Most of these domain name extensions were made up for InfoZone user groupings and were not otherwise available.)

.edu: Student Mentorship Program; Teachers' In-Service Training; Curriculum and Educational Materials Development; Schools Internetworking and Educational Exchange Projects; Regional Higher Education and Lifelong Learning Programs.

.gov: Town and County Government Agendas, Meetings, Documents and Archives; "Meet the Candidates"; Straw polling and E-Voting; Regional Master Planning; State Social Services Network; State and Federal Government Information Services.

.med: REACH (Regional Environment and Community Health) Project; Residents and Visitors' Healthcare Directory and Hot Line.
.mus: Telluride Historic Museum Archives and Histories; Lone Pine Cemetery.
.lib: Online Connections to other Libraries; MARMOT, ACLIN, and CARL; "BookTalk."
.art: Cultural Master Planning; Regional Arts Consortium; Galleries, Portfolios, and Salons.
.org: Nonprofit Organizations Online (Telluride Repertory Theater, Telluride Academy, Tomboy House Crisis Center, Telluride Education Foundation, Summer Festivals, etc.).
.net: Inter-connections with regional ISPs; Web Development; Wired and Wireless Connectivity and Services Provision; Colorado Region 10 Telecommunications Projects.
.com: KOTO FM Public Radio; TCTV (Telluride Community Television); Telluride Times Journal and Daily Planet.
.eco: San Miguel River Watershed Project; Smart Growth GIS Project.
.nap: Native American Writers, Arts, & Education Projects.
.bus: Banking and Financial Services; Business and Commerce Support Services; Regional Economic Development & Diversification; Teleconferencing Services, Local Business.
.rec: Seasonal Recreation and Sports Activities.
.kid: Voodoo Lounge Teen Center; Stories and Games.

Special Projects, Support, and Outreach

From 1993 to 1997, InfoZone staff designed and conducted research and demonstration projects, and sought support to help substantiate the nature, quality, and benefits of community networking. Following are brief notes on some of these projects.

Tele-Community '93
In July 1993, the Telluride Institute hosted its annual Ideas Festival on *"Tele-Community,"* bringing together leading thinkers and doers to discuss issues of "community" in the emergent Internetworked society. *"Tele-Community '93"* was also conducted live on the Internet via The WELL and on the local InfoZone. Participants included Lee Felsenstein, Howard Rheingold, Anne and Lewis Branscomb, John Naisbitt, Dave Hughes, Gene Youngblood, Judy Malloy, Lewis Rossetto, Steve Cisler, and many more. *"Tele-Community '93"* served as an important public debut of the InfoZone.

Telluride Summer Research Center
Scientists love the Rocky Mountain West. In addition to the Aspen Center for Physics and the Santa Fe Institute, the Telluride Summer Research Center has held annual

workshops since 1985. Beginning in 1993, the InfoZone provided science researchers with critically required Internet access and the ability to stay and work in Telluride. For instance, attendees of the *Workshop on Neuromorphic Engineering* were now able to collaborate with colleagues at the Salk Institute, Cal Tech, and Oxford University.

CATI–CRTP
The Colorado Advanced Technology Institute's Colorado Rural Telecommunications Project (CATI–CRTP) provided early technical support and multiyear funding for realization of the InfoZone. CATI–CRTP support also encouraged US West to extend the local calling area (5,000 square miles) to include a number of neighboring rural towns. I served on the CRTP Board for four years and helped plan the first three annual Rural Telecommunications Congress (RTC) programs that were held at the Aspen Institute.

CATI–EDA
The InfoZone was a key partner on the EDA-funded *Telecommunications and Rural Economic Development* project in 1994–1995, providing technical assistance and documenting (12) rural Colorado communities with Internetwork planning and strategies. Additionally, the InfoZone was contracted to web author and design the *Rural Telecommunications Investment Guide*. I first published the term "First Mile" in the Guide.[7]

CATI–NTIA
The InfoZone was a partnering node on the CATI: *Maps for People* project, promoting open, shared online access to data, mapping, and geographic information for public decision support, funded by the U.S. Department of Commerce's NTIA Technology Opportunities Program in 1995–1996. A direct early result of this project was the mitigation of contentious land-use conflicts between private developers and the county, minimizing the need for time-consuming, costly litigation.

Project Shooting Star
The InfoZone and staff were active participants in this mid-1990s Arizona, New Mexico, Utah, and Colorado Internetworking initiative, working with Navajo, Ute, and other Native groups, the towns of Durango, Farmington, Monticello, and Gallop, colleges, and businesses in the Four Corners region.

REACH for Health
The InfoZone was the recipient of a Colorado Trust and National Civic League's Healthy Communities Initiative grant, to facilitate a community health project, resulting in the 1995 book and web publication, *REACH for Health*, a community healthcare guide.

C2C-'94

The Telluride Institute's annual Composer-to-Composer Program in 1994, subtitled *Interactive Music, Media and Performance*, brought together (10) invited music and digital media artists who used the InfoZone to collaborate, perform, and from this high mountain community, turning the information revolution into a cultural revolution.

Arts Online

The InfoZone was the first recipient of a special National Endowment for the Arts and Benton Foundation: Arts Online grant in 1995–1996 to demonstrate arts, education, and cultural applications for and among Internetworked communities. The InfoZone served as platform for numerous formal and informal creative exchanges, including:

- *Tele-Comm-Unity* was a Rocky Mountain regional arts networking project series, funded by the Colorado Arts Council and an NEA/Western States Arts Federation "New Forms" grant.
- E-Café *Telebrations* involved Telluride creatives in global "picture-phone" and networked arts interactions, coordinated with Electronic Café International in Santa Monica, CA.
- Arizona State University's *Deep Creek Arts Program*, conducting summer classes 7 miles down-valley from Telluride, was able to provide students and visiting faculty with needed Internet access and online arts capabilities, quickly put to exceptional use.

Info/Eco

Durango-based economist, Dale Lehman, PhD, and I collaborated on research and writing on the information economy from an ecological economics perspective: *Info/Eco* presentations were made at the Aspen Institute, MIT, and NTT Data, Tokyo. This work has continued.

Extended Community Networking

Representing the InfoZone as an early example to be learned from, other 'Zoners and I were often invited to participate in regional, national, and international projects and programs. InfoZone staff played an active role online and in many working meetings, conferences, policy hearings, and consultancies, on grounded example setting of best practices and experiential learning about the nature of our developing local-global Internetworked lives and livelihoods.

For example, representing the InfoZone, I was an invited participant in working meetings held at the Carnegie Institute in DC, sponsored and organized by the Center for Civic Networking (CivNet '92). This led to Telluride Institute's hosting of the "*Tele-Community*" Ideas Festival in 1993.

In 1994 and 1995, I served as a consultant and jury member on an international design competition for the planning/design of ParcBIT, Mallorca, a proposed residential, research, and networked eco-community on the Spanish Mediterranean Island of Mallorca. Invited by the Balearic governor, I provided applied ecological telecommunications concepts and served on the jury that ultimately selected a most ecologically considered plan for ParcBIT, presented by the British architect, Sir Richard Rogers.

In 1995, I was invited by Amerigo Marras, of the Storefront for Art and Architecture in New York City, to be a consulting artist/architect on the French Mediterranean Island of Corsica (for Eco-Tech Corsica); to do a series of presentations on the Telluride InfoZone project as a model for networked rural communities; and to provide assistance to Corsican communities and institutions on getting connected to the Internet with its potential local-global services and opportunities.

Returning from Corsica, I visited the tiny Italian hilltop village of Colletta di Castelbianco[8,9] in Liguria. Colletta was an old stone village ruin, being rebuilt and reinhabited, while maintaining its historic structural integrity. Valerio Saggini, the young co-developer, with fiancé Stephania (working for SGI), also took me to Milan to meet the project's renowned urban architect/planner, Giancarlo DiCarlo, and the "father of digital computing in Italy," Luigi Dadda. Colletta is now a tele-village, with Telecom Italia fiber optic networking throughout—an aerie for "lone eagles."

In 1997, I was invited by Global Village Network[10] founder, Franz Nahrada, to give the keynote presentation at the Tyrol Worldwide conference on connecting rural Alpine villages, held in Innsbruck and Kals, Austria, inspired by the Telluride InfoZone's high mountain example.

During my time directing the InfoZone, and the Davis Community Network thereafter, we received occasional Japanese visitors, and I was repeatedly invited to visit and consult in Japan, where legacy copper landline infrastructure was being replaced by nationwide fiber optic and wireless networks, and networked society interests were keen. Yamada Village[11] was an early rural community fiber network test-bed that posed adoption problems, as young people had moved to cities, and the remaining population was elderly and not yet computer-literate. Yamada Village and InfoZone shared visits and learning.

The Association for Community Networking (AFCN) was founded at the now annual community networking conference, hosted in 1996 by La Plaza in Taos, NM, after having been held at Apple Headquarters in Cupertino, CA, in the two prior years (the *"Ties that Bind"* workshops). Madeline Gonzales, an early InfoZone "co-conspirator," became the first executive director of the AFCN. I was a member and also served on the AFCN Board during its active life.

While on the Board of the AFCN, I was a representative to the first Global Community Networking Congress in Barcelona in 2000 and the second Congress in Buenos Aires in 2001. Representatives from North and South America, Europe, Africa, Asia,

Australia, and New Zealand gathered to share ideas, tools, and methods; to discuss difficult issues; and to work on action agendas for the forthcoming UN + ITU convened World Summits on the Information Society (WSIS), which would be held in Geneva in late 2003 and in Tunis in late 2005. These meetings were informed by but my involvements continued post-InfoZone.

Recognition and Awards

The InfoZone received its share of attention at a time of early Internet exuberance, with articles in *Business Week, Wall Street Journal, US News & World Report, The Economist,* AP, *WIRED, Communications Daily, Entrepreneur Magazine, Denver Business Journal, Rocky Mountain News, Focus Magazine* (Germany), *Tele-Worker Magazine,* and *Yomiuri News,* as well as Italian, Norwegian, and Corsican press, plus radio and TV reports on Voice of America, National Public Radio, PBS, ABC, NBC, CBS, Channel 4 (UK), Nippon, FUJI TV, and Austrian, Swedish, and Finnish TV.

Researchers visited, studied, and wrote about the InfoZone. Alice McInnes published her paper, "Agency of the InfoZone: Exploring the Effects of a Community Network," *First Monday*, 1997.[12]

At the September 1994 Tri-State Economic Development Conference (in Kansas, Nebraska, Colorado), the Docking Institute for Policy Studies at Ft. Hays State University, Kansas, presented the InfoZone with its *Telecommunications Success Story* Award.

Today, the Telluride Institute, after its first 30 years, continues to nurture right-livelihood example-setting efforts from its spectacular but fragile, high mountain, hometown community.

A Personal Note

Consideration of our emergent information revolution, the nature of information, and the electromagnetic information/energy environment as dynamic, integral aspects of needed whole ecosystems understandings has been my fundamental creative motivation and basis for overt, example-setting work since the late 1960s. The InfoZone was a significant learning and doing opportunity along a dedicated personal communication, design, and eco-arts path that led from studies in architecture, design, and film at Pratt Institute in the 1960s to interest in systems, cybernetics, and new ecological understandings; to playing in the early electronic media arts and performance sandbox; to informal collaborations with NASA on weightlessness, biotelemetry-dance studies, ARPANET, and satellite arts projects. Telluride offered unexpected opportunities to apply these convergent creative interests pragmatically, on the ground, for real community building, while being an inspiring setting for ongoing work in the arts and cultural development.

The small successes and lessons learned from the InfoZone gave me a wonderful foundation on which to build, as I subsequently continued along rural information byways to current community open-broadband planning, working to address rural networked society disparities and inequities with public-private, economic win-win solutions and opportunities, and conducting international eco-arts-sciences tele-collaborations from a home base in Santa Fe, NM.

Organizational Details

The Infozone Program Staff included Richard Lowenberg, InfoZone Program Director; Madeline Gonzales, Program Associate; Lee Taylor, Applications Coordinator and Online Publishing; Bill McKee, Systems Administrator; Catherine Sellman, Outreach, Training, and Support Services Coordinator; Jim Harness, Software and Hardware Systems Specialist; Chris Hazen, GIS and Project Grants Coordinator; John Lifton, Telluride Institute Founding Board Member, Computing and Community Planning; and Pamela Zoline, Telluride Institute Founding Board President, Writer/Painter, Creative Inspiration.

The InfoZone Advisory Board was *Jonathan Aronson, International Relations, USC; *Anne Wells Branscomb, Harvard University; *Lewis Branscomb, Kennedy School, Harvard, University; Kit Galloway, Electronic Café International; Heather Hudson, McLaren School, University of San Francisco; Dave Hughes, Old Colorado City Communications; *Kathleen Kennedy, Kennedy/Marshall Productions; Ken Klingenstein, Computer and Network Services, University of Colorado; *Jack Kuehler, former president of IBM; Dale Lehman, Economist, Ft. Lewis College; *Amory Lovins, Rocky Mountain Institute; *Frank Marshall, Kennedy/Marshall Productions; *John Naisbitt, Megatrends, Ltd.; Sherrie Rabinowitz, Electronic Café International; Jeff Richardson, director, CATI-Rural Telecommunications Program; and Richard Solomon, Research Program on Communications Policy, MIT. (*Telluride Institute Board Directors)

Notes

1. Telluride Institute, InfoZone Program. Available at http://www.tellurideinstitute.org/infozone.html.

2. First Class BBS. Available at http://www.firstclass.com/.

3. Marshall McLuhan, "The Global Village," a term popularized in his books *The Gutenberg Galaxy: The Making of Typographic Man* (Toronto: University of Toronto Press, 1962) and *Understanding Media* (New York: McGraw-Hill, 1964).

4. The WELL. Available at http://www.well.com.

5. Jasia Reichardt, ed. *Cybernetic Serendipity, the Computer and the Arts* (New York: Praeger, 1969).

6. Secret Life of Plants: bio-arts, "Data Garden interviews bio-art pioneer Richard Lowenberg." Available at http://datagarden.org/43/richard-lowenberg-interview/.

7. *The Applied Rural Telecommunications Investment Guide* (1995) was sponsored by the Colorado Advanced Technology Institute and funded by the Economic Development Administration of the U.S. Department of Commerce. Developed under contract by the Telluride Institute InfoZone Program; Project Director: Richard Lowenberg; with web publishing assistance provided by Lee Taylor, Telluride Woodcraft. It is no longer available online.

8. Colletta di Castelbianco. Available at http://www.colletta.it/.

9. Valerio Saggini, "'Colletta di Castelbianco' Project: Rediscovering Sense of Place in the Era of the Global Village," teleura.com, November 13, 1995. Available at http://www.teleura.com/article/articleview/90/1/14/.

10. Global Village Network. Available at http://globalvillages.org/.

11. The Yamada Village E-Project, Japan. Available at http://siteresources.worldbank.org/INTEMPOWERMENT/Resources/14874_YamadaJapan-web.pdf.

12. *Alice McInnes*, "Agency of the InfoZone: Exploring the Effects of a Community Network," *First Monday* 2:2 (February 3, 1997). Available at http://www.firstmonday.org/ojs/index.php/fm/article/view/511/432/.

20 Community Networking, an Evolution

Madeline Gonzalez Allen

Over the years, "community networking" has evolved and contributed to what has become known as "social media," with many exciting and novel ways that we can all be interconnected—sharing information, inspiration, and our own personal experiences. With so much that is now able to be shared by so many, what is it that is *meaningful* to share? How can the sharing lead to more creative collaborative solutions to the issues facing our human species? To paraphrase Barbara Marx Hubbard, what is the meaning in our power that is good? How can people come together to decide what "good" means for them? How do we develop our abilities to discern meaning and relevance? With so much that can be constantly vying for our attention, how do we sustain our ability to be consciously choosing what it is we give our attention *to*? In my own personal ongoing balancing of technology with awareness, I have felt excited about the potential of technology as a tool for helping us deepen and expand our abilities to communicate and collaborate with each other, and I am poignantly aware of a need to keep this in balance with all the many other ways in which we experience being most fully alive as humans—our core *Beingness*.

In this chapter, I document how I followed a vision for community networking, of communities coming together and deciding for themselves how they wanted to use the then-nascent public Internet for the benefit of their own communities, rather than this technology being shaped and driven solely by those privileged few who at the time were "in the know" (and would eventually lead to the whole ".com" phenomenon). At that critical point in time, as the Internet was becoming a public medium, I felt a calling to do all I could so that *everyone*—regardless of their educational background, income level, employment status, ethnicity, gender, or any other "classification"— could have the same opportunity to learn about, shape, and benefit from this emerging technology. I worked with people from communities across Colorado (e.g., Telluride, Boulder, Southern Ute Tribe) to develop innovative community applications of the then-nascent Internet technology, shared what we learned with people from other communities (e.g., from Ute Mountain Ute Tribe to Oita province, Japan) and organizations (e.g., Morino Institute, Colorado Advanced Technology Institute, Ft. Lewis

College), and eventually co-led the creation of an international Association for Community Networking.

In early 1992, I was working as an international systems engineer for AT&T Bell Labs. In a sense, this had been a culmination of everything I had thought I wanted, had worked so hard for, and had earned my degrees for (in order to be able to make it through the "majoring" in Computer Engineering, I "minored" in French). I was working on developing and defining requirements for new international "user interfaces" (how people would interact with new computer systems), including traveling to Italy to work with people from the Italian phone company. I was loving the international travel, using the foreign languages I knew, and starting to learn new languages. Back in Holmdel, New Jersey, most lunch times would find me biking or running among the lovely rolling hills outside the AT&T Bell Labs facility. Then, more and more, I started to find myself searching for somewhere I could just sit. Just be with the strong emotions that were surfacing. What was going on, I'd ask myself. Didn't I have all I'd ever wanted? Hadn't I gotten the degrees I'd thought mattered? Hadn't I been hired by the company I'd most wanted to work for and now had the international job with the travel to Europe I'd always dreamed of? Yet as much as it didn't "make sense," I knew the time had come for me to leave. One day I found myself walking over to my manager's office. I heard myself saying how much I appreciated all I had learned, all I had contributed to and been a part of, and everyone I had met and worked with. But I had to leave. I had to find a greater sense of personal meaning. I had to find how I could be of greater service to this planet. My manager did not want to accept my resignation; instead he offered me something I hadn't known would be an option—a leave of absence. Without pay, but with the option to return if I wanted. I accepted, gratefully. I sold or gave away most of my belongings. I packed the few remaining things into my Honda Civic and drove from New Jersey back to Florida.

I spent time with my family. I allowed myself to have time with one of my lifelong passions—scuba diving in the beloved tropical coral reef ecosystem. I became a volunteer with the nascent Florida Keys National Marine Sanctuary, shared with visitors about how the reef was alive and how we could be protecting it. Then I began to hear a call of the terrestrial wild, wanting to experience and be of service to land-based wilderness. I learned more about backpacking, taught myself how to set it all up on my parent's living room floor, and then packed it up and headed out West to be a volunteer at Utah's Canyonlands National Park. I arrived in Grand Junction, Colorado, and set up my tent in the dark. I awoke the next day, opened the tent flap, and—there I was, surrounded by vibrant gold aspen leaves in full resplendent autumn glory. Colorado's mountains and Utah's canyonlands wilderness touched my core.

While backpacking through the San Juan Mountains, I came to a high-alpine jewel of a town surrounded by majestic peaks—Telluride. I met Richard Lowenberg and others from the Telluride Institute, and I shared with them ideas about the Internet. I wondered, could this emerging Internet technology be applied in ways that could be of real benefit to communities? This would become the central driving question at the heart of my vision for community networking. I stayed in Telluride. Back then the Internet was virtually unheard of, and few people outside of certain research settings/some large corporations had ever "sent an email message." I shared my personal experiences, the ways we had used the technology at AT&T Bell Labs, some of the ways it could potentially be used by people in communities to transcend geographical limitations. As a core member of the team, I helped to create the Telluride InfoZone, one of the first rural community Internet-based information systems in the world. The "InfoZone" system was a kind of electronic "bulletin board," providing community information and serving as a type of community repository. For me it was the education and collaboration that happened around the creating of the system, rather than necessarily just "the system" itself, that seemed to be of the greatest value.

In 1993, I participated in the *Telluride Ideas Festival*, and what a festival of ideas it was! Participants came from all over the country. I met a variety of remarkable people, including Steve Cisler, in charge of an innovative program called Apple's Library of Tomorrow organization; Ken Klingenstein, Director of Information Technology Services at the University of Colorado; Jeff Richardson, with the Colorado Advanced Technology Institute; Howard Rheingold, Eric Theise, and Judy Malloy, from San Francisco's The WELL; Randy Ross, American Indian Telecommunications; and Richard Civille, with the Center for Civic Networking in Washington, DC. The NSF had lifted commercial restrictions on the Internet. We talked about how we could help preserve the values that characterized this medium up until now—the sharing of information, by people sharing what mattered to them, collaborating in learning and creating. We talked about the possibilities the Internet and the newly emerging World Wide Web could offer communities.

That fall I accepted a position with the University of Colorado (CU) as the Coordinator for a new Boulder Community Network (BCN), through a partnering between the CU Journalism School and CU Information Technology Services (ITS). Ken Klingenstein was the CU ITS director who initiated the program and brought us together to form a management team. Bruce Kirshner was offering the Boulder County Civic Center from his home, a free BBS that provided some community information and discussions about local civic affairs. Neal McBurnett had been active at AT&T Bell Labs and served on the local Boulder Library Commission. Oliver McBryan was a CU computer science professor and researcher.

From the beginning, the aspect that most energized me was the opportunity to really bring in the community, to explore what it could mean for everyone in a community to have the opportunity to learn about some of the capabilities of this emerging communication medium, and really shape and direct it in ways that would be relevant and be of service *to* that community. In order for this to be authentic, to be real, I reached out to as many people, from as many community groups, as possible; people with a variety of perspectives and experiences who otherwise may not have heard about this technology in time to have an opportunity to help shape it. I reached out to people from the United Way, social services, the library, the public radio station, K–12 education, senior centers, the local employment & training coalition, local nonprofits, and local small businesses.

At the time, most people had never heard of the Internet. I had never been an extrovert, so all this "outreach" never came easy, but I was fueled by a need to have this be relevant to the community. I made cold calls, introduced myself, and introduced the whole notion of a new communications media. I shared about what the capabilities could be and why this could be relevant. I invited and deeply listened to them sharing about what their challenges were. Then I shared with them some of what we could explore working together toward relevant applications of this technology. I invited participation in the shaping of this program by all who wanted to help shape it, through what became a defining characteristic of our project, a thriving volunteer program. I worked with our team on writing grant applications. We received support from Steve Cisler through the Apple Library of Tomorrow program, and we received funding from the newly established National Telecommunications and Information Administration's Telecommunications and Information Infrastructure Assistance Program (TIIAP).

I stretched in ways I hadn't known were possible—with public speaking, with developing ways of facilitating group discussions and brainstorming that could lead (eventually) to decisions and the collaborative creating of something new, the Boulder Community Network (BCN). Its official launch was March 15, 1994. BCN was one of the first World Wide Web-based community networks in the world. Within two months, it had been visited by more than 2,600 computers and at least 14 foreign countries.[1] The BCN continues to this day as one of the longest-lived community networks,[2] with a wide range of topics ranging from Arts, Education, and Emergency Services to Health, Housing, Human Services, and Volunteering.

After launching BCN and leaving its evolution in capable hands, I accepted a position with Apple's Advanced Concepts Group, where I conceived and prototyped possible future services, authored papers, and gave presentations on web publishing and design. I loved the opportunity for full immersion in creativity. However, the pull toward *meaningful* use of technology remained strong. I continued to be active with community networking, including co-authoring "Boulder Community Network: a Publishing Overview" and presenting at the *1995 Ties That Bind: International Community Networking Conference* in Cupertino, CA. Eventually, I left Apple and returned full time to community networking.

I became a consultant focusing on working with educational, cultural, nonprofit organizations, and small businesses. I worked with the Southern Ute Indian Tribe through Fort Lewis College. I met with Tribal leaders, representatives from Tribal enterprises and organizations, K–12 teachers and school administrators, and members of their economic development organization. We created a collaborative, community-wide project and Tribal web presence (and soon heard how this was being used by kids in a classroom in Germany to learn about Native Americans directly from the Native Americans). I assisted with developing and delivering a curriculum for Internet training, and I worked on possibilities for stimulating more diverse local economic opportunities, including creating a telecommuting pilot project concept, to create new employment opportunities for the Tribe and neighboring towns.

In 1996, I co-led the creation of the Association for Community Networking (AFCN), became the executive director, and facilitated the interactions of the geographically distributed 5-person Board of Directors, 80-person Advisory Group, and members. The AFCN's mission was to promote the viability and vitality of Community Networking, linking and serving the many emerging community networks around the country, building public awareness, identifying best practices, encouraging research, and developing products and services. I had wonderful opportunities to work with a variety of people from diverse backgrounds and settings, and to participate in conferences nationally and internationally. As a speaker at the *1997 Hypernetwork Society Conference* in Oita, Japan, I was one of the few women on stage. After my talk, Mieko Nagano, a lovely Japanese woman who had been in the audience, animatedly thanked me through a translator for what I had shared and what I had shown was possible for a woman to do. She reminded me of my own mother. It was one of the most moving moments of my life.

While participating in the W. K. Kellogg Foundation's Managing Information with Rural America (MIRA) funding initiative, I provided consulting to W. K. Kellogg staff and grantees, participated in the MIRA proposal review process, and assisted with the determination of funding. What an invaluable insightful experience—to be on the *other* side of the grant process, as a reviewer, having to judge between so many worthwhile and potentially beneficial projects.

Through this process, I became increasingly aware of the need for community networking initiatives to become more self-sustaining, so as not to be competing for the scarce grant funding that was needed to meet other much more basic human/community needs (e.g., health, food, and housing). I became aware that community network's funds should ideally be used within their own community and not in support of an "association." In 1998, I accepted a position as a consultant with IBM.

To this day, I continue to love my client-centered consulting professional work as an IBM consultant—always changing and evolving, as business needs and technology capabilities continue to evolve. I have found my balance with offering service work in a variety of ways in my community—including offering summer wilderness survival camp for adolescents based on the Wilderness Awareness School approach, facilitating collaborative creative problem solving as enrichment for middle school students through the Destination Imagination program, and offering a mindfulness practice based on Mindful Schools in our local school district.

Through working with communities in creating and sustaining their community networks and working with clients in nonprofit and corporate settings, a crucial common aspect has been facilitating and sustaining authentic collaborative creativity among a group of people with often diverse and sometimes conflicting perspectives. Some frameworks and tools for facilitating effective communication and conflict resolution can be very beneficial, including those offered through Nonviolent Communication (NVC).[3] The collaborative process can be greatly enhanced if each person has a practice for sustaining his or her own presence, his or her own conscious awareness. Personally, I have been most inspired and sustained by the mindfulness practices offered by Thich Nhat Hanh, Jon Kabat Zinn, and Eckart Tolle, among others.

When we are able to come together consciously, in community, this is what can allow for the most relevant appropriate uses of technology, and for something much bigger than "the sum of the parts" to manifest. When *that* kind of collaborative creation manifests, it can be of service not only to that specific group of people, but to our humanity as a whole, in support of our conscious evolution.

Notes

1. Paul Tiger, "BCN's Early History," Boulder Community Network. Available at http://bcn.boulder.co.us/bcn/history.html.

2. Boulder Community Network Homepage. Available at http://bcn.boulder.co.us/.

3. Center for Nonviolent Communication. Available at http://www.cnvc.org/.

21 Cultures in Cyberspace: Communications System Design as Social Sculpture

Anna Couey

I am a weaver. But only in the beginning did weaving mean the process of intertwining threads to make cloth. Mostly I have practiced it by connecting people, ways of thinking, and forms of knowledge to shape culture.

Cultures in Cyberspace was a communications sculpture I organized in 1992 for the Third International Symposium on Electronic Art held in Sydney, Australia, in collaboration with George Baldwin, Phillip Bannigan, Anne Fallis (now Anne Fines), Susan (Sue) Harris, Judy Malloy, Joe Matuzak, John S. Quarterman, Randy Ross, and Eric S. Theise.[1] Structurally, the project involved group conversations on five online systems: American Indian Telecommunications/Dakota BBS (South Dakota, U.S.), ArtsNet (Australia), Arts Wire (U.S.), Usenet (international), and The WELL (U.S.). These five conversations were to be fed into one shared communications channel that all participating systems would be able to read and respond to. The five systems represented different cultural and geographic communities and communities of practice. Each was asked to discuss the cultural implications of computer-based communications systems. Participation was open to all members of these online communities. I hoped that the structure of the project would allow each community to articulate its distinct cultural perspective and that each of the perspectives would be woven into a larger whole.

I envisioned *Cultures in Cyberspace* as a telecommunications art work and a social intervention. My idea was that by engaging diverse communities in conversation, we could shape online communications infrastructure in a way that would facilitate power sharing and cultural exchange and that would resolve and heal our legacies of cultural oppression.

Concepts

I came to online communications as an artist. My entry point was the Art Com Electronic Network (ACEN), conceived by Carl Loeffler and Fred Truck, where I was exposed to experiments in participatory telecommunications art and networked art that connected people across online systems. The practices of artists using telecommunications in the 1970s–1980s, including those of Judy Malloy, Roy Ascott, and Electronic Cafe International (Kit Galloway and Sherrie Rabinowitz), signaled a radical departure from the hierarchical relationship between creator and audience that dominated cultural production (mass media and art) at the time. They inspired me to explore ways of using emerging telecommunications technologies to co-create new cultural forms. Carl Loeffler's systems-as-art approach to ACEN facilitated my engagement with cultural and political issues concerning the structure of online systems. In the early 1990s, I was beginning to consider how histories of racism and other forms of ideological supremacy are carried into contemporary culture. A process of reweaving—bringing more voices into the story and facilitating connections between them—could, I thought, instigate a shift in consciousness and practice.

The Third International Symposium on Electronic Art (TISEA), for which *Cultures in Cyberspace* was designed, was the first to be held outside Europe. TISEA organizers had a keen sense of Australia's geopolitical position on the Pacific Rim, and "Whither Cultural Diversity in the Global Village?" was a primary theme. While supporting the vision that computer networks were enabling the emergence of a many-to-many global communications territory, they also recognized the cultural inequities of the process. Artists were much more likely to be online in the United States than elsewhere. Language bias was deeply embedded in online communication tools. Although by the early 1990s Internet access and usage had proliferated far beyond the military industrial complex of its Cold War origins, its architecture still reflected U.S. geopolitics. My conversations with Australian artists and TISEA organizers Gary Warner and Tim Gruchy, while I was at Art Com, helped concretize the need to share networking strategies across national boundaries and to challenge hegemony in the development of networked cultures. In turn, Tim Gruchy recently noted, "It is impossible to underestimate how much our conversations with the Art Com crew in San Francisco influenced our understanding and thinking about the importance of networking and just how important a role it was going to be in the future. We were able to apply this thinking to the whole undertaking and approach to TISEA, striving and I believe succeeding in making it the internationally networked event it was. Its reverberations still faintly resound."[2]

The cultural politics of cyberspace were heightened by historical timing. 1992 marked the quincentennial of the European invasion/"discovery" of the Americas. In the United States, it was a time of looking back at the destructive and transformative changes wrought by the confluence of cultural convergence and the quest for wealth and power. *Cultures in Cyberspace* was a response to this pivotal moment—of considering the impact of colonial history and imagining the possibility of building a new cultural story into the infrastructure of global communication networks.

Structure

The five communication systems that participated in *Cultures in Cyberspace* represented different nations, cultural groups, and modes of thought engaged in shaping cyberspace prior to the commercialization of the Internet: tribal, regional, national, international, art, technology. ArtsNet was a national online service for artists and arts organizations in Australia "crossing the borders between art and commerce,"[3] established by Susan Harris and Phillip Bannigan in 1986. American Indian Telecommunications was an initiative to develop tribal information policy and networking infrastructure formed by Dr. George Baldwin, Randy Ross, Dr. Jim May, and Anne Fines; Dakota BBS was a bulletin board system for Native Americans located in South Dakota and operated by Anne Fines. Arts Wire was an online conferencing system for the arts in the United States initially led by Anne Focke, with a vision of building a national community of artists and arts organizations across disciplines and cultural identity, and a goal of developing a robust arts advocacy infrastructure during a time of right-wing attacks. The WELL is an online conferencing system in the San Francisco Bay Area founded by Steward Brand and Larry Brilliant in 1985, with countercultural roots and an interest in social experimentation; its members include virtual-culture visionaries and leaders of the community networking movement. Founded in 1979, Usenet was one of the oldest, largest, and most decentralized cooperative networks, with more than 265,000 users and 9,700 hosts on five continents in the early 1990s.[4]

Coordinators—artists, computer scientists, and activists involved in a diverse range of cultural issues related to the development of computer networks—facilitated *Cultures in Cyberspace* conversations on each system. Some, as founders of online systems for artists and arts organizations, were engaged with issues of inclusivity across cultural and geographical boundaries. In Australia, artists Susan Harris and Phillip Bannigan were connecting "small theatres from Alice Springs in the outback to major theatres like the Sydney Opera House"[5] to ArtsNet, motivated by "a strong sense of carving a cultural community."[6] Arts Wire director and poet, Joe Matuzak, was leading the development of a similar national model in the United States that was grappling with questions of cultural equity and participation. Randy Ross, George Baldwin, and Anne Fines were engaged with the interconnected issues of culture, information policy, and development for tribal nations—"how to survive in an even more rapidly growing environment of multiple threats from water rights, environmental exploitation, loss of lands, loss of language, loss of culture, among the many concerns of that time and still a concern today."[7] Dakota BBS was one vehicle for their work; Randy was also on Arts Wire's steering committee.

Other coordinators were deeply immersed in shaping the culture of online systems through the practices of electronic art and computer science. Judy Malloy was an early producer of electronic literature and designer of authoring systems, documenting interactive art as editor of the online *Leonardo Electronic News*.[8] She had produced a number of works on the Art Com Electronic Network and was a respected innovator within The WELL community. Eric Theise was well versed in the cultures of cyberspace at the time as an active participant on The WELL, Arts Wire, and Usenet, and as an academic "teaching about computer networks in the context of human-computer interaction."[9] Judy and Eric co-coordinated the *Cultures in Cyberspace* conversation on The WELL. John Quarterman was a recognized expert on the development of computer networks; his 1990 book *The Matrix: Computer Networks and Conferencing Systems Worldwide* was a leading source on the history, protocols, and connectivity of online systems. At the time of the project, he was focused on "documenting the growth of the Internet."[10] I asked John to coordinate the Usenet conversation in alt.cyberspace.

To connect conversations across systems, the project required a shared communications "channel." In consultation with TISEA organizers, Sue Harris, Phillip Bannigan, and I opted for creating a Usenet newsgroup that would originate in Australia to serve this purpose. The proposed newsgroup, alt.isea, would position the project outside the internationally dominant electronic arts centers, shifting cultural power, and it would provide a continuing platform for cross-cultural exploration of the issues through the auspices of the International Symposium on Electronic Art. For the first time in its history, Pegasus, the online conferencing system that housed ArtsNet, initiated the process to create a newsgroup.

Communication

My opening text for *Cultures in Cyberspace*, which coordinators posted on each site, raised questions about the unequal access to and use of online communications systems, and the role of cultural activity and identity in technological and social development:

Computer networks provide access to an internationally linked electronic communications territory—cyberspace. In cyberspace, communities form out of interest and choice, more than geography. As with multi-national corporations, computer networks are drawing new lines of social organization.

In contrast to the telephone or television, computer networks are a many-to-many communications medium—the virtual communities that inhabit them exist through active participation amongst their members. This technology would seem to incur a new social order—one based on reciprocity and interaction, rather than imperialist domination.

The catch is that computer networks are not accessible to everyone. Cyberspace is being colonized primarily by countries with access to a high level of technology; and within those countries, largely by the current power elite. At the same time, historically and now there is significant and effective international grassroots computer networking constituency, as well as public and local access BBS, at least in the U.S.

What is culture in cyberspace? In Australia, Canada, the U.S., and parts of Europe, some artists have gained access to computer networks and are using them to make and distribute art. Those working in a Euro-American artistic tradition often experiment with conferencing software to create works that evolve from a process of participatory, interactive communication. Native American artists in the U.S. have developed online graphic share-art, through which they represent their distinct cultures, and provide another source of income for tribal communities. In many 3rd world countries where poverty is high, and computers and phone lines are rare, networking projects are generally operated by non-governmental organizations or educational institutions, and tend to focus on economic or social development, not cultural preservation or participation.

How will cybercultures evolve? Is it important for cultural participation in cyberspace? And if so, how can and is equitable access made available to all cultural groups? What will happen to cultural groups that remain offline? Will cultural groups that do access cyberspace lose their distinct identities through a process of interaction? And, if so, is such an occurrence cultural evolution or homogenization—something to explore or something to avoid at all costs? What is the role of cybercultural activity in cyberspace itself; what is its role in the offline culture that initiated it?[11]

Instructions on how to participate accompanied my opening text:

Connect to one of the following systems between Nov 2–13, 1992, or post to the Usenet newsgroup, alt.isea, directly. Local discussions on the following systems will be distributed to other sites via alt.isea.

American Indian Telecommunication/Dakota BBS
coordinators: George Baldwin (baldwin@backbone.hsu.arknet.edu)
Anne Fallis (anne.fallis%dbbs@oldcolo.com)
Randy Ross (randy.ross%dbbs@oldcolo.com)
location: Dakota BBS, 406.341.4552 (8N1)

ArtsNet
coordinators: Sue Harris and Phillip Bannigan (suephil@peg.apc.org)

Arts Wire
coordinator: Joe Matuzak (jmatuzak@tmn.com)
location: artswirehub

Usenet
coordinator: John S. Quarterman (jsq@tic.com)
location: alt.cyberspace

The WELL
coordinators: Judy Malloy (jmalloy@well.sf.ca.us) and Eric S. Theise
  (estheise@well.sf.ca.us)
location: virtual communities (vc)

For further information contact:
Anna Couey
Telecommunications Subcommittee/TISEA
Arts Wire Network Coordinator
1077 Treat Ave.
San Francisco, CA 94110
tel: 415.826.6743
email: couey@well.sf.ca.us
couey@tmn.com

The ensuing responses varied across systems,[12] reflecting technology gaps and failures; existing community identities, culture, and history; and, perhaps, the challenges of using a temporary communications structure to connect communities whose members have not sought the connection. The shared communications channel, alt.isea, was not created according to protocol, and uunet, the main carrier for Usenet, refused to distribute it, rendering communications between systems ad-hoc and reliant on project coordinators who had accounts on multiple systems. Bridges between communities that formed through the project—between American Indian Telecommunications, Arts Wire, and The WELL—built on preexisting relationships and affinities.

On Arts Wire, people shared ontological reflections on the medium, creating a rich and lyrical tapestry. They discussed how to convey the performative aspect of a poetry slam in an online environment; how the history of other modern technologies might portend the development path of computer networks toward corporate consolidation; the qualities, culture, sensibility, and limitations of online text-based communication; how Native American languages convey meaning and the search for online communication models for reservations; the role of silence in conversation and how to represent different kinds of silence online; online anonymity and identity switching like "some kind of soul shopping center";[13] the likely development of a video-based Internet; the impact of cyberspace becoming a vehicle for popular culture; and how future telecommunications technologies would support human relationships. Arts Wire participants were Joe Matuzak, Anne Focke, Sarah Lutman, Jan Zita Grover, Eric Theise, Randy Ross, Frank Burns, and Jeff Gates. Randy Ross also posted excerpts of Dr. George Baldwin's writings on Arts Wire.

On The WELL, people stopped by the *Cultures in Cyberspace* topic in Howard Rheingold's Virtual Communities conference to introduce themselves and describe their work: Judy Malloy on electronic literature and *Leonardo Electronic News*; Brian Andreas on virtual opera; Fred Truck on Art Com Electronic Network; Eric Theise on human–system interface, Cinematheque, San Francisco experimental film and videomakers organizing to stop their marginalization, and Arts Wire; Robert Campanell on online/

offline community for rave enthusiasts; Jon Lebkowsky on bOING bOING magazine and Electronic Frontier Foundation-Austin; Steve Cisler on Apple Library of Tomorrow and using technology for artistic and cultural preservation; Craig Harris on computer music and Leonardo, the International Society for the Arts, Sciences and Technology; John Coate on networking artists in San Francisco; Barry Kort on Micro-Muse; Art McGee on online information sources about African, African-American, African-Caribbean, and African-Latin people, cultures, and issues, and Black/African-run or Black/African-oriented BBS in North America; and Ralph Melcher on community development and Internet access in Santa Fe. Participants did not all know each other; as Judy Malloy noted, the project also served "an unanticipated purpose of letting us here on the WELL know of parallel cultural developments in our community."[14]

Responses on other systems were quieter, less sustained, and/or lost, reflecting some of the very issues the project sought to discuss: access, relevance, differences in voice and forms of expression, and at TISEA, I believe, the overwhelming draw for people to focus on being together in person.

Weaving Across Time

Many of the people who invented the ARPANET and the Internet modeled both after academic interchange, which is why it's both so open and so insecure, and more like arts than like traditional business.
—John Quarterman, 2014[15]

But what about online communication *itself*? Initially it was used mostly for email and file transmission related to the net and to the physical sciences/engineering that the net supported. Like literacy three hundred years ago—a relatively closed system in which producers (writers) wrote for other formally educated people not unlike themselves—the net was initially a relatively closed loop whose members could speak in abbreviated ways because of their shared experience. ... Now that the online base of users has widened so greatly through both commercial and community access, the net and its individual nodes have widened their services and focuses. Popular culture—games, dating, sex, car-talk, horoscopes, 12-Step meetings in cyberspace—crowd the matrix. Losses and gains, hmmm? Shapes no longer very clear.
—Jan Zita Grover, 1992[16]

Of course the sheer raw power of economy invaded and occupied.
—Susan Harris and Phillip Bannigan, 2014[17]

Many of the technological developments envisioned by *Cultures in Cyberspace* participants more than 20 years ago have been realized. The near ubiquity of the Internet and social media has normalized new methods for information sharing and being in relationships, and these are continuing to evolve. George Baldwin notes, our early conception of online communities were the precursors to what happened with social networking ... the idea that anyone can seek out those of like mind and belief to associate with.[18]

Using contemporary social media, Baldwin has facilitated development of Native American communities in diaspora:

the Northern California Osage group I helped form has become a powerful lobby group that impacts the politics and elections of our tribe in Oklahoma. Without the online community the RL (real life) community would not be able to communicate ... the two go together and as such describe the communication patterns of our intentional Osage community in California ... not grounded in euclidic geography, but founded on the historical sense of belonging associated with our tribal heritage.[19]

The continuing participatory structure of social media enables a multitude of voices, no longer as constrained by oligarchic media channels. As Judy Malloy writes, "the voices are more diverse,"[20] and sometimes, as John Quarterman points out, an individual can "get good ideas to go viral."[21] The dream that participatory media would facilitate social change has borne fruit: mass movements have used social media spectacularly to challenge and transform oppressive power structures.

Yet the issues I sought to address with *Cultures in Cyberspace* are not resolved. In some cases, our language for them has changed, as Eric Theise notes—"digital divide," "paywall."[22] Along with increasing diversity and connection come new challenges, as Randy Ross describes, "Because of the video capability and multi-language applications, it [social media] has become a place of cultural exchange and extension. As more culture, through song, language, and art are shared across social networking, the adaptations or acquisition or adoption of Native culture then advances to new levels of 'Pan Indianism.' But on the other hand, it is possible that this social media can serve to advance particular tribal cultural experience as Tribes begin to harness language learning and experience to suit their own teaching and educational efforts to preserve, restore, and revitalize."[23]

The idea that decentralization of mass communications would usher in multicultural social equity seems insufficient today. Our histories of marginalization and oppression of particular groups of people have followed us in our fused media/real-life existence, and economic inequality has increased. The Internet and social media have helped normalize mass surveillance and fostered new forms of oligarchic power: our identities and actions have become capitalist commodities belonging to those who control the systems we use, not to us.

As I reflect back over our gains and losses, and my own path from building bridges between virtual communities to contemporary hybrids that continue to be shaped by differences in history, geography, race, class, gender, and so many other human identities, I still question: how will we shift our propensity for oppression? How will we move our cultures toward reconciliation and collaboration, equitable distribution of power and resources, and the ability to "see out of the me into us"?[24]

Notes

1. I would like to thank the collaborators for their insights on the evolution of networked cultures from *Cultures in Cyberspace* to social media, which informed my writing of this chapter.

2. Tim Gruchy, personal correspondence, 9/9/2014.

3. Susan Harris and Phillip Bannigan, personal correspondence, 8/24/2014.

4. John S. Quarterman, *The Matrix: Computer Networks and Conferencing Systems Worldwide* (Bedford, MA: Digital Press, 1990), p. 235.

5. Susan Harris, personal correspondence, 10/18/2014.

6. Susan Harris and Phillip Bannigan, personal correspondence, 8/24/2014.

7. Randy Ross, personal correspondence, 8/14/2014.

8. Judy Malloy, personal correspondence, 8/8/2014.

9. Eric Theise, personal correspondence, 8/15/2014.

10. John S. Quarterman, personal correspondence, 8/4/2014.

11. Anna Couey, *Cultures in Cyberspace* background, 1992. This and the following participation instructions are available at http://www.well.com/~couey/cultures/cultures2.html.

12. Cultures in Cyberspace conversations are available at http://www.well.com/~couey/cultures/culturesmenu.html.

13. Joe Matuzak, posted on Arts Wire (Artswirehub, Item 113:19: Cultures in Cyberspace), 11/10/1992. Available at http://www.well.com/~couey/cultures/cultures3.html.

14. Judy Malloy, posted on The WELL (Virtual Communities, Topic 83: Cultures in Cyberspace; A Virtual Panel, #34 of 69), 11/10/1992. Available at http://www.well.com/~couey/cultures/cultures4.html.

15. John S. Quarterman, personal correspondence, 8/4/2014. Quarterman further notes that this is his opinion.

16. Jan Zita Grover, posted on Arts Wire (Artswirehub, Item 113:19: Cultures in Cyberspace), 11/11/1992. Available at http://www.well.com/~couey/cultures/cultures3.html.

17. Susan Harris and Phillip Bannigan, personal correspondence, 8/24/2014.

18. George Baldwin, personal correspondence, 8/20/2014.

19. Baldwin, 2014.

20. Judy Malloy, personal correspondence, 8/8/2014.

21. John S. Quarterman, personal correspondence, 8/4/2014.

22. Eric Theise, personal correspondence, 8/15/2014.

23. Randy Ross, personal correspondence, 8/14/2014.

24. Adapted from Anna Deavere Smith, quoted in Sara Boxer, "Enter, the Audience: Trying to Gather Everyone In," *New York Times*, August 29, 1998, p. A13.

VI Social Media Poetics

22 Crossing-Over of Art History and Media History in the Times of the Early Internet—with Special Regard to THE THING NYC

Susanne Gerber

Crossing-over is a term used in genetics to describe a chromosomal event, which leads to new combinations of genes. In this chapter, the term is used for a cultural phenomenon. In the early days of the Internet, art history crossed the path of media history, and, subsequently, both disciplines conveyed characteristics of each other.

In the years after 1990, in an era of the rise and fall of the new economy, the phenomenon of "net art" emerged. Yet about a decade later, net art submerged in the art world. Although promising artists had created widely accepted work, net art as such was exceeding the concept of "art." In an interview from around this time, Wolfgang Staehle, artist and founder of THE THING NYC, posed the questions, "Where and how should art projects exist on the Internet?" and "How in 'real life' ought art projects reflect the conditions under which they are created?" His answer was, "Network features include immateriality, immediateness of transmission, and global access. These are interesting factors for artistic endeavor (intervention)."[1]

Art works examining the Internet did not lead to regaining the utopian potential of art, but its social, aesthetic, and conceptual approach referenced the future role of digital communication—crossing boundaries and thus directly impacting the social reality of life.

This chapter documents the role of communication in the art network THE THING and how it was realized.

The Founding of THE THING NYC: Theory and Practice

To begin, we look at the mathematician, artist, and theorist, Peter Weibel. In his theories, he uses the term "info-sphere" to summarize media-based communication between humans, communication between humans and machines, and communication between machines. Weibel considers the present and future development of the info-sphere as the logical and necessary consequence of the evolution of life forms.[2]

"What is the info-sphere?", Peter Sloterdijk asks in his "Spheres Trilogy" of *Bubbles, Globes*, and *Foams*.[3,4] The ancient doctrine of the Sphärik (Spheres) describes a shell-like, all-encompassing, rich in associations, self-referential, linked Cosmos. So the info-sphere has to be imagined as an exuberant universe of (technically) mediated human extracorporeal communication—which Weibel denotes as exo-evolution. The history of religions, the sciences, the arts, and the media all can be read as the progress of exo-evolution. The latest chapter in the history of the info-sphere is the history of the Internet.

The Internet, although born from the spirit of the military, in the early years developed as the bearer of a new idea of freedom and equality. The credo was, "We want no Kings, Presidents and elections. We believe in rough consensus and an executable code,"[5] which resulted in the formation of a community of network researchers who believed that cooperation among researchers was more powerful than any competition.

A social public domain was built, where everyone was free to read and write, to immerse and create oneself in the emerging World Wide Web and its predecessors—for example, multi-user dungeons (MUDs) that allowed several users to trundle along together on purely text-based message boards, to kill the dragon, to solve puzzles, and to chat with each other.

Initially, game environments, MUDs, and MUDS Object Oriented (MOOs) were subsequently also used for educational and discussion purposes. In 1988, another synchronous communication format was added with Internet Relay Chat (IRC). Bulletin board systems (BBSs) also existed, initially as stand-alone PCs with one or more dial-up connections.

In 1985, Stewart Brand and Larry Brilliant founded the legendary BBS whole earth 'Lectronic link (The WELL) in the San Francisco Bay area. Commercial online services such as CompuServe and AOL also began, and in the late 1980s, these separate networks became gateways to the Internet, allowing the exchange of email and news. To enable access to the Internet for people outside of universities, a series of so-called free nets that served communities and nonprofit endeavors were created. In the transition between the 1980s and 1990s, a rising stream of ads and spam were deployed in the network ("The September That Never Ended," when AOL expanded its membership, was a turning point). In 1990, targeted efforts were undertaken to get commercial and noncommercial information service providers in the network, with commercial Internet providers (ISPs) emerging beginning in 1989 and 1990.

That was the key moment that inspired Wolfgang Staehle to found THE THING in 1991. One of the first and most important art platforms—which entered the new space of data networks to communicate, distribute, and produce—THE THING was a forum for a decentralized, global exchange of information, discussion, and net-based art projects.

We knew this would change everything.
—Wolfgang Staehle

THE THING began in New York City. Nodes followed in Cologne, Berlin, Vienna, London, and Stockholm (view timeline for details). Staehle states, "This development would have a big impact; a transaction would happen much faster, it would have impact socially, politically, in the financial world, in the military, everything would be accelerated; machines would make decisions for us. We knew this would change everything."

During the mid-1990s, the Internet began to grow faster and faster, and a large part of the general population had grasped its importance. With the increasing speed of modems, the Internet flourished and gained popularity. It was also economically more interesting. Many large companies began to use the Internet to present their products on home pages. As the company name and domain (which for commercial providers usually ends with ".com"), thrived in this boom period, it became known as the dot-com boom.

Wolfgang Staehle, the founder of THE THING NYC, was not only at the forefront in this period of the rapid development of media technology, but he was also important in connecting and fostering the possibilities of telecommunications and the idea of net art (net-based art), which was taking hold in the art world.

Indeed, the common assumption that the Internet boom triggered the idea of net art should be reconsidered. Artists had embraced the idea of the network and the network itself before the emergence of electronic networks. Thus, artistic actions were ready and able to recognize the new technological possibilities and dimensions and to further push the boundaries of digital telecommunications before it reached the general public.

Dieter Daniels refers to the group of artists who had the strength and imagination to fuse art and technology together as the "net pioneers." Daniels observes that at the time of the breaking point in art and media history, "a rapidly developing international art found itself racing a fast changing techno-sociological context."[6]

In Daniels' words, the net pioneers investigated the following approaches on the stage of art and Internet:

A critique of the "bourgeois" concept of art, the commercialization and institutionalization of art ...

A kind of "art for all" that would reach its audience directly, bypassing the gatekeepers of the art context ...

Collective authorship, or anonymous works, as a critique of the idea of "genius" ...

The transition from art to life and politics ...

Art that does not want to be recognized as art ...

Art as an effect or shock of the real ...

Internationalism or non-nationalism ...

Reflection on the medium in the medium and the deconstruction of its materialism ...

The revision of formalist approaches, as regards the network medium ...[7]

The community of telematic artists was first an insider group of worldwide active, technological-oriented artists, and the new free spaces were acknowledged as liberties from the regulatory mechanisms of the art scene. The ideas of networking, sharing, and collaboration energized a desire for a new art, which was more direct, transparent, varied, faster, smarter, and fun; more about communication than ever before; and with the intention of exceeding the scope of the meaning of what constituted art in the future. In Staehle's words,

In those days there was a movement, an institutional critique ... the ironic thing about that was that the institutions very rapidly caught up with it, framed it, and then re-institutionalized it themselves. So I thought, someone needs to actually try to do that again outside of this institutional framework ... rather, we chose very deliberately to take an outsider position, simply to create a discourse that might possibly be independent of the constraints that impose institutions always ... to be able for once, to talk freely about the whole phenomenon of art.[8]

Examining the individual elements that comprised THE THING, it becomes clear how it had already anticipated and realized the entire spectrum of communication potentials that the Internet offers.

THE THING's menu includes features such as threads, thing.review, audio, video, projects, and editions:

THREADS were moderated message boards with topics ranging from aesthetics to politics; from net activism to issues of net-specific art.

RADAR was an exhibition calendar with moderated events.

FOG-CHAT was an anonymized chatroom.

THING.REVIEW presented critical essays regarding exhibitions, books, films, new media, and other cultural phenomena. Readers could add their own comments, in effect opening the various essays up for discussion.

VIDEO offered video-on-demand.

AUDIO offered nonstop "RealAudio" music by leading groups in the avant-garde music scene, samples by invited DJs, and artists' sound works—ranging from radio plays to sound collages.

PROJECTS was THE THING's virtual exhibition space. Since the BBS days, THE THING collaborated with artists to develop specific online works.

EDITIONS was an e-commerce shop that offered art editions—electronic as well as object editions—published by THE THING in collaboration with artists. The edition series was pioneered by THE THING in 1993 to provide a financial venue for artists and to benefit the nonprofit THING.

Interview with Wolfgang Staehle /Susanne Gerber, September 2014

SG: Wolfgang, when you dream about THE THING, now with the distance of some years, what comes to mind?

WS: Nice question, ok, let's daydream. What I see now is my old Amiga computer and my first modem, back then, something most people had no idea what to use it for. Then I see the spaces in which THE THING came into being. First there was the basement space at 44 White Street in Tribeca. You went downstairs to a different world. Monitors, cables, modems, a fan—quite a science fiction ambience, a kind of cyber cave. Then in 1995 a very different feeling when we moved into a loft on the 16th floor of the Starrett Lehigh Building in Chelsea. Nineteen thirties industrial charm, an Internet company office with a vintage seating corner and a panoramic view of the Hudson—very cinematic. Then a few years later, we moved onto the fourth floor in the same building, a bit closer to earth and more affordable during the real estate boom. But the much larger space which connected and encompassed everything was of course the virtual space of THE THING. This electronic network in which and with which we lived, for which we worked, which grew bigger, which gave us headaches and an infinite amount of fun. The network of THE THING was worldwide and very much alive.

SG: What kind of role had your friends, the users, co-workers, artists, and other interested people?

WS: There was this contagious enthusiasm for what was suddenly possible, combined with a sense of play and a curiosity for what would happen next. There was this inspiring community feeling that we were able to reframe the discourse, that we had the means to realize our ideas independently of the traditional art system. It was a never-ending current of discussions, talks, and debates around the globe. People from all over would come to our office in Manhattan to work with us for a while or realize some specific project. And then, of course, there were the parties. When I see the whole project in its totality today, I see a huge sculpture with concentrations in some parts and countless lines of flow of people and information. The whole thing was a social and technological work of art—a creation and a creature. Did we dream enough?

SG: Can you ever dream enough? Can you mention one art project that stands out?

WS: To pick one is of course difficult, there were so many interesting projects. But one that was really amazing was the "Toywar." A project that we did not initiate, but kind of fell into our lap, since we were asked to host and manage its technical infrastructure. It was a legal fight for the domain etoy.com in 1999/2000. The antagonists were the European art group etoy and the American online toy retailer eToys. The name Toywar originated from the name of the website www.toywar.com which we hosted in New York. The artists were accused of cybersquatting, even though they had registered their domain long before the company eToys was founded. The campaign at its most active phase had about 1,700 international participants and led to all kinds of curious online happenings. The interface with all those funny Lego warriors was not only beautiful, but also functioned remarkably well as a tool for social and political mobilization. As Beuys said, "Democracy should be fun!" Within weeks this crazy propaganda machine led to a dramatic drop of the stock value of eToys. Soon they were trading as penny stock and eToys eventually folded and the artists were able to keep their etoy.com domain. This was an example of a very successful and very creative defense in an unjustified civil suit. One can still draw lessons from it.

SG: Toywar, yes, what an example! The belligerent, masculine language of THE THING has often been noted and frequently criticized.

WS: I remember that the tone in some of the debates was sometimes a bit rough—a friend of mine once called me jokingly "Blade runner on the Hudson"—and it was not always easy for the sysop or moderator to tamp down the rhetoric. Some groups totally resisted moderation. So yes, maybe in the beginning you could say it was a bit male dominated, but already the establishment of the Cologne node was initiated by a woman, namely Andrea Wagener, who brought in the whole crew of "Friesenwall 116a," an interesting artists/activists collective. In Cologne there was almost parity between male and female participants. Later on, many women artists participated with various projects: Mariko Mori, Eva Grubinger, Vanessa Beecroft, Coco Fusco, just to name a few. Also, the toywar infrastructure was almost single-handedly managed by Andrea Mayr, one of the few women at the time interested in programming and who came to us from Vienna. And since the guys in New York—myself included—were unable to keep any records, we would not have survived without Gisela Ehrenfried, our indispensable office manager and archivist.

SG: Let's jump into the here and now, the year 2014. You just had a show in Italy where you showed images of the social networks of your friends using data from Google or Facebook. The images consist of lines connecting individual nodes or various affinity groups. Are you trying to show the beauty of communication, of social networking?

WS: Ah, now it's getting a bit complicated. Beauty yes, but also much more. I was always fascinated by the aesthetics of networks, and therein certainly lies a hidden, but also quite visible, motivation for my work. There can be structural beauty even in a Facebook social profile. But that doesn't mean that I condone the development of mass communication via PC or mobile as we have it today, the way some mega corporations now control and exploit our communications. What I am missing is the desire for autonomy and self-determination that propelled us to do what we did 20 years ago. Where is the courage for criticism and change? Where is the will to escape or defy the control and manipulation of the data suckers, be they state agencies or commercial entities, or increasingly all of them together? Why do people voluntarily allow them access into the deepest corners of their private lives? You see, now I'm asking questions.

SG: Questions which I ask myself. Questions for which I do not have any answers either. In the early years of the Internet, I thought that the digital avant-garde would always have a conceptual and technical advantage. I believed new technologies would be a vehicle for creative autonomous zones, democratic decision making, and better distribution of resources—to make it short, more discourse, more problem solving, and less aggression and dominance. This is not the world we have today. What is your personal attitude in the face of this situation?

WS: Hell, life goes on. My work goes on. There are still cells of resistance. There is no need to be on Facebook, to twitter constantly, or even use Google. We can still communicate on our own terms, with our own tools, even if it is a bit more complicated. I certainly do so with my very close friends and confidants. I don't really know how to solve the big political questions, but I know we could solve a lot of problems if we would use our technological potential creatively. One good thing could be to decentralize the net, to let millions of independent servers bloom. Instead, what's happening is quite the opposite. We have a development where our data are being co-opted by corporate assholes and government spooks and then used to turn us into better consumers and gullible, manipulated citizens. My fear is that our "digital doubles" one day will be more "real" than our actual selves. So for now, the smaller the footprint you leave out there, the better.

THE THING NYC Exhibited a Playful Approach and Exemplified Far Ahead of Its Time What Is Capable of a Networked Communication

In *Net Pioneers*, Dieter Daniels documents the programmatic objectives of THE THING NYC, as a whole, in three points:

Construction of an independent, partly self-designed technological infrastructure

Formation of a self-organized, net community and the collective design and testing of a corresponding model of discourse

Development of a form of art specific to the network, exploring the medium's potential in an experimental, self-reflective way.[9]

Daniels notes that "this development took place in an autonomous situation, as unusual for the media as it was for the art world at that time; the frameworks were not only independent of any art institution, but also existed outside of state or commercial media control."[10]

Idea and execution were in close proximity to each other at the particular moment of the meeting of art history and media history in the mode of a crossing-over, as described in the opening of this chapter. So THE THING provided a blueprint for everything that was offered on the Internet in the following years, an early, perhaps also innocent blooming period, of the Internet-based phase of the communication age.

This spirit of the time was encountered in those years in different fields. For instance, within the framework of their respective art & science projects, the microbiologist Regine Hakenbeck and the artist Susanne Gerber collaborated to create genomics biotopics and a short film—about the microbiological communication of bacterial strains, recently discovered as an essential engine of evolutionary processes—called "communication is evolution is communication."[11]

Returning, at this point, to Peter Weibel and his analysis—in that the development of the info-sphere creates and extends the environment for humans on the earth—so that an intelligent networked humanity can survive in far greater numbers on the planet than it could without these means. Thus, a collective responsibility for the fair growth of exo-evolution (which, yes, is an intentional process) exists, and future potential should not be squandered. Weibel sees a misused and therefore counterproductive use of resources, for example, in the construction of excessive surveillance and monitoring equipment from governments and manufacturers—with the result that the info-sphere becomes an instrument of control and dominance and loses the potential for growth and development.

Here it is inspirational to look back at the early era of the Internet. THE THING NYC exhibited a playful approach and, far ahead of its time, exemplified what is capable in networked communication.

Appendix: THE THING—Timeline and History

Created by Gisela Ehrenfried.

November 1991
THE THING starts operating out of a basement at 44 White Street (Tribeca), New York City.

March 1992
THE THING Cologne node opens.

June 1992
"Wochenschau," THE THING's first online symposium, organized by THE THING New York and THE THING Cologne, focuses on new modes of art production and exhibition.

July 1992
"Pressure on the Public," THE THING's online symposium is part of a project on the art public, organized by Mitchell Kane, The Hirsch Farm Project, Northbrook, IL (catalog).

October 1992
"Remaking Civilization: Rethinking Evolution, Intentionality, Time, and Identity," online discussion group. Project is a collaboration with Blast.

November 1992
THE THING Dusseldorf node opens. Public terminal of THE THING at Friesenwall 116a Project Space, Cologne, Germany, during Cologne UnFair Event. Publication of the *Yellow Reader*," print publication with excerpts from online discussions.
Public terminal of THE THING at Daniel Buchholz Gallery booth at Cologne Art Fair.
Public terminal of THE THING at "F.A.R. Bazaar," Foundation for Art Resources, Los Angeles. Live demonstration of the network by artist Kelly Hashimoto.
"Manifesto," first visual art project online THE THING, curated by Benjamin Weil, with works by artists Henry Bond, Gavin Brown, Angela Bulloch, Laura Emrick, Sylvie Fleury, Liam Gillick, Dominique Gonzalez-Foerster, Happier Days, Yasuma Morimura, Marco Mazzuconi, Julia Scher, and Wolfgang Staehle (available online; show also exists in poster format).

December 1992
Benefit Art Auction for THE THING at Nathalie Karg Gallery, New York. Public terminal of THE THING to demonstrate the network in function.

April 1993
"1916," limited electronic art edition by Olivier Mosset, produced and published by the artist and THE THING (available online). First edition online THE THING.

July 1993
THE THING Berlin node opens. "Accrochage," online art exhibition curated by Wolfgang Staehle.

September 1993
"Superdream Mutation," unlimited, numbered electronic art edition by Peter Halley, published by the artist and THE THING (available online).

October 1993
Introduction of electronic dissemination of art magazines and journals via THE THING, including the *Journal of Contemporary Art* (published by Klaus Ottmann), *Lusitania Magazine* (published by Martim Avillez), *Lacanian Ink* (published by Josefina Ayerza), and *Texte zur Kunst* (published by Isabelle Graw and Dr. Stefan Germer).

November 1993
"Transactivism," online symposium organized by Jordan Crandall and THE THING. The panel discusses the production and circulation of art and sociality in transactional space. Invited panelists include artists, critics, and curators (archived in the File Area; also available on disk).

December 1993
"Building Process," online art project by John F. Simon. The artist creates "Line Drawings" from a paint program he designed after Paul Klee's concept of "active lines, passive lines, and mobility agents."
THE THING Frankfurt node opens.
THE THING Vienna node opens.

March 1994
"Bioinformatics," moderated online forum. Part of a larger project for the Kunstverein Cologne. The project functions as a map or guide with which readers can situate themselves as living biological systems within many informational systems to orient themselves as bioinformatic entities.
"Julio," limited electronic art edition by Rudi Molacek, published by the artist and THE THING (available online).
"9 Sculptures, New York," online art project by Helene von Oldenburg with floor maps of nine New York Museums and a legend indicating dimensions and location of nine imaginative sculptures in these spaces.
"nOn Television—THE THING," television documentary by artist Aki Fujiyoshi, broadcast on public access Channel 16, New York, March 23, 1994 (videotape available). Part of a television series on collaborative art projects in New York.

"Copy," limited electronic art edition by Rainer Ganahl, produced and published by the artist and THE THING (available online).

"Snap to Grid," first online interview (W. Staehle/R.Ganahl) from a series of online interviews in one of the public fora of THE THING. Each interview was archived in the File Area for later retrieval (publication in the print media was encouraged).

"Provisional," online art exhibition by Felix Stefan Huber. The interactive artwork is designed by the artist to bring together your world with that of homeless people and refugees—right into your living quarters.

THE THING introduces Internet email and newsgroups.

April 1994

"virusheaRtbeAt,"limited electronic art edition by Joseph Nechvatal, published by the artist and THE THING (available online).

"No Cover, No Minimum," online interview with artist Dike Blair in The Thing's Talkshow forum.

"Artists in the Information Ghetto—A Way Out," workshop introducing artists to computer technologies. Participants: Wolfgang Staehle, THE THING ("Networking"); Gretchen Bender, artist; Stephania Serena, Charles Warren, consultants; Rainer Ganahl, artist; Marshall Blonsky, writer, professor of Semiology at The New School for Social Research, New York, April 30, 1994.

"Mean Things," limited electronic art edition by David Diao, published and produced by the artist and THE THING (available online).

May 1994

"Basic English, Basic Japanese," online art project by conceptual artist Rainer Ganahl. Includes video, image, sound (interview R. Ganahl/Sagawa), and text files (online interview W. Staehle/R. Ganahl). This On Show project dealt with the study of a new language as a non-object-oriented, but personality and social relationship altering cultural exchange. In October 1994, the project was also presented by THE THING Vienna (additional interview F. Rakuschan/ R. Ganahl).

October 1994

"What Is In Your Mind," group exhibition curated by Frederick Harleman at the National Museum of Science and Technology, Stockholm, Sweden, with participation of THE THING (organized by Wolfgang Staehle/THE THING New York). Other participants included Bigert & Bergstroem, Klaus vom Bruch, Aris Fioretos, Rainer Ganahl, Michael Joaquin Grey, Michael Joo, Laurel Katz, Jon Kessler, Mikael Lindgren, Matthew McCaslin, Nam June Paik, Jean Tinguely, Teddy the Artist, Dan Wolgers, and Fredrik Wretman/Mats Hjelm. October 21–November 30, 1994.

"Cybersphere," Symposium at Kulturhuset, Stockholm, Sweden, October 22 and 23. Wolfgang Staehle participant as the founder of THE THING. Other speakers included Michael Benedikt, Dir. Center for American Architecture and Design at University of Texas at Austin, TX; Donna Haraway, History of Consciousness Board at the University of California, Santa Cruz, CA; Allucquere R. Stone, Dir. of ActLab, University of Texas, Austin, TX; Rob Tow, Researcher at Interval Research Corporation, Palo Alto, CA; Peter Weibel, Dir. Inst. for New Media at Frankfurt Art Academy, Frankfurt, Germany; Norbert Bolz, Prof. Communication Theory at University of Essen, Germany; Amy Bruckman, Researcher at MIT Media Lab, Cambridge, MA; Brenda Laurel, Researcher at Interval Research Corp., Palo Alto, CA; Elisabeth List, Assoc. Prof. Department of Philosophy, University of Graz, Austria; Marcos Novak, Dir. Advanced Design Research Program, School of Architecture, University of Texas, Austin, TX; and Jeffrey Shaw, Dir. Inst. for Image Media at ZKM Karlsruhe, Germany.
THE THING Stockholm node opens.
"Pain," online video show curated by Shauna Sampson and Steven Overman. Project includes an online catalog with image and text files. Project produced and published by the curators, the artists, and THE THING. Artists include: Laura Parnes, Kelly Parr, Jason Fox, David Weeks, M.M. Serra, Skip Arnold, Richard Karnatz, Ricardo DeOliveira, Cheryl Donegan, Rainer Ganahl, Wayne Gonzales, Mike Wodkowksi, Janine Gordon, Michael McAuliffe, Andrew Perret, Joshua Singer, Wolfgang Staehle, John Tremblay, Jane Duncan, Kenneth Goldsmith, Meghan Gerety, Ken Goldberg, Bryan Gonzales, J. R. Gross, Charles Labele, Peter Lunengeld, Julia Parker, Crystal Reiss, Trudie Reiss, John Tipton, and others. Opening reception at "Here," a nonprofit space in Soho, New York, October 22, 1994.
"jon.tower@thing.nyc.ny.us," an online project by conceptual artist Jon Tower. The artist opens an interactive office for consultation. The project also includes video, sound, and text files by the artist. Opening reception at I. C. Editions, Soho, New York, November 1994.

November 1994

"Altwien Neuzeit," group exhibition curated by Warren Niesluchowski, New York, and Hubert Winter, Vienna, with participation of THE THING.
"The Laws of Humans," online project by Noritoshi Hirakawa in the On Show project area of THE THING (avi, gif, and wav files). The project was also to be presented as part of a solo exhibition by the artist in Amsterdam, The Netherlands, February 1995.
Media Alliance, lectures on "Art-related computer networks," organized by David Green, New York Foundation for the Arts, with participation of THE THING representative.
Limited electronic art edition (portfolio) by James Nares, produced and published by the artist and THE THING New York (available online). Production of electronic art edi-

tion by Peter Schuyff, produced and published by THE THING New York (print version was be available via Pace Gallery/Pace Editions, New York).

December 1994
"Informatics: The Electronic Frontier and You," seminar by Jordan Crandall and Wolfgang Staehle, THE THING, at White Columns, New York, December 12 and 19. The seminar was part of the seminar series "Theoretical Studies in Art" at White Columns.
"Freaks Online," online art project by Claire Jervert (gif files).

January 1995
"Dagegen/Dabei—Production and Strategy in Art Projects Since 1969," a six-part exhibition series at Kunstverein Hamburg, Germany, curated by Bettina Sefkow and Ulrich Doerrie. One of the projects presented there is THE THING (organized by Michael Krome, TT Cologne).
"(inter)ACTIVE electronic art channels," group exhibition.
Presentation of THE THING by W. Staehle, January 23, Trenton State College, Department of Art, College Art Gallery, Trenton, NJ, January 23–February 15, 1995.
Online art project by Chris Kramer (gif files, text collages).

February 1995
"Art, Identity, and Boundaries: Limits and Their Transgression," a cycle of four conferences organized by Carolyn Christov-Bakargiev and Ludovici Pratesi at Palazzo delle Esposizioni, Rome, Italy, February 26–March 19. Wolfgang Staehle is invited, as artist and founder of THE THING, to participate on the February 26 panel, "Beyond Physical Boundaries: New Cybernetic Communications." Other participants include Antonio Muntadas, artist, USA; Catherine David, organizer of Documenta X, Germany; Jimmie Durham, artist, USA; Michelangelo Pistoletto, artist, Italy; Renee Green, artist; USA; and Hermann Nitsch, artist, Austria.
"Blast 4: Bioinformatica," exhibition at Kunstverein Cologne, Germany, with participation of THE THING. February 4–March 19, 1995.

March 1995
"Quicktime Movies," by John Baldessari, produced for THE THING, by David Platzker. Concurrently, an exhibition of Baldessari's "Books and Ephemera," also curated by David Platzker, takes place at Printed Matter at Dia Art Foundation, New York, March 9–April 28.

April 1995

The *Journal of Contemporary Art*'s WWW pages are incorporated in THE THING NYC's Web site. James Nares portfolio of JPEG images, produced by the artist and THE THING, presented online TTNY BBS.

May 1995

Springer, a new Viennese magazine "focusing on investigations of the broad terrain of history, theory, and criticism of the visual arts, while concurrently inquiring other scholarly fields such as new media and pop culture, opens online forums on THE THING BBS with writing access for their contributing editors based in various European and American cities and reading access, as well as a feedback channel, for the public." *Springer Magazine* also resides on TTVienna WWW and in the form of a bimonthly print publication.

June 1995

Ars Electronica (June 20–23), Linz, Austria. presentation of THE THING, premiering multipage World Wide Web sites of THE THING NYC, Vienna, and Basel, with telnet function into THE THING BBS's message forums and live conferencing area. The website functioned as an ever-changing exhibition and publishing area. Taking advantage of the HTML programming language, this included hypertexts, still images, video clips, and sound files. The event included a symposium with a lecture by Wolfgang Staehle, THE THING NYC. THE THING NYC's WWW projects produced for Ars Electronica include: Image Files, an interactive stock image bank project by Wolfgang Staehle and Franz Stauffenberg; Alter Stats, an interactive visualization of user access of the website, by John Simon; video clips, sound, and image projects by Mariko Mori, Sam Samore, Noritoshi Hirakawa, Beat Streuli, Felix Huber, and Christian Marclay; The *Journal of Contemporary Art Magazine* web project; THE THING Archive, a selection of texts from THE THING NYC BBS (including Transactivism/online Symposium; Gray Goo Lounge/ Interview with Dike Blair; The Twist Thread); and telnet function into THE THING BBS. Catalog available, including texts on THE THING by Andreas Kallfelz, Jordan Crandall, and Klaus Ottmann.

"Art and Telecommunication: Universality—Balance/A Pancultural Project," Civitella d'Agliano, Venice, Italy. Internet art project with participation of THE THING (coordinated by THE THING Vienna). Other participants include Geert Lovink, Digital City, Amsterdam, The Netherlands; Pit Schultz, Museum for the Future, Berlin, Germany; and Derrick de Kerkove, McLuhan Institute, University of Toronto, Canada.

July 1995

THE THING NYC moves to a loft space on the 16th floor of the Starrett-Lehigh Building in Chelsea (601 W. 26 St., New York City, NY 10001). A T-1 leased line circuit to provide full Internet connectivity is installed and an SGI Web Server connected to our LAN. The setup allows combining the global access, multimedia capabilities of the WWW with the interactive/discursive qualities of the message-based information system (THE THING BBS). A telnet link from THE THING WWW site into THE THING BBS allows for real-time conferencing and participation in THE THING's local and international discussion groups.

August 1995
"Arctic Circle," an exhibition on the Internet by Felix Huber and Philip Pocock via THE THING NYC BBS and THE THING website. The project is a "double travel," a physical journey over the Arctic Circle to the least populated, last remote wilderness on earth (Klondike, Yukon, Northwest Territories, Alaska) and, concurrently, over the globe-blanketing infobahn. "Arctic Circle" represents an investigation of contemporary loneliness in a natural wilderness and in front of the computer screen. A series of short performance video and sound loops were produced on and above the 66.67th Parallel, the Arctic Circle, as well as additional text, image, sound, and video files relating to the travel-as-art-as information. The project was also presented at "Photography after Photography—Defining Photography Through Digitality," a traveling exhibition sponsored by Siemens Cultural Program, Munich, Germany; and "Telepolis," a Luxembourg Goethe-Institute Exhibition, Luxembourg, and other locales.

September 1995
Martin Kippenberger, Achim Kubinski "Beuys," audiovisual piece in THE THING WWW ("Art Stuff").

October 1995
"Cyber Soho" Arts Festival, Soho, New York City. 3-day public presentation of WWW and CD-ROM projects such as THE THING, Laurie Anderson/Voyager, Whitney Museum, Dia Center for the Arts, adaweb, Tractor, and ArtnetWeb. With a series of talks moderated by Janine Cirincione (Microsoft) and Timothy Druckrey (New York University).

November 1995
"THINGreviews" is an ongoing art review project on THE THING website, as well as on THE THING BBS. Edited by artists/writers Susan Goldman and Craig Kalpakjian, "THINGreviews" publishes on-the-spot reviews by an international group of contributing art critics and artists/writers reporting on exhibitions and art events in the United States and abroad. The intention was to further communication within the art community on the global level and to offer translated versions of reviews (i.e., English-language reviews translated into other languages and vice versa).

"A Garden Project," by artist Alyson Shotz. The multimedia project was presented in THE THING WWW "Artstuff" Section (http://www.thing.net).

December 1995

"Bulletin Board," exhibition/investigation at Spot 71, New York City (participants included Mike Ballou and Four Walls, Devon Dikeou/Zing Magazine, Jackie McAllister, Printed Matter Bulletin Board, and Alexis Rockman, THE THING International BBS).

"Aliased Father," a website by artist Stefan Beck, produced for THE THING NYC WWW's Art Stuff section.

January 1996

"Out A Site," a web project premiering on THE THING NYC website, produced by artist Steven Pollack with multimedia projects including an unpublished interview by author Paul Bowles, a previously unreleased music video of David Byrne, unpublished photographs of Brancusi's studio, and more.

"From the Arctic-Circle to the Tropic of Cancer," a continuation of the website road movie by Felix Huber and Philip Pocock at THE THING website (an autovisual diary from trips to the Arctic Circle, Summer 1995, followed by reports from Mexico, January/February 1996).

THE THING introduces THING WORLD, "an exciting new chat application with a specially designed fantasy art world tour, where visitors can explore, alter and play in galleries, studios and rooms familiar to many in the New York art world, talk with friends and strangers in the back room of an art dealer or just have unlimited martinis, which are always at hand, while viewing a gallery show. Almost anything can happen in THING WORLD. Check it out!"

February 1996

"follow, follow the yellow brick road" exhibition at the New York Kunsthalle with presentation of THINGWorld, the multimedia live chatrooms on THE THING's website. Other participants include Felix S. Huber/Philip Pocock's "From the Arctic Circle to the Tropic of Cancer," a web project also residing on THE THING's website, and works by Warren Neidich (opening February 23, 1996).

THINGReviews is reviewed by The McKinley Group's professional editorial team of international publishers, technologists, and information specialists and rated a "4-Star" site, the highest rating an Internet site can achieve in Magellan, McKinley's Internet navigational and informational directory.

FAT Magazine, a New York bi-annual print publication, which mixes fiction, commentary, and art in an enigmatic tabloid format, with each issue loosely organized around a theme, such as "Good and Evil" (Issue #1) and "Surrender" (Issue #2). FAT

Magazine's website, designed by THE THING/John Rabasa, premiered this month on THE THING.

"Schnittstelle Netzhaut," a project by The Swiss THING as part of the project series "Sprechende Koerper" at the Skulpturhalle, Basel, Switzerland, February 29, 1996.

March 1996

THE THING introduces its "WWW Discussion Board" open to the public.

THE THING Amsterdam node (mirror site of THE THING NYC website with telnet function into THE THING BBS).

"Cyber Stars" award for THE THING NYC by Virtual City's First Annual Awards (sponsored by IBM) featuring a link to THE THING website (Virtual City, NYC).

April 1996

"ThingReviews" Party at THE THING NYC headquarters (4/9/96).

"Reading Seminar: Deleuze" is a public discussion forum moderated by artist Rainer Ganahl on THE THINGS WWW Discussion Board (with invited participants).

May 1996

Participation at "Version 2.2," a cycle of conferences ("Communication Internet") at the Museum of Contemporary Art, Saint-Gervais Geneve, Geneva, organized by artist Barbara Strebel and Andre Iten, Art and Electronic Media Director, Saint-Gervais Geneve, Geneva, Switzerland. Other participants include Max Kossatz, Felix Stefan Huber/Philip Pocock, Joachim Blank, Walter van der Cruissen, and Herve Graumann.

"Version Box," 22 editions by 25 artists (including Wolfgang Staehle/Ricardo Dominguez/The Thing New York) coordinated by Simon Lamuniere, published by Saint-Gervias, Geneve, Switzerland in the context of the "Version 2.2" conference.

Exhibition participation at "Departure Lounge," a group exhibition at Clocktower Gallery/The Institute of Contemporary Art (PS1) organized by Arfus Greenwood (PS1), artist Franz Stauffenberg, and writer Deborah Drier. "Departure Lounge" was conceived as a gathering space where visitors moved through a network of projects, objects, sound, moving and still images, performances, and cocktail parties. This network was developed by encouraging artists to introduce other artists to the project, create "links" between their work, or combine the works of other artists with their own.

"Quick Times," a group show in the Artstuff section of THE THING's website. Curated by Wolfgang Staehle, the show included videos by John Baldessari, Klaus vom Bruch, Cheryl Donegan, Rainer Ganahl, Herve Graumann, Felix Huber and Phillip Pocock, Rudi Molacek, Daniel Pflumm, Steven Pollack and Renate Sturmer, Christopher Roth and Franz Stauffenberg, Wolfgang Staehle, and Michael Smith. The presentation marked increased commitment to present original multimedia arts content on THE THING's website. The clips ranged in style from the "cool" Techno loops of Daniel

Pflumm to the ideosyncratic animation of Rudi Molacek. "Quick Times" focused on artists who not only produced outstanding video works but also fully understood the concept and mastered the requirements of networked computing.

THE THING workshop at "The Space of Information at the Rotunda," a program of informational workshops, talks, and an installation on the intersection of art with electronic environments, curated by Laura Trippi, May 4–30. Other participants included Echo, Word, artnetweb, adaweb, and New York Online. Party to celebrate the "Out-A-Site" project on THE THING website, May 1, 1996.

June/July 1996

"Super," a fictive "faux Hollywood-style" movie by artist Danny Hobart produced for THE THING's website. The movie trailer was "generated" by way of a series of original video clips, soundtracks, scripts, and still images periodically updated and "promoted" by collectibles such as posters (screen savers) and trading cards (online editions).

October 1996

"In the Flow: Alternate Authoring Strategies," exhibition curated by Daniel Georges at Franklin Furnace, New York City, October 1996 (the exhibition was accompanied by a website). Other participants included titok (Robin Silverberg and Secret Providers), Planet and Eies Texts (Frank Gillette with Teleconferees), Phantasmagorium/Blast 5 (X-Art Foundations and participants), Posters (Group Material), Mail Art (Beattie and Davidson), Photographs (Louise Lawler), Bus Poster (Group Material), GoGo Drawings and Ink Blots (Laura Parnas and GoGo Customers and Dancers), Mail Art (from 1984 FF exhibition), Especially for You (Gabriel Martinez and Interactors), we both belong (Ben Kinmont and participants), PS 217 Sites Mural (Sylvia Benitez and Students of PS 217), and Wall Drawing (Sol Lewitt and Drafters).

"In the Flow," new project series in the ArtStuff section of THE THING including projects by Susan Goldman "uniCity," Franz Stauffenberg "Happier Days," Zhang Gu "Untitled," Eva Grubinger "Bikini Project," Christine Meierhofer "Order a Theft," and Ursula Endlicher "Live Performance."

November 1996

Party at THE THING to celebrate the "In the Flow" web project series.

December 1996

"Thing World," a project by Wolfgang Staehle for the back cover of *Lusitania*, a bilingual art magazine published by Martim Avillez (Vol. #8 "Being On-Line—Net Subjectivity," guest editor: Alan Sondheim), New York City.

January 1997
"Digital Eros," organized by Ricardo Dominguez at THE THING, with readings by Doll Yoko/Gashgirl of VNS Matrix, Petrol Head, Shelly Marlow, and Robert Kylee; a new web project by Zhang Gu; and online videos by Prema Murty and Wolfgang Staehle.

March 1997
"Floating Thing," two evenings of CuSeeMe projects via THE THING website based on a live performance by Floating Point Unit at THE THING office (organized by Ricardo Dominguez).
"Future's Memory," a digital soap opera in 13 scenes. Screenplay by Ricardo Dominguez and Diane Ludin; CuSeeMe project by Floating Point Unit; produced at THE THING and broadcast as a weekly cross-media Internet/cable TV show on Channel 16 (public access) starting March 30 at 11:30 p.m.

April 1997
Complete redesign of THE THING website interface and "Maintenance/Web" (the uglier side of technology) by Kevin McCoy, Jennifer McCoy, and Torsten Zenus Burns (organized by Ricardo Dominguez). Publication of new online projects, including new features such as "TT TV" (Real Audio/Video) and "WTTR" (The Thing Radio), and custom-designed messaging and live chat applications, as well as the launch of a new "Spotlights" series of individual art projects, and new publications of Thing Editions.

May 1997
Panel participation at "Transmedia" cycle of conferences, organized by Internationale Stadt, Berlin (May 26–June 1).
THE THING is featured in "evelmachines," a multimedia kiosk that is an "ambient interactive commingling" between Zing Magazine (published by Devon Dikeou) and cyberNY (produced by Mike Brown) with its premiere version launched with a party at Club Void, May 22.

June 1997
Launch party on June 21 with a performance of the band "Blood Necklace" to celebrate THE THINGS's participation at the documenta website showcasing the "new" THING site, a new series of art projects and programs, such as WTTR (The Thing Radio). "Blood Necklace" is a New York City TechNoCore band with Steven Parrino, Trudie Reiss, and Jennifer Syrie.

Summer 1997

THE THING website is selected by the curatorial committee of the quintannual international "documenta X" exhibition, Kassel, Germany, to be presented via their official website, June–August 1997.

Fall 1997

Publication of new web projects by Bullseye Art, Franz Stauffenberg (second version/"Happier Days"/co-produced by THE THING), Rainer Ganahl ("Basic Korean," co-produced by The Thing and also presented at the Kanju Biennale of Art, Kanju, Korea), Susan Goldman ("second version/uniCity"), Max Kossatz/Holger Friese ("antworten.de"), and Paul Devautour ("Sowana").

January 1998

Publication of new web project by Vanessa Beecroft, co-produced by THE THING (ad announcement in Index Magazine).

February 1998

The Nettime mailing list, focusing on net theory and criticism and moderated by Geert Lovinck, Diana McCarthy, and Pit Schultz, is archived on the web exclusively by THE THING ("Threads" section). Inaugural launch party on February 22, 1998 (with presentation by nettimers Pit Schultz and Diana McCarthy).
Launch of THE THING's new interface and of in-house developed "community server/messaging" software (code by Max Kossatz) providing features like paging, user profiles, threaded messaging, and much more. It is based on an SQL database.
Opening reception for Vanessa Beecroft.

March 1998

The "Threads" section of THE THING website is expanded to include the following discussion and announcement boards: "Bulletin" for general announcements, "Thingist" moderated by Arfus Greenwood and Wolfgang Staehle, "Infowar" moderated by Rick Dominguez, "Rainer's Reading Seminar" moderated by Rainer Ganahl, "Almost (A)live from LA" moderated by Stephan Pascher, "Nettime," and "Guestbook" for comments on the website.

June 1998

Reception for Sawad Brooks and Yoshi Sodeoka, June 25, 1998.

May 1998

Publication of new online project by Sawad Brooks "[sous rature ...] A Reflection on Digitial Media (As Drawing)" (THE THING "projects" section).

April 1998
Publication of new online project by Yoshi Sodeoka "Prototype #22," 5 products from OPT Technologies, Inc. (THE THING "projects" section).
Autonomedia and THE THING book launching party for "Media Archive" by Adilkno (The Foundation for the Advancement of Illegal Knowledge) published by Autonomedia, with presentation by Geert Lovinck (Adilkno) and introduction by Jim Fleming (Autonomedia) at THE THING space, May 11, 1998.

September 1998
"The Telegraph Wired 50," online project by Heath Bunting. Another Heath Bunting accolade. Click on it. Own, be owned or whatever...
"Do You Like Mathematics?" online project by Nicholas Frespech.
"Collider," weekly live web broadcast. Live streaming audio/video program moderated by Gerard Hovagimyan: Live interview with artist Wolfgang Staehle.

October 1998
"New Media Art: The Artists, The Market, The Politics," seminar organized by United Digital Artists (UDA) and Rhizome at UDA, New York, October 22, 1998. Panel participation by Wolfgang Staehle/THE THING; other participants include Maciej Wisniewski, Natalie Jeremijenko, Beth Stryker, Vivien Selbo, Tamas Banovitch, John Ippolito, Barbara London, and Rachel Greene.
"local.language" by Rainer Ganahl; publication of online web project with discussion board in connection with solo exhibition at Kunsthaus Bregenz, Austria.
"Collider," weekly live web broadcast. Live streaming audio/video program moderated by Gerard Hovagimyan:
Live interview with artist Stephan Pascher, moderator of "Almost (A)live From LA" web discussion board published by THE THING.
Live interview with Paul Garrin, artist and founder of PG Media, Inc.
Live interview with Miltos Manetas, artist.

November 1998
"Collider," a weekly live web broadcast online THE THING. Live streaming audio/video program moderated by Gerard Hovagimyan.
_Interview with Marisa Bowe, editor in chief of the online magazine "Word."
In [audio]: Bob Dodds "Bob's Media Ecology."

December 1998
Publication of web project "The History of Moving Images" by Vuk Cosic. From the Official History of Net.art, volume III: Watch films. Star Trek, Blow Up, Deep Throat ...

In Collider #12, GH talks with Ricardo Dominguez from the Electronic Disturbance Theater about recent FloodNet actions.

In Collider #11, GH talks with Peter Fend, the internationally renowned eco-artist.

January 1999

In [video]: Momoyo Torimitsu's "Miyata Jiro." Three videos, taking the format of "commercials," featuring her Japanese businessman robot Miyata Jiro.

"Web Performer" by Ursula Endlicher, produced for and published by THE THING (in "projects"). "Web Performer" is a web project that introduces six different characters. The characters are based on some of her video/live performances. As the piece develops, new images are downloaded directly from the web based on a search-engine result for each character.

Opening reception (January 22) for artist Ursula Endlicher on the occasion of the inauguration of "Web Performer" online THE THING.

In [audio]: The Electronic Disturbance Theater interviews Manuel De Landa. This section opens with his view of strategies versus tactics under the flows of neomaterialism and the Left.

"Collider," a weekly live web broadcast online THE THING. Live streaming audio/video program moderated by Gerard Hovagimyan (GH).

In "Collider" #14, GH features an interview with Rainer Ganahl, conceptual artist and photographer of academic superstars.

In "Collider" #13 GH and writer/media theorist Peter "Blackhawk" von Brandenburg discusses socioculture and media theory.

February 1999

THE THING is 1 of 13 large (web) communities selected for presentation at ArcoElectronico (electronic media arts festival in Madrid, Spain, entitled "the post-media era"). "Hosted by aleph, and organized for ArcoElectronico99, 'the post-media era' introduces itself as a system that tries to facilitate a critical approach to the contemporary transformations of the public sphere—those induced by the emergence of new media, especially the internet—focusing the analysis on the role that concerns all cultural and artistic practices in that context." [The post-media era]—A constellation of (web) communities of media producers: The constellation of selected webs are: [alt-X], [betacast], [blast], [convex tv], ::eco::, [gallery 9 / Walker art center], [nettime], [nirvanet], [P.A.R.K. 4DTV], [raveface radio], [rhizome], [The Thing], [Xchange].

Musee d'Art Contemporain, "Musique en Scene," exhibition participation of THE THING (represented by Wolfgang Staehle), with a presentation of "office radio" (see below). The show is focused on electronic sound projects by selected internationally known artists and arts organizations prominent for their work in this field.

In [audio]: officeradio [the mix]. The story of the cut T1-line. An audio collage by THE THING crew produced for "Musique en Scene."
"GraphicJam," a web artwork by digital artists Andy Deck and Mark Napier, connects visitors into a live, online collaborative drawing. A collage of creative impulses, GraphicJam is a live mix of doodles, drawings, and color created entirely by those who visit the website.

March 1999
"CyberArt99," mailing list hosted by and exclusively web-archived on THE THING website. Moderated by Cynthia Pannucci/ASCI with invited participants, including Max Anderson, Director of the Whitney Museum; John Ippolito, Guggenheim Museum; Martha Wilson, Franklin Furnace; Steve Dietz, Dir./New Media Initiatives, Walker Art Center; Bill Jones, Editor/Artbyte Magazine; Randall Packer, UC Berkeley; Robert Atkins, art critic; Kevin Teixeira, Intel Corp.; Doree Duncan Seligman, Bell Labs Software Research Department; Mark Napier, artist; Wolfgang Staehle, artist/Dir. THE THING; and others.
New features in THE THING [video]:
"Collider #19," live TV webcast show moderated by GH Hovagimyan: Interview with artist Prema Murthy, whose new web project for The Thing site will be launched in May.
"Hood Ornament," video by artist Skip Arnold produced for THE THING.
"Circle's Short Circuit," film by artist Caspar Stracke.

May 1999
Web casting of "Five29Ninety9," a one-day art symposium with 24 lectures, an exhibition, and a SoundLab performance; at St. Ann's Church, Brooklyn, New York, May 29, 1999.
"CyberArt99," conference organized by Cynthia Pannucci/ASCI with invited panelists, including Wolfgang Staehle of THE THING. At Cooper Union School for Art and Architecture, New York.
Bindi, web project by Prema Murthy for THE THING [project] section. Bindigirl is a character or Murthy's avatar. She is a construct of fe/male desire, created out of what is deemed "exotic" and "erotic." Murthy takes Bindigirl pictures of herself and juxtaposes them with ancient Indian texts excerpted from Hindu Deity mythologies and The Kama Sutra as translated by Sir Richard F. Burton.

Fall/Winter 1999
Book/CD ROM publication "THE THING Itself," edited by Wolfgang Staehle (artist, founder/director of THE THING New York) and Jordan Crandall (artist, publisher of Blast Editions, founder/director of the X-Art Foundation). With essays (original

writing) by contributing writers, excerpts from THE THING's online forums, as well as art work reproductions originally published by THE THING, covering history of THE THING from 1991 to present.

Publisher: Autonomedia/Semiotext(e), Fall/Winter 1999.

Series of four curated video group shows (curators included Johan Grimonprez, Florian Wuest, and others).

Notes

1. Private archive Wolfgang Staehle.

2. Peter Weibel discussed the info-sphere in a lecture on the conference republica 2014 in Berlin. Available at http://www.youtube.com/watch?v=64GzkfWaBu8/.

3. Peter Sloterdijk, *Bubbles* (Cambridge, MA: MIT Press, 2011).

4. Peter Sloterdijk, *Globes* (Cambridge, MA: MIT Press, 2014).

5. Slogan probably originally from David Clark, "A Cloudy Crystal Ball/Apocalypse Now," IETF Conference, 1992, where he wrote, "We reject: kings, presidents and voting. We believe in: rough consensus and running code."

6. Dieter Daniels. "Reverse Engineering Modernism with the Last Avant Garde," in Dieter Daniels and Gunther Reisinger, eds. *Net Pioneers 1.0: Contextualizing Early Net-Based Art* (Berlin: Sternberg Press, 2010).

7. Ibid., 44–46.

8. Private archive Wolfgang Staehle.

9. Daniels, 2010, 26–27.

10. Ibid., 28.

11. Regine Hakenbeck and the artist Susanne Gerber. Available at http://kuukuk.de/communication_is_evolution_is/. (The animation was on the referenced website until 2013, when Professor Hakenbeck was head of the Department of Mikrobiology. She is emerita since 2013.)

23 Arts Wire: The Nonprofit Arts Online

Judy Malloy

Creating online community and presence for the nonprofit arts from 1992 to 2002, Arts Wire, a program of the New York Foundation for the Arts (NYFA), was a social media platform and Internet presence provider that bridged the eras of text-based conferencing and the World Wide Web, beginning in 1992 with a text interface, moving to the web in 1996. With eventually more than 80 different online discussion platforms, Arts Wire fostered online discussion among artists and arts workers, worked with artists and arts organizations to provide Internet presence, created *Arts Wire Current* (later *NYFA Current*), a weekly electronic newsletter about issues in the arts, and worked to expand diversity and collaboration in the cultural Internet environment.

Before Internet usage was ubiquitous, in an era when every time the technology was mastered, it changed, Arts Wire's strong presence on the Internet, its emphasis on bringing the nonprofit arts community online, greatly contributed to the early presence of the arts on the Internet. Many artists and arts organizations, who continue to be central information providers on the World Wide Web, got their start on Arts Wire.

"Interest groups," such as the ARTISTS conference and NewMusNet, were gathering places for artists. Interest groups—from NEW YORK STATE, which brought local New York State arts councils online and collaborated with the Buffalo Free-Net to exchange arts-related information, to ABBNET (Art Beyond Boundaries), a conference for the arts in Montana, Nebraska, North Dakota, South Dakota, and Wyoming—were gathering places for arts organizations. Additionally, groups such as the Native Arts Network Association (NANA) and the Alliance of Artists Communities hosted private online conferences to bring participants across the country together on a shared online platform.

The vitality, diversity, and cultural significance of its individual artist and nonprofit organization members were at the core of Arts Wire's collective vision. Among Arts Wire members were writers, visual artists, musicians, dancers, and theater artists. Arts Wire members also included critics, arts administrators, arts funders such as the National Endowment for the Arts and the Andy Warhol Foundation, and arts organizations—from Out North in Anchorage, Alaska to DiverseWorks in Houston, Texas, to the Frank Silvera Writers' Workshop Foundation in New York City; from American Indian Telecommunications in Rapid City, South Dakota, to Opera America in New York City. Creating a diverse cultural presence on the Internet, Arts Wire members also included the National Association of Latino Arts and Cultures, the Asian Cultural Council, the Kitchen, and PS122, as well as the Association of Independent Video and Filmmakers, the American Music Council, The Joyce Theater, and Americans for the Arts, among many others.[1]

Eventually hosting more than 100 websites for artists and arts organizations, Arts Wire provided training for online access and presence. With links to more than 400 artists and arts organizations on its MAP, Arts Wire also worked to provide web visibility, not only for members whose websites were hosted on its server, but also for members whose websites were hosted elsewhere,

Arts Wire Goes Online in 1992

In 1988, at Orcas Island, off the coast of Washington State, a group of about 200 artists and arts organizations, including the National Association of Artists' Organizations (NAAO), State Arts Agencies, and private funders, gathered at *The Orcas Conference*. Organized by the NYFA, a leading institution in the support of artists and arts organizations, the conference explored new ways of directly supporting artists.

After the conference, many participants were interested in finding a way to continue talking about the ideas that the conference generated. In response, Anne Focke, the founder of the Seattle-based nonprofit And/Or Gallery, a creator of the public art program at the Seattle Arts Commission, and a founder of Artist Trust, originated the idea of Arts Wire, an online communications system where the arts community could virtually gather. She recollects that she was riding a ferry and thinking about names when she came up with name of Arts Wire.

Anne's vision encompassed a virtual place for the arts of all kinds. At this time, when the culture wars were beginning, she was also greatly aware of the need for arts advocacy. Even in this era before the World Wide Web, there was excitement about bringing together artists' communities on online systems, where ideas, information, and work could be shared by artists and arts organizations of all kinds.

Established in 1971 as an independent organization to serve individual artists throughout New York, NYFA works to empower artists and arts organizations in their creative lives. The idea of Arts Wire was shared by NYFA executive director Ted Berger, whose broad knowledge of artists and the arts, vision that encompassed the importance of arts advocacy for the nonprofit arts, and support of Arts Wire over the years were invaluable. NYFA Director of Programs, Penelope Dannenberg, was also involved with the development of Arts Wire from the beginning, as was Director of Communications, David Green.

At its inception, Arts Wire's mission was expressed as follows:

The mission of Arts Wire is to provide the arts community a communications network that has, at its core, a strong composite voice of artists and community-based cultural groups. With this foundation, Arts Wire intends to develop a broad and inclusive online community that allows distinct communities to establish their own standards and patterns of use within a system that reinforces democratic values and encourages interaction among its users.

Arts Wire also provides education and technical support in the use of this tool, generates and maintains online resources that meet information needs of its constituency, and works to create a place for artists and the arts community in the national communications infrastructure as part of a larger effort to develop for artists a more integral place in society as a whole.

With funding from a coalition of arts organizations and state arts agencies, as well as additional support from NYFA, Pew Charitable Trusts, the John S. and James L. Knight Foundation, the Glen Eagles Foundation, Metropolitan Life Foundation, The Rockefeller Foundation, and the Andy Warhol Foundation for the Visual Arts, among others,[2] Arts Wire went online in 1992. The initial staff included Anne Focke, Director; Anna Couey, Network Coordinator; David Green, Director of Communications; and Dan Talley, Editor, Arts Wire News.

Initially led by Focke and then by Michigan-based poet, Joe Matuzak, Arts Wire was run from a virtual online "office," a staff conference where from all over the country the staff met to discuss work. The presence of staff from many parts of the country created a welcoming atmosphere for artists of all kinds and from all places. For instance, in early 1996, in addition to Joe Matuzak in Michigan, the staff included Barry Lasky and David Green at the New York Foundation for the Arts;[3] flutist and nonprofit administrator Beth Kanter in Norfolk, MA; graphic artist Tommer Peterson in Seattle; I (Judy Malloy) in El Sobrante, California; and arts advocate Kim Adams in Detroit, MI.

The Steering Committee also "met" online. Among others, over the years, they included Jane Bello, Association of Hispanic Arts; Ted Berger, NYFA; Nancy Clarke, American Music Center; Steve Durland, artist/writer, *High Performance*; Pauline Oliveros, composer; Randy Ross, American Indian Telecommunications; Gary O. Larson, National Endowment for the Arts; Sarah Lutman, arts advocate and cultural innovator; Louis LeRoy, artist, Association of American Cultures; Bill Pratt, Montana Arts Council; and Dan Martin, Director of the Master of Arts Management Program at Carnegie Mellon University.[4] Other people involved with Arts Wire's beginnings included artist Jim Pomeroy, until his death in 1992 of an acute subdural hematoma that was undiagnosed after a fall.

Joe Matuzak, who became Arts Wire Co-Director and then Arts Wire Director in 1995, is a poet who brought to Arts Wire a coherent technical vision, experience in both Arts Management and technology, a broad knowledge of arts advocacy and the importance of advocacy, and face-to-face networking ability. He also believed in the need for an online communications system for the arts and worked tirelessly to make Arts Wire sustainable.

The Arts Wire Conferencing System

Beginning in 1992 with a text interface and moving to the web in 1996, the Arts Wire conferencing system brought artists, arts organizations, and arts funders together on the same platform. An incubator for online activity for many artists and arts organization, Arts Wire's conferences were an unprecedented attempt to bring diverse segments of the U.S. arts community together at a time when national arts support was in crisis. Beth Kantor, who was Arts Wire's second Network Coordinator, has emphasized that at that time on Arts Wire, there was a secure place online where both artists and people who work for arts organizations could have professional conversations about their work.[5]

In 1992, when Arts Wire went online, Anna Couey, Arts Wire's first Network Coordinator, brought her experience working with Art Com Electronic Network to Arts Wire. Anna coordinated the development of the first iteration of Arts Wire's conferencing system on The Meta Net, which utilized Caucus, a text-based conferencing system developed by Camber-Roth, Inc. It offered users online "Interest Groups" where they could share news, advocacy, ideas, art making, and programs. She had the difficult job of not only bringing the Arts Wire system online but also signing up new members, encouraging participation, and producing *Hotwire* (the initial name for *Arts Wire Current*). At that time, going online meant getting a computer (not everyone had one), acquiring a modem (they weren't standard with computers), and then getting it to work with your computer (which wasn't always easy). A connection (such as SprintNet) was needed to get to the system; then to negotiate the system using text-based commands, a certain amount of no-longer-necessary technical skill was essential. In addition to requiring a lot of effort, pre-web online communication was *very* slow.

As a network artist, Couey brought an understanding of the dynamics of art networks and a community-centered approach to Arts Wire's online environment. In her words:

Arts Wire was envisioned as a national arts news and information-sharing network. By the time I became Arts Wire's first Network Coordinator in November 1991, Director Anne Focke had already established a national steering committee to guide development of the project and identified an extensive group of artists, arts organizations, associations and funders who were prospective participants. Anne's broad vision of connecting the culturally diverse arts community as a whole—from alternative artists' spaces to the symphony—infused the project. My role was to coordinate development of the online system and build its resources and membership.

Arts Wire was developed in response to the needs of the arts community. We undertook an extensive and inclusive process of collecting input from a broad range of prospective participants to determine the project's technical needs. The system host we selected, The Meta Net, was a small conferencing system that was willing to work with us to design an arts community space with online databases.

We envisioned Arts Wire as a collection of conferences, each managed by an arts organization or group serving a specific sector of the arts community. However, we also wanted to encourage communication and information sharing across communities. We designed an opening conference, ARTSWIREHUB, that all users would enter first. ARTSWIREHUB was a place for sharing news and information relevant to everyone.

I assisted an initial group of arts organizations to develop their online services with the help of volunteer computer mentors provided by CompuMentor, who worked with each organization onsite. With technical infrastructure and initial content in place, Arts Wire opened to subscribers in the summer of 1992 and continued to grow.[6]

Arts Wire's conferencing, coordinated initially by Anna Couey and later by Beth Kantor and Judy Malloy, was an unprecedented foray into bringing the nonprofit arts online in a community networking forum. Arts Wire members were people whose voices supported the arts community, people who were committed to a place for the arts in contemporary culture. The sense of community and the opportunity to communicate with the national arts community were important. There was enough interest in discussion so that eventually Arts Wire had 80 interest groups.

In a review of Arts Wire for *Internet World*, Kenny Greenberg observed that:

Another interesting Arts Wire feature is its organization of groups into conferences. When you enter a group, you can simply browse, or you can see who is in conference, or who has visited the conference and what they have seen. You can view each participant's self-introduction and contribute your own introduction and thoughts. With a little practice you will be able to jump in at any number of points and contribute your thread.

He also observed that:

As with art—Gophers, SIGs, and HTTP sites notwithstanding—it is the human spirit that makes Arts Wire special. The voices behind the information and the personal reactions to the data make Arts Wire a lively place.[7]

Interest Groups as Central Forums of Discussion

Central Interest Groups included AWNEWS, a copyright-free area for posting news; AWHUB, an area where members could talk about issues in the arts; and AWTECH, a place for information and discussion about the technical information needed to negotiate the online climate at that time. AWNEWS was copyright free, but discussion in

AWHUB was protected in that anyone who wanted to repost it needed to ask the permission of the person who wrote the words. Among Arts Wire's other Interest Groups and interactive resources were:

AIDSWIRE, an online information resource maintained by AIDS and LGBT activist, Michael Tidmus.

ARTISTS, a conference collaboratively hosted by artists active on Arts Wire, including among many others dancer/choreographer Linda Austin, musician/composer Doug Cohen, and CMU Studio for Creative Inquiry-based environmental artist Tim Collins.

CENTER FOR SAFETY IN THE ARTS, an online source of information on art hazards, maintained by Michael McCann. Information provided by The Center for Safety in the Arts was also available to the general public around the world via the Gopher, a text-based infoserver that was in wide use on the Internet before the advent of the Web.

INTERACTIVE, an online laboratory for focused discussion and production of interactive art, founded by Anna Couey and Judy Malloy in 1993. Defining interactive art as "involv[ing] exchange between is originator, work, and participants," the Interactive Art Conference hosted a virtual artist-in-residence program that provided new media artists a forum to discuss their work. Discussion topics covered a broad range of interactive art media, including art telematics, electronic literature, interactive digital art, artists books, public art, social sculpture, installation, and performance. The resulting conversations were informal, providing a snapshot in time of the approaches of new media artists to their work. Almost 20 years later, two interviews that initially took place on the Interactive Art Conference have been published in print.[8,9,10]

LITNET. Hosted by Lisa Cooley, LITNET was a collaboration between the Council of Literary Magazines and Presses and Poets & Writers—in support of freedom of expression and advocacy for the literary arts.

NATIVE ARTS NETWORK ASSOCIATION (NANA), a group of Native Arts Organizations, from New York City to California. Participants included Joanna Osborne Bigfeather, American Indian Community House, NYC; Janeen Antoine, American Indian Contemporary Arts; E. Donald Two Rivers, American Indian Economic Development Association, Chicago, IL; Jennifer Baxter, Guilford Native American Association, Greensboro, NC; Pat Petrivelli, Institute of Alaska Native Arts; and Susan Stewart, Montana Indian Contemporary Arts, Bozeman, MT—with core involvement from Atlatl in Phoenix, AZ (Carla Roberts and Wendy Weston-Ben); and technical advice from Randy Ross, American Indian Telecommunications. NANA was funded by the Nathan Cummings Foundation.

NEA INFO. Hosted by the National Endowment for the Arts, NEA INFO contained a comprehensive listing of grant recipients.

NEWMUSNET. One of Arts Wire's first online Interest Groups, NewMusNet,[11] a virtual place for information, discussion, and publication of new music and issues about/for composers, performers and presenters of experimental music, began in the summer of 1992. NewMusNet was coordinated by musician/composers Pauline Oliveros, Douglas Cohen, and David Mahler. Others active in NewMusNet were electronic music composer Carl Stone, musician/arts advocate Gary O. Larson, and composer/musicians Thomas Bickley, Matthew Ross Davis, and Joseph Zitt. After meeting and collaborating on NewMusnet, Bickley, Davis, and Zitt formed Comma, an ensemble that—using techniques including extended uses of the voice as well as electro-acoustic environments—recorded newly composed musical works, performance poetry, chant from various early traditions, and group improvisations.

NPN. The National Performance Network received funding to bring its primary sponsors online to conduct NPN business and talk about the field. NPN conferences included a public conference and two private conferences for Sponsors and Steering Committee. Among the people who kept the public NPN Conference lively was Mark Russell, then artistic director of PS122.

PROJECTARTNET. Created in 1993 by Aida Mancillas and Lynn Susholtz, PROJECTARTNET was a San Diego-based community arts networking project that brought children from schools in immigrant neighborhoods online to create a community history.

WESTAF. The Western State Arts Foundation is a regional arts organization serving the Western states. On Arts Wire, WESTAF initiated a subscriber-based online version of ARTJOB, their comprehensive job listing service.

In a column in the *Village Voice*, Robert Atkins observed that:

The part that's most difficult for noncomputer aficionados to imagine is the sense of community that some subscribers feel. Composer Pauline Oliveros says she's made new friends; one of them helped her set up an electronic mail-order catalogue that's the best distribution system for her avant-garde music she yet found.[12]

However, in hindsight, the continuing creation of many online conferences (as previously noted, eventually there were more than 80 interest groups) weakened the idea of Arts Wire as a central place where the arts community could gather as a whole. It also meant that in some interest groups, there was not enough participation to maintain vitality. Additionally, with the advent of the web, much arts community activity on Arts Wire migrated to website building.

Online Components of Live Conferences

Perhaps because they were created in an initial burst of energy that did not need to be sustained, and also because they offered a way for those who could not attend the face-to-face meeting to participate online, some of the most successful Interest Groups on Arts Wire were set up as online components of live conferences. For instance, in 1996, Arts Wire hosted an online component of the *Fourth National Black Writers Conference*, held at Medgar Evers College of the City University of New York on March 21–24. The theme was "Black Literature in the 90's: A Renaissance to End all Renaissances?" A publicly accessible interactive version of the conference was available via Arts Wire on the World Wide Web, implemented on what was to become Arts Wire's new virtual home—the server at The Heinz School of Arts Management at Carnegie Melon University.

Keynote speakers for the Black Writers Conference were Paule Marshall and Amiri Baraka. Marita Golden, Terry McMillan, Bebe Moore Campbell, Walter Mosley, Arthur Flowers, Thulani Davis, and others participated in a series of panels that included "Black Literature: Who Are the Readers?" and "Black Literature: The Politics of Publishing." Texts of many of the speeches and online components for many of the panels were available on Arts Wire. The face-to-face event received funding from the National Endowment for the Humanities. The web version was funded by the Reed Foundation and implemented by Arts Wire's Beth Kanter, Barry Lasky, and Tommer Peterson. The work was coordinated by Elizabeth Nunez, Director of the National Black Writers Conference.

Other face-to-face conferences that had simultaneous Internet presence on Arts Wire included *Art-21*, a conference convened by the National Endowment for the Arts (NEA) and held in Chicago in April 1994. Arts Wire created an online conference for *Art-21* and uploaded presented information.

Technology Transfer for the Arts

Arts Wire began working to bring the arts community online before the era of Facebook and Twitter, at a time when many people did not see the need or have the experience to go online and "talk" with words and/or at that time were not comfortable representing their organizations online in a situation where their words remained visible—even though Arts Wire had a "you own your own words" policy. Additionally, conferences were not available to the general public (much more privacy than is afforded in most contemporary social media).

Also, in the early years, although some members were enthusiastic about sharing ideas and action online, others were daunted by the time and expertise needed to negotiate the system. To combat the difficulties, Director Joe Matuzak worked to set up ways to help new users. He confronted the difficulties of everything from hooking up modems to frustration with the technical limitations of what was available at the time to the intricacies of online etiquette. He created strategies, including a user manual, to deal with difficulties, such as the arcane methods of uploading and downloading to the text-based interfaces. When I (Judy Malloy) came aboard in 1993, initially as Front

Desk Coordinator, much staff energy was dedicated to helping new users to negotiate the technologies. In addition to staff support, there was also a Mentor Program, in which we matched new subscribers with volunteer mentors to provide assistance and support as needed. When the web took hold, Arts Wire Education Coordinator Beth Kanter created SpiderSchool to help users negotiate the creation of their own websites. In her words:

> Arts Wire's SpiderSchool provides information for the arts community on how to integrate technology into their work. SpiderSchool was developed in 1995, initially as a resource for Arts Wire members who were building Web sites. About the same time, Arts Wire worked with the Arts Challenge Fund of New Jersey to create a series of workshops and print manual (written by Beth Kanter) titled BUILDING ARTS AUDIENCES AND COMMUNITIES ON THE WEB. Soon after, SpiderSchool became the home of the digital version of the manual, as well as web-based courseware for the face-to-face workshops conducted by Arts Wire staff and a cadre of trainers.[13]

Moving to the World Wide Web

The web began taking off in 1994, and by the fall of 1994, Arts Wire had its own home page, significantly enhanced by Tommer Peterson's elegant interface graphics. We also acquired our own domain name.[14]

Migrating our text-based conferencing to web-based conferencing was an intensive project. Working—under the direction of Director Joe Matuzak—with Charles Roth, the developer of Caucus software, TMN, and the web design team of Interconnect, and with the financial support of the New York State Council on the Arts, Arts Wire staff, including Barry Lasky, Beth Kanter, Tommer Peterson and Judy Malloy, created an interface that merged the graphic ease and linking power of the Web with the online discussion affordances of Arts Wire's original conferencing software.

The process was intense, requiring much discussion of the interface needs of our user base, but in October 1995, Arts Wire unveiled a new graphic interface that made our interactive conferencing system accessible via the World Wide Web. The new web-based conferencing system utilized a specially designed server and Common Gateway Interface (CGI)-compliant scripts that ported the underlying caucus conferencing database onto the web. Because at the time many Art Wire members did not have web access, the conferencing system also remained accessible via the text-based system.

Websites Created by Arts Wire Members

A part of Arts Wire's mission was to generate and maintain online resources that met information needs of its constituency and to work "to create a place for artists and the arts community in the national communications infrastructure as part of a larger effort to develop for artists a more integral place in society as a whole." Thus, in the mid-1990s, Arts Wire also began to offer website building and hosting as part of its membership. It should be noted, however, that if the addition of helping artists and arts organizations build and host websites was important in the changing Internet climate, it also diffused the community dialog focus of our original mission. Nevertheless, the increasing presence of the arts on the rapidly growing World Wide Web was thrilling. Then Network Coordinator Beth Kanter was instrumental in providing information and guidance in the process. As Content Coordinator and editor of *Arts Wire Current*, I worked to promote arts community websites.

Memorable early websites by Arts Wire organizational members included among many others: American Composers Orchestra; Art in the Public Interest; the Association of Hispanic Art; DanceUSA; Dia Center for the Arts; DiverseWorks (Texas); Frank Silvera Writers' Workshop; Junebug Productions (New Orleans); The Kitchen; The Joyce Theater; Manhattan Theater Club; The National Black Writers Conference; TAAC, The Association of American Cultures; the New Jersey Theater Alliance; Opera America; Out North (Alaska); Poets & Writers; Randolph Street Gallery (Chicago); Side Street Projects (Santa Monica, CA); and Space One Eleven (Alabama). To attract visitors to member websites, "The Map," initially maintained by Beth Kanter and later by Judy Malloy, was a list of links to Arts Wire members' websites that provided an extensive portal to arts presence on the World Wide Web. The Map of member websites was complimented by Arts Wire's WEBBASE (later renamed ARTQUARRY), a searchable database of cultural resources to which everyone in the cultural sector was invited to contribute. Because WEBBASE was not curated, initially it served the purpose of an open resource for artists and arts organizations. Eventually the need for a more curated approach was apparent. However, this issue was partially addressed by profiling entries of interest in various issues of *Arts Wire Current*. Profiled entries ranged from Anderson Ranch Arts Center, a nonprofit art center located on the slopes of the Rocky Mountains, to the Polish Culture Page/Polskie Wydarzenia Kulturalne, a guide to what's new and noteworthy in Polish culture.

In an article about Director Joe Matuzak and Arts Wire, the *Detroit Free Press* wrote:

It could be the poster child for Internet theory 101; that's the one that says those who focus their attention on content will rule the Web. Arts Wire's pages are an overflowing cyber-font of content. Indeed, it is one of the richest sources on the Internet for information about the arts and arts education, and it earned a 1997 *ComputerWorld* designation as one of the 100 most effective organizations at using Internet technology.[15]

Among the many Arts Wire artist members who built websites in the early years were Steven Durland, John Maxwell Hobbs, Akua Lezli Hope, Takahiro Iimura, Meredith Monk, Pauline Oliveros, Shirley Sneve (Rosebud [Sicangu Lakota] Sioux), Carl Stone, Troika Ranch, and Wendel White. Their early web-based projects included Shirley Sneve's *A Story of Mothers and Daughters in South Dakota*, Judy Malloy and Cathy

Marshall's *Forward Anywhere* (under the auspices of Xerox PARC and eventually published by Eastgate), John Maxwell Hobbs' award winning *Web Phases*, and Wendel White's evocative photos and texts about African American towns in New Jersey, which are still available online as *Small Towns, Black Lives*.[16]

In an interview on the Interactive Art Conference on Arts Wire, artist/web designer/teacher Jeff Gates pointed out the unique opportunities that artists have in the Information Age, noting, in particular, the opportunity that artists have to participate in defining this new medium, as well as the opportunity to explore "the nature of our work and our roles in society."[17]

Working with CMU's Master of Arts Management Program

As a result of the need for more server space and support for web-based applications, in June 1996, NYFA and Carnegie Mellon University (CMU) announced an ongoing collaboration between Arts Wire and CMU's Master of Arts Management (MAM) Program.

"If not a match made in Heaven, this is at least a match made in cyberspace," Dan J. Martin, Director of the MAM Program at CMU, observed in a press release. "Both Arts Wire and my program have a keen interest in discovering and sharing ways the emerging and evolving information technology can be used by artists and arts organizations in the creation of their work and the fulfillment of their missions. In working together on this front, we hope to truly serve the arts community by providing advanced tools for leadership and management, as well as tools for the development of new and larger audiences and stronger constituencies."

In collaboration with CMU MAM, Arts Wire made RealAudio available for musicians and others who wanted to incorporate sound files on their websites. Additionally, an Excite search engine for *Arts Wire Current* allowed anyone interested in arts news to find information in the *Arts Wire Current* archives as far back as 1995.

By 1997, Arts Wire had a solid membership; significant web presence as an arts portal; web-based conferencing; respected arts news, available both by email and on the Web; and a training program that focused on bringing the arts community online.

Arts Wire Current

One of Arts Wire's initial goals was to provide current arts news that was not widely available and was vital to the work and life of Arts Wire's members. The original idea was that online news feeds from members would be posted in a copyright-free area of Arts Wire (AWNEWS) and then be published weekly in an email delivered newsletter. Later this was supplemented with in-depth features on issues of interest to the community.

The first *Arts Wire Current* editors were, when it was *Hotwire*, Anna Couey and Penny Boyer, followed by David Green, who renamed it *Arts Wire Current*.

After David Green left NYFA in 1996 (to become the director of the National Initiative for a Networked Cultural Heritage), I (Judy Malloy) was appointed editor of *Arts Wire Current* (renamed *NYFA Current* in mid-November 2002).

On the Internet, it was possible to reach a large niche audience by covering arts information not available on the major media. Because the arts community was vitally interested, it was possible to present news with collaborative input[18,19] that was a hybrid between information and journalism. Under my editorship, *Arts Wire Current* went from a circulation that was mainly Arts Wire members to a circulation in 2003 of more than 5,000 via the email mailing list and about 25,000 web hits for each (weekly) issue (at that time a substantial audience for a niche publication).

For eight years, from 1996 to 2004, I worked to create a weekly publication that covered arts information not carried by the major media—such as alternative art spaces, performance art, censorship, diversity, arts advocacy, small dance and theater companies, issues of interest to composer and musicians, artists' housing, and artists' websites. In producing each issue, my editorial approach was to weave together information in a series of related themes, focusing on both individual voices and a coherent whole. For instance, an issue with a focus on artists' housing included the following articles: "Cultural District Planned for the Brooklyn Academy of Music"; "Cultural Coalition Works to Buy Buildings in Boston"; "Artspace Housing Projects Planned in Poughkeepsie, NY and Bridgeport, CT"; "Chicago's Roentgen School to Convert to Artist Housing"; "Riverside Hotel in Reno Begins New Life as Artists Lofts"; "SF to vote on Propositions to Alleviate Dot Com Encroachment"; "Housing Planned for Endangered Artists in Seattle's Pioneer Square"; "Creating an Arts District on Mid-Market (San Francisco)"; and "Open Home: Westbeth."[20]

Arts Wire Current's Coverage of 9/11

One of *Art Wire Current*'s most important roles was documenting damage to the arts community in New York City in the aftermath of 9/11.

Arts Wire Current headlines tell the story: "In the Wake of the Terrorist Attack, Artists and Arts Organizations Rally to Help Those in Need; Jamaican Sculptor Michael Richards Missing; LMCC Offices Obliterated"; "Call for Artists to Work with the Families of WTC Destruction Victims"; "Reports of Arts Community Losses Accumulate as Artists and Arts Administrators Cope with the Aftermath of Disaster"; "Stories of Survival and of Helping Hands Bring Light in the Darkness"; In Union Square, Shrines Express a Collective Grief"; "As Americans Gather Together to End Terrorist Attacks, Freedom of Expression is a Vitally Important Liberty: Bipartisan Coalition Urges Protection of America's Constitutionally Guaranteed Freedoms"; "Help for Artists and Organizations Impacted by WTC Attack"; "CERF Provides Disaster Relief to Craftspeople"; "Santa Fe Art Institute Offers Free Residencies to Artists Whose Spaces Have Been Compromised by the Terrorism"; and "Michael Richards, 1963–2001."[21]

"Each night that the lights are turned on, the tickets sold, the programs printed, is a miracle." —Mark Morris

Taken as a whole, *Arts Wire Current*'s coverage also provides a record of many battles fought during the height of the Culture Wars.

In 1999 alone, among many other things, *Arts Wire Current* covered issues including how a Charlotte, North Carolina, student's winning play wasn't performed because of lesbian characters; how Texas rescinded funding for the Texas Shakespeare Festival after a production of *Angels in America*; how The Esperanza Center legally challenged the climate of censorship in San Antonio; the extent of the "Digital Divide" for minorities, despite soaring Internet participation; and a report on the decline in hiring of women and minority directors.

As reported in *Arts Wire Current*, the intensity of culture war battles in 1999 clearly illuminates the power of the arts to focus attention on contemporary issues. "The intersection of the artistic imagination and the civic realm offers a fertile ground for aesthetic innovation and civic participation," said Americans for the Arts President and CEO, Robert L. Lynch, to introduce "Arts Animating Democracy," an Americans for the Arts initiative that fostered artistic activity that encouraged civic dialogue on contemporary issues.[22]

Additionally, as recorded in *Arts Wire Current*, many friends of the arts and diversity stepped in to offer advocacy and safe havens. A coalition of Hispanic organizations staged a boycott of major television networks as the first step in an effort to convince the networks to cast more Latino actors. The National Association for the Advancement of Colored People convened a public hearing to examine discriminatory treatment by the television industry against African Americans, Asian Americans, Latinos, Native Americans, and other minorities. When an exhibition of work by artist Tanya Batura, which was scheduled to open at a city-run gallery in Hartford, CT, was canceled by the city because the work was considered sexually suggestive, Real Art Ways exhibited the work in their new Arts Center.

Among many other artworks and art actions and new platforms for arts coverage documented in *Current* in 1999 are the 11th year of commemorating Day Without Art, a day of action and mourning in response to the AIDS crisis; the *Gender Identity in New Media* panel, which I developed and hosted on Arts Wire for the *Invencao Conference* in Brazil; *Central Park*, a New York City Opera Trilogy; the American Music Center's *NewMusicBox*; *2 Weeks in August* at the Northwest Asian American Theater in Seattle; the *Middle East Film Festival* at Bates College in Maine; and in Oklahoma City *Through the Looking Glass: Fresh Perspectives by Artists with Disabilities*.

In the 1999 year-end summary of the News as reported by *Art Wire Current*, I emphasized that:

in 1999, all around the nation, there were places where the arts flourished—from Florida where a study showed that the Miami Beach economy gets a $100 million boost from the arts ... to Boston where the ICA will build a new museum on the waterfront; to New York where construction of the Mark Morris Dance Group's $5 million dance center in Fort Greene, Brooklyn is underway.

For music education advocacy, a 1999 highlight reported by *Current* was the *Concert of the Century for VH1 Save the Music*, where President Clinton remembered the role of his school music teachers. The then president said that in school, "Music taught me how to mix practice and patience with creativity. Music taught me how to be both an individual performer and a good member of the team." The lesson for everyone, he continued, is that "we are stronger when we are playing in harmony based on our common humanity."[23]

Dancer/choreographer Mark Morris concluded a keynote speech at the Midwest Arts Conference by saying,

And I'm here because I want to tell you how important it is what you do, what we do. It's hard. Conditions are worsening. Live performance is being pushed farther and farther to the fringes of our national culture.... Each night that you open your theatres is a miracle. Each night that the lights are turned on, the tickets sold, the programs printed, is a miracle. It is a miracle each night that your community is invited to gather in your buildings and hear music or see theatre or dance.[24]

More Extraordinary Arts Wire Staff

Arts Wire staff—from Barry Lasky, Arts Wire's second Technical Coordinator, whose skill and expertise were essential in our move to a web interface; to Resource Director Kim Adams, who brought a love of the arts and a knowledge of the nonprofit art world to the difficult job of fundraising; to Doug Cohen, Arts Wire's Systems Coordinator, beginning in 1997, who not only created system enhancements and helped many users negotiate Arts Wire but also was instrumental in the initial creation of the NewMusNet Conference—brought an extraordinary range of skills that were needed in the creation and maintenance of this online system for the arts.

With a background in graphic design and communications, Tommer Peterson, who was Arts Wire Publications/Marketing Coordinator for about five years, brought the ability to produce interesting and informative issues of *Arts Wire News* (Arts Wire's print publication initially edited by Dan Talley) and later created the graphics that gave Arts Wire's website a distinctive look and feel.

Beginning in the spring of 1994, Beth Kanter brought her connections with the music community, her management experience, and her tireless energy to Arts Wire as Network Coordinator. Beth, who was trained as a classical flutist but decided to go into orchestra management (and is currently a nonprofits on the Internet guru), created

SpiderSchool to provide guidance on website building, as well as an associated print publication, *Building Arts Audiences and Communities on the Web*, a "hands-on" step-by-step guide for arts organizations to plan, develop, implement, and promote their websites.

I came to Arts Wire with a background that included experimental artists books, pioneering work in electronic literature, initial coordinator of *Leonardo*'s fledgling electronic publications, consultant in the document of the future at Xerox PARC, and artist-in-residence at Deep Creek School in Telluride, where I had worked to teach MFA students how to document their work online.

Unlike email, which is difficult to access on an archival basis, AWSTAFF, the Arts Wire staff conference, provided a place where information about the making and running of Arts Wire was documented as it evolved. Because the conferencing system was additive and organized by topic with the list of "items" providing an overview index, it was possible to go back and look at information on topics such as setting up conferences, bringing in new members, and conducting training sessions. Sharing details of the work, problems, ideas for the future—in one virtual place accessed from all around the country—Arts Wire was a model for an organization run virtually via online conferencing.

In many ways Arts Wire led the way. That meant we made the mistakes, but it also meant there were a lot of times we mapped out new terrain. —Joe Matuzak

Arts Wire was initially founded as a national program, sponsored by NYFA. However, as web presence became a central part of organizational missions because there was greater demand for online multimedia capability in the arts, and at the same time a greater emphasis on NYFA's primary mission to support New York artists, in November 2000, NYFA understandably took Arts Wire completely in house. By mid-2001, all that remained of Arts Wire was *Arts Wire Current*. In mid-November 2002, *Arts Wire Current* became *NYFA Current*, and all the associated Arts Wire pages, including the Jobs and Opportunities, were gradually assumed by NYFA. However, as the last Arts Wire staff member, I was happy to remain as editor of *NYFA Current* until the spring of 2004.

In summary, from 1992 to 2001, through an online portal for artists, arts administrators, and arts funders from all around the country and potentially globally, Arts Wire offered a forum to share ideas, work, programs, and core advocacy concerns.

If Arts Wire was ahead of its time, nevertheless participation on Arts Wire provided confidence and experience in working online that greatly contributed to the rich and diverse presence of the arts in the contemporary Internet.

Because of the many important issues facing the arts community, there is still a need for a central online gathering place. In today's web-based Internet, this would not be as difficult to negotiate as it was in Arts Wire's early years. The challenge would be to create a focus on interesting issues that are important to participants and to create an environment that would sustain such dialogue.

Over the years of his tenure as director, Joe Matuzak emphasized Arts Wire's successes, including accelerating the online presence and expertise of arts service organizations and state arts agencies and working with technology transfer for artists and arts organization's not only online but also through workshops and conference presence—as well as Arts Wire's groundbreaking efforts in putting Conferences such as the NEA's *Art-21;* and the *National Black Writers' Caucus* online, and in creating and disseminating core sources of online arts information, such as *Arts Wire Current*, SpiderSchool, and Arts Wire's conferencing system.

In his words, "The bottom line is that in many ways Arts Wire led the way. That meant we made the mistakes, but it also meant there were a lot of times we mapped out new terrain."[25]

Notes

1. Arts Wire was based on a membership model. The membership fee partially supported the network. The sliding scale membership fee system ranged from $60/year for individual artists and small organizations to an Institutional Membership category that enabled large organizations to support Arts Wire as well as bring all of their interested staff online.

Among Arts Wire member organizations were: Academy of American Poets, American Music Center, Alaska State Council on the Arts, Albuquerque Arts Alliance, Alliance of Artists' Communities, Alliance of New York State Arts Councils, American Arts Alliance, American Music Conference, Andy Warhol Foundation; Appalachian State University Office of Cultural Affairs, Arlington, VA Cultural Affairs Division, Arlington Arts, Art in the Public Interest, Artist Trust, Arts Council of Greater Kalamazoo, Arts Council of New Orleans, Arts Midwest, Association of Hispanic Arts, Audio Description Home Page, Bronx Arts Council, Chicago Theater Homepage, California Arts Council, Chicago Department of Cultural Affairs, City of San Antonio Department of Arts and Cultural Affairs, City Lights Youth Theater, The Clay Studio, Colorado Council on the Arts, The Corning Museum of Glass, Cranbrook Dance Theater Workshop, Illinois Arts Council Legion Arts, Maine Sponsors Association, Manhattan New Music Project, Media Alliance, Mid Atlantic Arts Foundation, Missoula Cultural Council, Mobius, National Association of Artists' Organizations, National Arts & Disabilities Center, National Assembly of State Arts Agencies, The National Campaign for Freedom of Expression, The National Black Writers Conference, The National Endowment for the Arts, Nebraska Arts Council, Newton Television Foundation, New Jersey State Council on the Arts; New Jersey Theater Alliance, New York State Alliance for Local Arts Agencies/Rural Program, The New York State Council on the Arts, North Carolina Arts Council, Ohio Arts Council, On the Boards, Opera America, Oregon Arts Commission, Out North, P.S. 122, Paint Creek Center for the Arts, Pacific Northwest Ballet, Peace Troupe, People for the American Way,

Photography New York, Pittsburgh Dance Council Homepage, Poets & Writers; Rheedlen's Rise & Shine Productions, Rhode Island State Council on the Arts, Rochester Civic Music, Seattle Art Museum, School of the Museum of Fine Arts, Artist's Resource Center/Career Services, Society for Photographic Education, Soundout, South Carolina Arts Commission, The South Dakota Arts Council, State Arts Council of Oklahoma, Third World Newsreel, Virginia Commission for the Arts, Visiting Artists Program, School of the Art Institute of Chicago, Walker Art Center Performing Arts, Webster's World of Cultural Democracy, Women's Studio Workshop, and The Writer's Voice of the West Side Y.

2. David Green, "Arts Wire Archive: Funders with Vision," *ArtsWire News* 2 (Fall 1995): 4–5. Over the years, other Arts Wire funders included The Walter & Elise Haas Fund, the Robert Sterling Clark Foundation, the LEF Foundation, the Nathan Cummings Foundation, and the List Foundation.

3. NYFA staff who interfaced with Arts Wire and *Current* over the years and are not listed elsewhere in this chapter included Amy Hufnagel, Matthew Deleget, Waddy Thompson, Sarah Pierce, Jeff Fischer, Ann Wagar, Kevin Duggan, Alex Burke, Randi Goldman, Jennifer Hickman, Lisa Pue, Kate Wilson, and Carla Mapelli. FYI Editors Joe Hannan and later Alan Gilbert also interfaced with Arts Wire and *Current*.

4. Other Arts Wire Executive Committee members included over the years Theodore S. Berger, NYFA; Susan Dickson, National Council for the Ceramic Arts; Anna Couey, network artist; Nathan Lyons, photographer, Visual Studies Workshop; Rose Parisi, Illinois Arts Council; Dan Talley, artist, The Forum Gallery; Peter Taub, Randolph Street Gallery; and Barbara Thomas, artist.

5. Information about Beth Kanter is available at http://www.bethkanter.org/.

6. Personal Correspondence with Anna Couey, December 12, 2014.

7. Kenny Greenberg, "On-line Salon," *Internet World* (October 1994): 102–103.

8. Anna Couey and Judy Malloy, *Interactive Art Conference*. Available at http://www.well.com/~couey/interactive/.

9. Anna Couey and Judy Malloy, "A Conversation with Jim Rosenberg (on the Interactive Conference on Arts Wire)," in Jim Rosenberg, *Word Space Multiplicities*, edited by Sandy Baldwin (Morgantown, WV: WVU Press, 2015).

10. Anna Couey and Judy Malloy, "A Conversation with Sonya Rapoport (on the Interactive Conference on Arts Wire)," in Terri Cohn, ed., *Pairing of Polarities* (Berkeley, CA: Heyday Press, 2012).

11. Information about NewMusNet is available at http://www.newmus.net/.

12. Robert Atkins, "Scene & Heard," *Village Voice* (April 27, 1993).

13. Beth Kanter, "*Arts Wire SpiderSchool and Arts Wire Workshops,*" Arts Wire Current, April 28, 1999.

14. Letting your domain name lapse can cause identity problems even if your organization is no longer operational. Sadly, Arts Wire's domain name—http://www.artswire.org—was allowed to lapse and is now owned by a very different organization that has no connection with Arts Wire.

15. David Lyman, "Mixed Media: The Man Who Runs a Web Site for Artists Says Computers and Art Needn't Be Strange Bedfellows," *Detroit Free Press* (August 23, 1998).

16. A web history of Wendel White's *Small Towns, Black Lives* is available at http://www.blacktowns.org/.

17. Anna Couey and Judy Malloy, "A Conversation with Jeff Gates on the Interactive Art Conference on Arts Wire," January 1997. Available at http://www.well.com/~couey/interactive/jgates.html.

18. An example of collaboratively created content on *Arts Wire Current* is "Arts Community Urges Government to Restore Individual Artists Fellowships," the lead story in the May 28, 2002 issue, which is available at http://www.well.com/user/jmalloy/awcurrent/Arts_Wire_Current_May28_2002.pdf.

19. Among the many Arts Wire members who provided news were the NEA; The National Campaign for Freedom of Expression (NCFE); The National Association of Artists' Organizations (NAAO); the ACLU Arts Censorship Program; People for the American Way; Americans for the Arts; the National Assembly of State Arts (NASAA); and many state and local Arts Councils, including The New York State Council on the Arts (NYSCA); New Jersey State Council on the Arts; Illinois Arts Alliance; Arts Foundation of Michigan; Arts Midwest; Bronx Council on the Arts; Virginia Commission for the Arts; and the North Carolina Arts Council. Among other Arts Wire members who were regular news providers were the American Music Center, Garland Thompson, The Frank Silvera Writers' Workshop; and Visual AIDS. Artists who provided news included San Diego artist, Aida Mancillas; Washington, DC-based artist, Jeff Gates; Texas artist, Louis LeRoy, who provided information about The Association of American Cultures; and Colorado-based stage designer and arts advocate, Richard Finkelstein, who also initially hosted the email mailing list version of *Arts Wire Current* via the Department of Theater and Dance at the University of Colorado. Additionally, Robert Lederman, the President of A.R.T.I.S.T. (Artists' Response To Illegal State Tactics), provided information about A.R.T.I.S.T.'s battle to have visual art accorded the same first amendment rights as literature. Playwright Cindy Cooper provided leads and commentary on stories of interest to the feminist and theater communities.

20. "Solutions Sought as Real Estate Boom Impacts Artists Nationwide," Arts *Wire Current* (November 7, 2000).

21. Selected titles of articles from *Arts Wire Current*, September 18, 2001 and *Arts Wire Current*, September 25, 2001. *Arts Wire Current*, September 18, 2001, is available online at http://www

.well.com/user/jmalloy/awcurrent/Arts_Wire_Current_Sept18_2001.pdf. An index to Arts Wire's 9/11 coverage is available on the Internet Archive at https://web.archive.org/web/20011212020910/http://www.artswire.org/current/911.html.

22. "The Arts Year in Review," *Arts Wire Current* (December 28, 1999).

23. Ibid.

24. Ibid.

25. The quote from Joe Matuzak is from the consultant's report on Arts Wire that I created for NYFA in 2004.

24 Electronic Literature Organization Chats on LinguaMOO

Deena Larsen

It was the year 2000—not all that long ago in real time, but a millennia ago in Internet time. With well over a million web domains and nearly four million users in the world, we were dancing at the apex of the dot.com; we thought everything and anything was possible online. Over the previous few decades, we had all flocked to electronic literature, each of us attracted to the creative possibilities outside of print literature.

Yet we were scattered across the globe, each a lone voice in a wilderness of universities or linear writer communities, without a place we could talk about our passion and be understood. Hypertext is just choose-your-own adventure books, right? All those fancy links and images and sound—you really don't employ them to convey content, do you? To combat these misperceptions, we needed our own community. In this new Internet world, we could connect. You might be the only person in Denver to like pink cotton-candy poodles with sunglasses or to be addicted to hypertext/electronic literature writing. But in 2000, now you could be part of a true community of your peers by simply turning on your computer.

A Home of Our Own with the Electronic Literature Organization

Throughout the 1990s, we'd meet face to face, but mostly at conferences with many agendas other than simply electronic literature. We had figured out some ways to answer the crying need to "facilitate and promote the writing, publishing, and reading of literature in electronic media."[1] By 1999, I'd been hosting/conducting hypertext/electronic literature workshops at these conferences, hosting online chats, and creating online writers workshops for a few years. At CyberMountain 1999, I hosted 30-odd electronic literature writers in a mansion for an intense four days of work. Brown University had also hosted one of the first conferences for "print and electronic writers" that year.[2]

That same year, Scott Rettberg, Jeff Ballowe, and Robert Coover formed the Electronic Literature Organization (ELO) and assembled an impressive board of directors, comprising most of the electronic literature luminaries of the day. ELO launched projects, including an eye-catching $10,000 prize for electronic literature. Marjorie Luesebrink, the second ELO president, sponsored a symposium and a scintillating electronic literature gallery, laying the foundation for a reading series and conferences. I moved the chats I had been conducting independently under the auspices of ELO, and for four years (2000–2004), we convened biweekly or monthly online.

An Algonquin Hotel in Text and Screen

I'd invite a lecturer or pose a topic—most of the luminaries in the field flitted through our screens. Then I'd advertise in every place I could dream of, cajoling people to join in the fun. The chats were worldwide, so some would have to get up at 5 a.m. and others stay up until 9 p.m. I'd fire up the computer, flex my fingers, and start typing at about 200 words per minute.

I had been using multiobject-oriented space (MOOs) because not everyone could afford the bandwidth of more in-depth conferencing tools. Some people still used Telnet, a text-only precursor to browsers, so MOOs provided access for as many people as possible. An academic space, LinguaMOO offered us "a text-based synchronous learning environment."[3] Recording was relatively easy, and the learning curve was not too high.

In the MOO, I'd invite our audience in, ask questions, answer questions, and direct traffic—all from the keyboard. (Robert Kendall arrives at door. "Hi Rob, just type @go ELO chat and join us.") With nothing but text to limit us, we could afford to meet in grand ballrooms, under the sea, or perched on Saturn's rings. I'd serve virtual brandywine with elephant ears drenched in raspberry swirl icing as I introduced our guests of honor with a prewritten little spiel: "Jane Yellowlees Douglas is our guest today." Everyone would commence filling the screen with ludic, insightful text. I'd follow at least four conversational threads at once, eliciting thoughts: "So, Jane, how do these secrets affect the work even if readers don't happen to stumble across them?"

Afterward, I'd clean up the chat for the archive: fixing spelling here, moving conversation threads there so it was easier to follow the flow of one of the discussions—the conversations were already desultory, flitting from chaos theory to type fonts in two seconds flat—no need to keep the fragmentation of the lag time. Slower typers often came in on a thought that had already scrolled out of sight.

Writers throughout the ages have needed a place to munch on delicate prisms of possibilities, don omni-spectacles to see all the points of view, and talk about the Issues That Matter. For us, this was even more vital. We were exploring new frontiers—new dimensions of writing never before possible. Our readers had to traverse an ever-changing landscape of browsers or platforms, figure out mice and screens, and understand navigation and image before they could read one of our works (filled with links and sounds, videos and games) or even dream about discovering the concepts, the meanings we so carefully hid deeply locked away in our ergodic treasure chests.

The Not-So Guilty Parties: Who Wasn't Who

A few of these chats are still tucked away on the web at http://eliterature.org/chats/. As I look through the tattered remains of the chat archives, I see some familiar names, some who are still vibrantly contributing to our growing community and others who are now sorely missed.

I hung out an eternal welcome mat for everyone, and many crossed that threshold. We had brave publishers of web journals, gathering materials in one site for readers, such as Claire Dinsmore (*Cauldron and Net*), Talan Memmott (*Beehive*), Jennifer Ley (*Riding the Meridian*), and even Mark Bernstein of Eastgate Systems, publisher of disk-based literary hypertexts. These were talented writers in their own right as well. Electronic literature writers such as Rob Kendall, Nick Montfort, Noah Wardrip-Fruin, Stuart Moulthrop, Marjorie Luesebrink (M. D. Coverley), Stephanie Strickland, Jim Rosenberg, John McDaid, Diana Slattery, Bill Bly, and more graced our textual rooms. Writing communities such as the trAce Writing Center (Helen Whitehead and Sue Thomas), Online Writing Center, and others actively participated. Theorists, academics, and teachers such as Jay David Bolter, David Kolb, and Katherine Hayles also mingled with the writers, all exploring this wonderful new world from various angles (or, as our frequent typos had it, angels). People came from all over the world, including Susana Tosca, Lucia Agra, Raine Koskimaa, Gonzalo Frasca, Lisbeth Klastrup, Inna Kooper, mez, and so many more. Our circle was not exclusive, for we reached out to people in libraries, digital humanities, art, and other areas, inviting them into the fray as well, and forming the basis for many partnerships.

But we had curious guests, for on the Internet, you did not have to introduce yourself as you truly were, but you could take on any persona you desired. As a good host, I did not press for details. Some of the others were not so subtle:

Janet Murray wonders if it is polite to ask if Salmon is a fish.
    Salmon surreptitiously picks at a scale or two.[4]

We had fictional guests drop by as well. In one chat,[5] I pulled that same game, dropping out as Deena the host and coming back as Rachel Cole, a bossy storekeeper from my work *Marble Springs*, circa 1890s.[6] (People dropped in and out all the time from rolling blackouts, battery problems, and unexplainable shutdowns. In those days, you didn't expect that your Internet connection or your computer would be stable enough to last for an entire hour online.) Yet I never quite knew the truth about Maddy, who said of herself:

Maddy says, "I KNOW I am fictional, but I live a real life!"[7]

This led us straight into identity problems:

Helen says, "It's a very sensitive issue: identity and multiple personalities!"
    Deena passes out party masks and masks under those masks to all and sundry.
    Salmon grins.[8]

Thus, amid suggestions that we all drop our names and quandaries about ascribing thoughts, we pondered who among us was real:

Deena says, "Was Maddy a fictional character masquerading as a person, or a nonfictional real life person masquerading as a character?"
    Dev smiles, as truth is just a constructed fiction itself, no?[9]

What We Touched On

In these talks, we kept circling around the issues of archiving, tools, reading, and conventions—the same conversations we have today. A few things had changed from when we first started to explore these ideas in the late 1980s and 1990s. Even by 2000, we were giving up the idea that one could be Lost in Hyperspace; few people still despaired of ever getting rid of "click here to go to the next page." We spent much more time discussing reader expectations, for at that point, the oldest of the "net generation" had just entered first grade and were just beginning to learn how to read at all, let alone read our works.

E-delivery versus E-literature versus Hypertext: What Are We Writing?

We obsessed over the differences between print literature and electronic literature. Fifteen years later, we are still arguing about what to call what we are writing, and we still have people conflating electronic literature with electronic publishing (which we defined as merely moving a book experience to the screen without engaging sound, imagery, links, navigation, etc.). The terms people used for the 2001 trAce New Media Writing Competition indicated interesting directions (hypermedia literature, revolutionary web-specific writing, hypermedia poetry, interactive literature, randomly created web narrative, interactive poetry, hyper-essay, informational sculpture, digital literature, cyberpoetry, and moving poetry, to name a few).[10] We tried and tried to define what we were talking about.

Scott Rettberg says, "The basic distinction I would make is that electronic literature is lit that couldn't be presented in a paper form that uses the capabilities of the computer to do things that authors couldn't do in a bound codex."[11]

Electronic Literature Organization Chats on LinguaMOO

We remonstrated with Greg Newby of Project Gutenberg (with a lofty goal of 10,000 e-books on every computer)[12] and Karen Weisner, author of Electronic Publishing,[13] that hypertext, that electronic literature was indeed something new, something more than just one word following another. Christian Crumlish, a literary agent for new media writers, said:

xian says, "I see hypertext as a natural extension of alternative canons, apocrypha, alternate folios of Shakespeare, and all the other things in writing that have always mitigated against the ultimate authority of the writer's voice, or version of events."[14]

Helen Whitehead also touched on this:

First films were just recorded stage plays—then the producers and writers realised that a new medium meant different possibilities. That's the case here too.[15]

Intrepid, Daring Readers: Who Were We Writing For?

Many of us just wrote for the love of writing, but others never forgot that we were writing for someone:

Deena says, "mez, actually, me too. I tend to just create for the work's sake, and forget who may be seeing it—or how they will see it."

mez says, "deena, i hope most audiencez can get something from my art, whether it B a resonance with the colours or the m-mages or the complex conceptual bent."[16]

We noted that our readers were evolving, and we talked about the expectations of engaging with this screen and rapid click:

Katherine says, "Maybe it's more accurate to say that the ways in which we read are changing—for frequent users of the Web, perhaps, becoming more impatient, wanting to read smaller blocks and move on."[17]

Of course, these views have changed dramatically with new generations who have embraced the Internet from babyhood. Thus, the era before the Internet is reflected in our archived comments:

A. C. Chapman says, "I think of the big problems with developing an audience for e-lit, art, etc. is that people expect to do certain things on a computer, and relax and enjoy is not usually one of them."

Andrew Stern says, "My parents (who actually are now using email) still would never dream of having a meaningful interactive narrative experience on a computer."[18]

One problem we touched on has eased quite a bit, at least in the United States, although not everywhere in the world yet. It used to be expensive to remain online for more than a few minutes, so we bemoaned that people would download work incorrectly or not engage with it for long enough. Do not get us started on disk-based work or the incompatibilities of browsers!

At the time, readers expected closure, and they expected to know how much further it was to the "end of the book." So a work that presented a world that did not need to be explored—or even explained thoroughly—was puzzling. Even though, as Nick Montfort pointed out, we had had interactive fiction for decades, with spatial organization and mind maps, readers still felt compelled to finish works:

Maggie says, "They feel the same frustration w/ hypertext sometimes, fearing they'll 'miss' something if they don't choose the best plan of action."[19]

We talked of how to teach and what to teach, of whether this writing should be confined to the academics or reach out to the world. We bemoaned the fact that there were gatekeepers, yet we could not imagine manning the floodgates of the amount of materials on the web without such gatekeepers.

Helen says, "I think academics will be quicker to accept electronic journals—it is very established now—but literary works are still seen as lesser than print."

Greg Newby says, "Web publishing is fine, but look at my comments above about authority. Anyone can make a Web page. That is the problem, even though it holds the seed of the solution!"[20]

Software and Hardware and Bugs, Oh My: Why Did We Feel Compelled to Write?

In our communities, we bolstered each other's faith in writing, despite the lack of tools, the paucity of readers willing to experiment with us, or the complete and total lack of any business model whatsoever.

minimusW says, "Every e-writer, I believe, has a conversion experience, when they realize there''s something they can do on the computer that they can't do in print. Would it be interesting to share these experiences?"

RobKendall says, "I think multimedia (flash, etc.) can engage on a more visceral level than just print on the page. For me, music and movement adds the excitement of live performance."

elizabethj says, "I like the fact that it encourages minimalism."[21]

Cleo says, "Thinking is so much more intimate with hypertext than I often find it to be with linear text, because I get deeper inside the creator, see how her/his mind works. It feeds curiosity."[22]

Oxymoron says, "I think that many of us recognise that hypertext is a way of representing the world that many of us feel 'fits' with our perception of multiplicity in ourselves and our world."[23]

We joyfully pushed the limits of the tools, exuberantly exploited bugs in code, and played:

cleo says, "for me it is the all-inclusiveness of the medium—I always felt so many limits before—here i can play with image, music and text, as well as have a kinetic element, sometimes of change, that makes it very alive for me, shall we say."[24]

scottrettberg says, "it's beautiful that so many different forms are being created, kind of smashing borders."[25]

Archiving: Could We Retrieve What Was Already Lost? Or Save Anything?

The silicon beneath our fingers shifted so rapidly we could barely keep up, so our archiving woes were intensified—Mosaic as a browser was vanishing, Netscape was everywhere, and Internet Explorer was just beginning. Most of us old hats, by that time, had lost at least one work if not more to incompatible browsers or dead software. If you haplessly choose the wrong software—the one that did not survive—you faced the waste of years of efforts.

ELO worked on PAD to recommend best practices for archiving,[26] and we discussed many of these ideas in the chats. We thought of keeping machines or just giving ourselves up to the gods of ephemera:

a.c.chapman says, "I think the argument, for me, for keeping those old technologies alive is that certain stories are explicitly tied to said technologies. Noah and I have a friend, Nick who has a large store of old machines specifically for this reason. However, I would argue that we need to start thinking of computer related stories and art in much the same way we classify 'Happenings' or 'performance art' in that there's an acknowledged temporality."[27]

We talked of good copies and bad:

Rob_Kendall says, "Another big problem with online works is that sometimes there are multiple versions of works floating around out there, some up to date and some old versions full of bugs. I'd suggest avoid having 'deprecated' versions of your works floating around."[28]

Of life and death:

Helen says, "It's particularly difficult for an elit author to ensure their work will continue after their death—manuscripts can be burned but aren't usually, whereas websites are routinely ignored by relatives. I had a call from a writer with a terminal disease and I wasn't able to advise him where he could go to be sure his website of poems would stay online."[29]

Prognostication: Where Will We Be in 2015?

We also speculated on the future, imagining the merging of web and television, of digital paper, and speech solely based on visual icons. Some thought it would take time:

Bob Zwick says, "I just think the evolution of dead tree books to electrons will be slow coming."[30]

Others looked for a bright future for both writers and readers within our lifetimes:

Jeffrey says, "10 year vision: 5 small publishers doing the most cutting edge work with a variety of new and evolving technical platforms for creating and distributing; 20 years: large publisher using one or two content creation and distribution platforms (heavily reliant on peer-to-peer) publishing mostly e-lit by established print writers, but also publishing a few works by new elit writers. ... Publishers are making gobs of money, writers are making some (a bit more than in print as a % of profits)."[31]

Sue says, "[M]y vision, if I have one at all, is that most media will be multimedia including multisensory, and that the legacy of this time will be interactivity—that is what has changed literature more than anything else and that is the key, I think, to the new reader experience."[32]

Afterwards

I would sum up those times and our talks as, it was the best of times, it was the worst of times. It was the most interesting of times to write and explore, and the most heartbreaking. Perhaps it still is—we may well say the same things 15 more years from now. Thank you, my friends, for sharing these times with me.

The housekeeper arrives to cart Deena and the rest of the crew off to bed.

Notes

1. Mission statement of the Electronic Literature Organization. Available at https://eliterature.org/.

2. Brown University, Technology Platforms for 21st Century Literature, April 7–9, 1999. Conference proceedings. Available at http://cds.library.brown.edu/conferences/TP21CL/.

3. C. Haynes and J. R. Holmevik, *High Wired: On the Design, Use, and Theory of Educational MOOs* (Ann Arbor, MI: University of Michigan Press, 2001).

4. ELO chat transcript, September 27, 2000. Available at http://eliterature.org/chats/092700.html.

5. ELO chat transcript, July 1, 2001. Available at http://eliterature.org/chats/070101.html.

6. Deena Larsen, *Marble Springs*, First Edition (Cambridge, MA: Eastgate Systems, 1993). Second edition available at http://marblesprings.wikidot.com/.

7. ELO chat transcript, September 27, 2000. Available at http://eliterature.org/chats/092700.html.

8. Ibid.

9. ELO chat transcript, July 1, 2001. Available at http://eliterature.org/chats/070101.html.

10. ELO chat transcript, February 4, 2001. Available at http://eliterature.org/chats/020401.html.

Electronic Literature Organization Chats on LinguaMOO

11. ELO chat transcript, October 15, 2000. Available at http://eliterature.org/chats/101500.html.

12. ELO chat transcript, September 16, 2001. Available at http://eliterature.org/chats/091601.html.

13. ELO chat transcript, August 30, 2000. Available at http://eliterature.org/chats/083000.html.

14. ELO chat transcript, July 8, 2000. Available at http://eliterature.org/chats/070800.html.

15. ELO chat transcript, May 19, 2002. Available at http://eliterature.org/chats/051902.html.

16. ELO chat transcript, January 16, 2000. Available at http://eliterature.org/chats/012600.html.

17. ELO chat transcript, August 5, 2001. Available at http://eliterature.org/chats/080501.html.

18. ELO chat transcript, July 15, 2001. Available at http://eliterature.org/chats/071501.html.

19. ELO chat transcript, August 19, 2000. Available at http://eliterature.org/chats/081900.html.

20. ELO chat transcript, September 16, 2000. Available at http://eliterature.org/chats/091601.html.

21. ELO chat transcript, May 19, 2002. Available at http://eliterature.org/chats/051902.html.

22. ELO chat transcript, April 21, 2002. Available at http://eliterature.org/chats/042102.html.

23. ELO chat transcript, December 3, 2000. Available at http://eliterature.org/chats/120300.html.

24. ELO chat transcript, February 23, 2000. Available at http://eliterature.org/chats/022300.html.

25. Ibid.

26. ELO PAD Project. Available at http://eliterature.org/programs/pad/.

27. ELO chat transcript, November 5, 2000. Available at http://eliterature.org/chats/110500.html.

28. ELO chat transcript, June 15, 2003. Available at http://eliterature.org/chats/061503.html.

29. Ibid.

30. ELO chat transcript, January 7, 2001. Available at http://eliterature.org/chats/010701.html.

31. ELO chat transcript, March 18, 2001. Available at http://eliterature.org/chats/031801.html.

32. Ibid.

25 trAce Online Writing Centre, Nottingham Trent University, UK

J. R. Carpenter

The trAce Online Writing Centre was founded under the short-lived name, The Cyber-Writing Project, by British author and academic Sue Thomas at Nottingham Trent University (NTU), in the United Kingdom (UK), in 1995.[1] For the next decade, trAce expanded organically and somewhat haphazardly into a vast interlinked local and international network created and populated by a multitude of multidisciplinary artists, authors, programmers, and researchers during a period of rapid social and technological change.

From 1995 to 2005, the trAce Online Writing Centre hosted and fostered a complex media ecology: an ever-expanding web site; an active 24-hour web forum; a local and international network of people; a host of virtual collaborations, online studios, and artist-in-residencies; a body of commissioned articles and artworks; a pedagogical program of workshops and online writing courses including *Kids on the Net*, run by Helen Whitehead, and the *trAce Online Writing School*; the *trAce/Alt-X International Hypertext Competition*; the *Incubation* conference series, which ran biannually from 2000 to 2004; and *frAme: the trAce Journal of Culture and Technology*, edited by Simon Mills, 1999–2004. As its name suggests, the trAce Online Writing Centre was more interested in writing than literature; more interested in the way people behave online together than in the writing they might eventually produce; more interested in the interactions than in the outcomes. As a result, trAce was always messy. It was a shifting morphing hybrid entity. It welcomed anybody and everybody—studio-based artists, career academics, educators, fiction writers, poets, computer programmers, chatroom lurkers, locals, kids, and so on. What all these people had in common was that they used the Internet as a medium and a meeting place, a site of exchange and a platform to generate, disseminate, and debate new media writing/writing practices.

Online Community

Building on Thomas' formative experiences within the text-based virtuality of LambdaMOO, a multi-user text-based online community founded in late 1990 by Pavel

Curtis at Xerox PARC, trAce experimented with a number of online community-building tools, including the Mailbase listserv and Lingua MOO, before finally settling on O'Reilly WebBoard. From late 1999 to 2003, the trAce WebBoard was used by 3,000 to 5,000 people from around the world, of which roughly 500 were regular participants. For many, these online conversations were the essence of trAce. Echoing the hubbub of a market town, the forums were sites of exchange—of ideas, information, links, tips, feedback, and debate. Award-winning UK-based author Christine Wilks credits the trAce Online Writing Centre with luring her into the world of digital literature:

I found the forum for giving and receiving feedback on work in progress particularly useful—so important when, most of the time, you're working in isolation. I really didn't have much confidence at all when I first started, so the course and the forums were invaluable. (correspondence, October 23, 2011)

The *Archive of the trAce Online Writing Centr*e contains a wealth of ForumLive[2] and Online Meeting Lingua MOO/Chat logs.[3] Invaluable to researchers not just as records of what was being said in them, these carefully transcribed conversations also evoke the syntactic context of their creation. A glance at the following excerpt of a transcript of a bit of chitchat that took place prior to a chat on Erotic Hypertext on Sunday, November 19, 2000, offers a glimpse into the visual, textual, grammatical lingua franca of Lingua MOO:

Chat Transcript:
Erotic Hypertext
Start log: Sunday, November 19, 2000 12:02:27 p.m. CST

wanderer arrives from Tower of Babble
mez quietly enters.
vika quietly enters.
vika says, "Hello, all."
mez says, "Hey vika"
vika says, "Ah, hello mez. I see you are 'all.'"
mez says, "Yeah, sorry about that:) negate that qs if possible, yeah?"
mez says, "So we r earli then?"
vika says, "Apparently."
mez says, "So vika, wots yr story?:)"
vika says, "Five more minutes, by my clock."
mez says, "k."
vika says, "Well, it's all in my 'bio.':)"
mez says, "Not b-ing a regular MOO boffin, can I ask how 2 check that please?"
vika is very new to the MOO interface and is 'sploring.[4]

In some cases, these spaces were used in overtly performative ways. In *Material Virtualitites: Approaching Online Textual Embodiment*, Jenny Sundén argues, "When text-based domains in cyberspace are seen as performance space, a literary format indissolubly intertwined with the electronic environment is created."[5] In 2001, Marjorie Luesebrink, under the pen name M. D. Coverley, invited members of the public to create online characters, interact with others, and find clues as part of a mysterious project called *M is for Nottingham?*[6]—presented as an interactive performance in the virtual haunts of the historical city of Nottingham, on the web, and as a drama enacted at the Incubation conference held by trAce at NTU in July 2002. The following excerpt of a chat transcript about the project reveals the playful interweaving of performative and informative writing modes that coexisted in Lingua MOO:

Chat Transcript:
M is for Nottingham?
Start log: Sunday, June 16, 2001, 21:04:03 p.m. CST

Margie, Margaret, Sue, cahoots, and the White Lady arrive.
Sue says, "Thanks all—now we have a recording in operation!"
cahoots introduces herself to margie: "Hello margie, I'm cahoots. Pleased to meet you."
margie says, "hi cahoots, do I know you by another name?"
Margaret says, "Cahoots—have you a character in M for Nottingham? yet?"
Jean_Smith arrives from Eliterature.
cahoots bows coquettishly, "perhaps, perhaps not:-)"
The_White_Lady scribbles on her slate.[7]

Life-long friendships formed in these forums. Sue Thomas recalls:

At some point in the late 1990s, Randy Adams turned up at the O'Reilly WebBoard, which then housed the *trAce Online Writing Community*. He never really left, and in the years that followed, he became a vital member of the trAce team. He would stay with us to the end—2005, when trAce was finally closed and archived. (Thomas 2014)

The end of trAce did not necessarily mean the end of these relationships. In 2006, Randy Adams began R3/\/\1X\/\/0RX,[8] a collaborative online creative space for the remixing of digital media, which continued after Adams' death in 2014, and now has more than 500 works. Most members were brought together, initially, by trAce.

Collaboration

A significant number of works supported by trAce were built in collaboration and depended heavily on contributions sent in from writers around the world. Perhaps the best known of these, *The Noon Quilt*,[9] was designed and maintained by the Australian artist and writer Teri Hoskin, based on an idea by Sue Thomas, using Perl scripts written by Ali Graham. *The Noon Quilt* was "an assemblage of patches submitted by writers from around the world. Together they formed a fabric of noon-time impressions. The quilts were stitched over a period of approximately five months 1998–1999."[10] A 112-page, full-color, wire-bound print book iteration of *The Noon Quilt* was made, and the source code was made available for anyone who wanted to create a new version of the web-based project, resulting in at least three new works: *The Eclipse Quilt*, commissioned by trAce in 1999, and The *Dawn Quilt* (2004) and *The Road Quilt* (2005), both commissioned by the British Council.[11] The influence, regardless of whether it is acknowledged or even conscious, of this early use of an online community to create a mass-participation art work may be seen in contemporary works such as #dawnchorus365,[12] an online artwork devised by Natasha Vicars that invites tweets that reinterpret dawn every day of the year, November 15, 2013 to November 14, 2014.

Publication Venues

Many of the names in the trAce archive are well known to us today in electronic literature, new media art, digital humanities circles, and beyond. For many of these authors, trAce provided a first publication venue at a time when there were few places for the public presentation of experimental web-based writing. The first trAce/Alt-X Hypertext Competition, held in 1998, helped spread awareness of trAce's activities into American networks. The competition was judged by Robert Coover, who wrote, "In the first awarding of a major new prize in this new emergent form of literature—if 'literature,' or skill in letters, is still what it is—it is natural that the range of forms and creative concepts among the entries should be almost bafflingly diverse."[13] Coover, along with Scott Rettberg, coauthor of one of the competition's two winning works,[14] went on to found the Electronic Literature Organization (ELO)[15] the following year. Many of ELO's early active members had been published by trAce and/or had attended trAce events. The Incubation conference series (2000, 2002, 2004) offered an opportunity for digital writers, readers, critics, and artists from around the world to meet in the flesh, and it provided a vital platform for the live presentation and/or performance of digital writing. From 1999 to 2004, *frAme, the trAce Journal of Culture and Technology*, published more than sixty "bafflingly diverse" essays and creative works, including early works by Randy Adams, Mark Amerika, M.D.Coverley, Deena Larsen, Geert Lovink, Talan Memmott, Netwurker Mez, Kate Pullinger, Scott Rettberg, Francesca da Rimini, Alan Sondheim, and Eugene Thacker.

From 1999 onward, trAce supported a number of writers-in-residencies, online studios, and research fellowships that provided authors a more sustained space to experiment with new forms. In return, these authors made significant contributions to the trAce community by "initiating online projects, attending conferences, and joining in WebBoard discussions."[16] During his residency (September 1999–March 2000), Alan Sondheim made 2,903 WebBoard posts, mostly in response to questions or creative work posted by other members.[17] In 2004, trAce writer-in-residence Tim Wright traveled throughout the UK working with people he met along the way to piece together *In Search Of Oldton*.[18] Other collaborative projects supported by trAce took a considerably more local focus. In 2000, Nottingham-based digital writer and editor Helen Whitehead staged a series of creative writing workshops in textile factories aimed at teasing out stories of Nottingham's industrial history and making them available online. The resulting work, *Web, Warp, and Weft*,[19] explored the resonances between the making of textiles and the making of the Web.

Prehistory

Although, or rather, specifically because *trAce* was predominantly an online community built and distributed through global digital communications networks, it is critical to consider the brick-and-mortar city within which this dispersed community was conceived, founded, and physically located. Nottingham is smack-dab in the middle of England. Neither a port town nor a major financial center, in the Middle Ages, Nottingham thrived as a market town—a hub for commerce, a meeting place, and a site of exchange of material goods and of information. During the Industrial Revolution, the city grew into a prosperous center for textile production. In "The Future Looms: Weaving Women and Cybernetics," Sadie Plant writes, "The computer emerges out of the history of weaving. ... The loom is the vanguard site of software development."[20] Charles Babbage's proposed (but never realized) Analytical Engine (first described in 1837) was to have employed the same punch card system that made Jacquard's looms capable of weaving complex patterns. The technology of textile production introduced memory, complexity, and abstraction to calculation. Ada Lovelace recognized that these punch cards would make Babbage's Engine programmable, in the modern sense of the word.[21] In "Material and the Promise of the Immaterial," Ingrid Bachmann underlines the central role that textile production played in the Industrial and Digital Revolutions of the 19th and 20th centuries, respectively: "The mechanization of textile mills involved the transfer of workers from the home to the factory, from the country to the city via the newly developed railways powered by the steam engine."[22] In one sense, long-distance communications technologies liberate people the from the specificity of location. Even as steam engines hastened migration to the city, steamer packet services aided and abetted the migration of people and information from the old world to the new. Since the 1960s, packet switching has enabled the transmission of data through networks, thereby simultaneously reducing the need to travel for work and making it possible to work from almost anywhere. Yet Bachmann cautions against the utopian rhetoric of progress: "The appalling work conditions of the multi-national electronics industry rival the deplorable conditions of the nineteenth-century textile mills."[23] As textile manufacture fell into decline in the UK after World War II, the appalling work conditions were outsourced, the infrastructure re-purposed. Nottingham's Lace Market is now a shopping mall. Other former factory buildings have been transformed into artists' studios and gallery spaces.

It is not my intention to suggest that trAce grew directly out of the textile mills of Nottingham. I wish to make a more general association between seemingly disparate yet similarly paradoxical technological eras. Just as new weaving technologies created affordable textiles *and* deplorable working conditions, so too new communication technologies offered writers great opportunities *and* threatened their way of life. trAce was born at a moment of great optimism *and* great anxiety, and it evolved over a period of profound technological change.

Humans have been attempting to communicate across long distances through the written word for thousands of years. The telegraph braced us for speed; the wireless for a wide range of address. But the Internet introduced a hitherto unprecedented degree of atemporality, asynchronicity, and multiplicity to long-distance communication, which radically and irrevocably altered a wide range of writing practices. Since the early 1980s, amateurs, artists, authors, academics, and activists had been exploring text-based virtuality through email, listservs, Telnet, Gopher, FTP, BBS, USENET, MOOs, MUDs, and other multi-user online communities. The British computer scientist Tim Berners-Lee first announced his invention of the World Wide Web to the alt.hypertext newsgroup on August 7, 1991. In *Feminism and Technoscience*, Donna Haraway observes, "Making connections is the essence of hypertext."[24] Hypertext radically altered relations between subjects and objects. The new digital writing spaces of the 1990s enabled people, women in particular, to circumvent existing power structures by building new virtual networks. By late 1994, the Netscape Navigator Web browser had provided a visual interface for hypertext, graphic, and sound file formats, paving the way for the graphic and increasingly proprietary Web we know today. trAce was born within the same year as the visual Web, but conceptually it was informed by an earlier, opener, more generous, and perhaps more naive text-based virtuality within which people strove to communicate solely through the written word.

Formation

In 1995, as Course Leader of a new MA in Creative Writing at NTU, Sue Thomas was keenly aware that these new text-based, nonlinear, multi-user, and multimedia spaces would challenge all current conventions of creative writing, and thus could be used to foster new forms of reading, writing, and interacting online. Originally trained as an artist and bookseller, Thomas had returned to university as a mature student in 1985 to take a degree in English and History. A minor option in Computer Studies inspired her first novel, *Correspondence*, published by The Women's Press in 1992, in which a woman turns herself into a computer. In 1995, Thomas attended the Virtual Futures conference held at Warwick University, UK. There her growing interest in online writing was peaked by a workshop on LambdaMOO led by the Australian artist, author, and performer Francesca da Rimini (also known as Gash Girl). Through LambdaMOO, Thomas sought out, got to know, and got advice from a wide range of international artists, writers, performers, academics, and other early adopters of online writing spaces, including: VNS Matrix, an Australian performance group of which da Rimini was a member; Amanda McDonald Crowley, then director of the Australian Network for Art and Technology (ANAT); and Dale Spender, author of the massively influential book, *Man Made Language*.[25] I highlight the influence of these Australian women in particular to underline the degree to which online writing spaces served to both bridge vast geographical distances and empower women to occupy and transform these spaces. In "Monstrous Agents: Cyberfeminist Media and Activism," Tully Barnett writes of this wave of Australian cyberfeminism that so influenced trAce as:

characterized by ideas for changing society, its structures, and perhaps even the human itself in ways that would bolster prospects for women's equality, relationships to new technologies and the movements and practices they inspired. It was a moment of sharp interest in the intersection of new technologies with notions of gender, sexuality, the body and social equality.[26]

Through LambdaMOO, Thomas also encountered a number of other early online communities based in the United States, which went on to have a major influence on the structure of trAce Online Writing Centre. These included Lingua MOO,[27] an educational online community created and run by Cynthia Haynes and Jan Rune Holmevik, a married couple who met online; Voice of the Shuttle,[28] a resource for humanities research founded as a series of static inter-linked web pages by Alan Liu in 1994; and Alt-X Online Network,[29] an online art and literary network, originally founded as a Gopher site by Mark Amerika in late 1992, which featured art, fiction, poetry, and critical theory from international contributors.

From the outset, the trAce Online Writing Centre was formed and informed by an active engagement with online writing, yet pragmatically trAce's first output was a word-processed photocopied booklet. *Select Internet Resources for Writers*[30] was compiled in the summer of 1995 by Simon Mills, who was then an MA student in Creative Writing at NTU. As its introduction announces, "This booklet is written to aid the newcomer in using the Net to access the resources it holds in relation to creative writing."[31] Among these resources were links to sites distributing journals and zines, a poignant reminder that trAce emerged just before the explosion in digital publishing, which was to forever alter the literary landscape at a time when print and digital enjoyed a more symbiotic relationship.

The trAce community embraced both camps, and some early chatlogs contain lively discussions about the use of mixed media in writing. ... The creative hypertexts and hypermedia in the trAce Archive can easily be compared to the multifarious pages of an artist's book.[32]

Fittingly, considering its humble beginnings with photocopy, among the last outputs of the trAce Online Writing Centre was also a booklet of sorts, edited by Randy Adams. *trAces: A Commemoration of Ten Years of Artistic Innovation at trAce* was published as a full-color wire-bound print document in 2006, and it is still available online as a PDF.[33]

Funding

With funding from the English & Media Studies Department at NTU, Mills was hired to build the first trAce website, which launched at the Virtual Futures Conference held at Warwick University in May 1996.[34] In 1997, trAce applied to the Arts Council of England (ACE) for funding from their lottery-funded Arts for Everyone scheme and was awarded a substantial three-year grant. This money was used in part to fund a one-day conference called "Writers & the Internet," which took place on Friday, October 16, 1998, at The Broadway Media Centre in Nottingham. This was the first conference in the UK to be dedicated to the topic of writing on the web. The conference website gives a clear indication of the optimism and anxieties of the time:

The internet offers great opportunities for writers. There are fascinating new forms of writing to be discovered; interesting people to meet, and swathes of research material to be mined. But it also brings concerns. Authors are worried about copyright and intellectual property. They are wondering how they can earn money from working online. They fear that The Book may be dying. This conference brings together an international group of professional authors and educators with extensive experience of the internet to address some of these anxieties and provide informed opinion about the potential of the net for the artistic community.[35]

Among the guest speakers were the aforementioned Australian feminist Dale Spender, the Americans Mark America and Cynthia Haynes, and the Norwegian Jan Rune Holmevik. Although many of the conference attendees had been in close online contact for a number of years, most had never met in person. Over the years, trAce remained committed to fostering an online community and, whenever possible, bringing people together in real time.

How an organization is funded has a profound effect on how it operates. That first big ACE grant turned into regular funding. That this funding came from outside academia was significant for two reasons. First, it indicated that the Arts Council had recognized the importance of online writing and that trAce was a leader in the field.

Second, it meant that Thomas could no longer be paid directly by the university. She was seconded from her teaching post but remained at NTU as the Artistic Director of trAce. From that point onward, trAce evolved as a nonacademic, transdisciplinary, transgressive space physically located in the middle of a suburban university campus, yet weaving an ever-widening global network of both human and digital resources, finding and funding opportunities for the creation, promotion, and study of online writing.

In 2003, trAce was awarded funding from National Endowment for Science, Technology, and the Arts (NESTA),[36] an independent charity supporting innovation in the UK, to manage the Writers for the Future project and to begin archiving the by then sprawling trAce website. trAce used a considerable amount of its funding to support artists and writers directly by commissioning new work, supporting practice-led research, subsidizing travel costs, and hiring artists and writers as course leaders and peer mentors. trAce also employed numerous people over the years, as administrators, forum monitors, editors, and educators, including, but by no means limited to, Randy Adams, Caroline Bamborough, Catherine Gillam, Jill Pollicott, Simon Mills, Helen Whitehead, and Kate Wilkinson. As the founder, principal fundraiser, Artistic Director, public face, and tireless ambassador of trAce, Thomas was among the organization's greatest assets. Yet as Thomas has always been the first to acknowledge, that the organization relied so heavily on the fundraising efforts of one individual was one of trAce's greatest weakness. In 2005, Thomas left NTU to take up a Professorship at de Montfort University in Leicester, UK. Gillam, Mills, Whitehead, and Wilkinson also moved on. All left with the hope that trAce would continue. But sadly, with no new money coming in, it did not.

Archive of trAce Online Writing Centre 1995–2005

To give a full account of the history, people, and output of this one-of-a-kind moment-in-time online community would require a multiplicity of perspectives and far more space than that afforded here. Although I was aware of trAce from the late 1990s onward, I was not an active participant in any of the web forums, conferences, or online projects hosted by trAce. In researching this chapter, I visited Nottingham Trent University, with the generous support of the Bonnington Art Gallery. I consulted press-clippings referencing trAce and print documents produced by trAce. I interviewed Sue Thomas and, less formally, a number of other writers who were involved with trAce in various ways. I have not attempted to present a linear history of trAce. Deleuze and Guattari urge, "Make a map not a tracing. ... What distinguishes the map from the tracing is that it is entirely oriented toward an experimentation in contact with the real."[37] The majority of my research for this chapter has been undertaken through direct interaction with the vast and varied digital outputs of trAce preserved in the large and fully searchable *Archive of trAce Online Writing Centre 1995–2005*.[38]

Throughout its existence, trAce supported experimentation with new literary forms, new software applications, and new hardware platforms, many of which are now antiquated. Many of the site's features no longer function as they once did. Sometimes you have to know what you're looking for some things just aren't there. The trAce archive does not include the *trAce Online Writing School* (2002–2005) because its contents were private to the tutors and students, nor does it include *Kids on the Net* because, although it shared server space with trAce, it had a different audience and remit.[39]

One of the many interesting things about the *Archive of the trAce Online Writing Centre 1995–2005* is how much it reflects the context of the creation and dissemination of its contents. Whereas archives held in museums or libraries generally contain artifacts created elsewhere—manuscripts illuminated in a monastery or photographs printed in a darkroom, for example—the digital artifacts contained in the online *Archive of the trAce Online Writing Centre* continue to exist in the medium within which they were first created—the web. That said, the web and the devices through which we view it have continued to change. Wolfgang Ernst observes, "If a radio from a museum collection is reactivated to play broadcast channels of the present, it changes its status: it is not a historical object anymore but actively generates sensual and informational presence."[40] Similarly, viewing old web pages in modern browsers, we are confronted with a temporal paradox. Layer upon layer of dated web-design aesthetics peel like wallpaper, revealing earlier versions beneath. Pages optimized for lower resolutions now take less than a third of the screen. Ghosts of browsers past mingle with occasional page errors, dead links, and missing images. Sound files play automatically. Warnings abound, issued from earlier eras, addressed to readers who are not us:

best viewed on a 800x600 screen with 32 bit color without the browser toolbar.
this piece is viewable only on Level 4 and 5 Microsoft Internet Explorer.
Netscape 6 will not support many of the features in this essay.
the Flash4 [or later version] plug-in is required to view this site.

The digital material traces of trAce present a number of opportunities for further research that have yet to be taken up. These are not artifacts of a dead web but, rather, signposts on a map of a living web pointing to a web as it once was, a web in progress, a web in the making.

Since this chapter was written, the archive held at Nottingham Trent University has been taken offline for technical reasons. It is being transferred to the permanent collection of the Electronic Literature Organization and will be available at http://eliterature.org/trace-online-writing-centre-archives/.

Notes

1. trAce was also known as the trAce Online Writing Community for a number of years.

2. Available at http://tracearchive.ntu.ac.uk/forums/.

3. Available at http://tracearchive.ntu.ac.uk/community/logs.cfm.

4. Available at http://tracearchive.ntu.ac.uk/community/19nov00log.cfm.

5. Jenny Sundén, *Material Virtualitites: Approaching Online Textual Embodiment* (New York: Peter Lang, 2003), 154.

6. M. D. Coverley, *M Is for Nottingham?* (Nottingham, UK: NTU: trAce Online Writing Centre, 2002). Available at http://tracearchive.ntu.ac.uk/misfornottingham/Mpart1/welcome.htm.

7. Available at http://tracearchive.ntu.ac.uk/community/16junlog.cfm.

8. R3/\/\1X\/\/0RX. Available at http://www.runran.net/remixworx/.

9. Teri Hoskin, *The Noon Quilt* (trAce), 1998–1999.

Available at http://tracearchive.ntu.ac.uk/quilt/index.html.

10. Ibid.

11. *trAces: A Commemoration of Ten Years of Artistic Innovation at trAce 1995–2005* (Nottingham, UK: NTU: trAce Online Writing Centre, 2006), 8 (www consulted October 27, 2014). Available at http://tracearchive.ntu.ac.uk/traces/traces.pdf.

12. Natasha Vicars, *#dawnchorus365*, Fermynwoods Contemporary Art, UK, November 15, 2013–November 14, 2014.

13. trAce/Alt-X Hypertext Competition, 1998. Available at http://tracearchive.ntu.ac.uk/hypertext/.

14. William Gillespie, Scott Rettberg, Dirk Stratton, and Frank Marquardt, *The Unknown*. Available at http://unknownhypertext.com/.

15. Electronic Literature Organization. Available at http://eliterature.org/.

16. *trAces, 2006*, 15.

17. Ibid, 18.

18. Time Wright, *In Search of Oldtown* (Nottingham, UK: NTU: trAce Online Writing Centre), 2004. Available at http://tracearchive.ntu.ac.uk/writersforthefuture/view/insearchofoldton.htm.

19. Helen Whitehead, *Web, Warp, and Weft* (Nottingham, UK: NTU: trAce Online Writing Centre, 2000). Available at http://webwarpweft.com/.

20. Sadie Plant, "The Future Looms: Weaving Women and Cybernetics," in Lynn Hershman Leeson, ed., *Clicking In: Hot Links to a Digital Culture* (Seattle: Bay Press, 1996), 123–135, 123.

21. I am indebted to Sue Thomas for pointing out to me that Ada Lovelace's father Lord Byron's ancestral home, Newstead Abby, is located in Nottinghamshire, and Lovelace herself is buried at the Church of St. Mary Magdalene, Hucknall, Nottingham.

22. Ingrid Bachmann, "Material and the Promise of the Immaterial," in Ingrid Bachmann and Ruth Scheuin,g eds., *Material Matters: The Art and Culture of Contemporary Textiles* (Toronto: YYZ Books, 1998), 23–34, 26.

23. Ibid., 26.

24. Donna Haraway, *Modest_Witness@Second_Millennium.FemaleMan©_Meets_OncoMouseTM: Feminism and Technoscience* (New York & London: Routledge, 1997), 126.

25. Dale Spender, *Man Made Language* (London: Routledge & Kegan Paul, 1980).

26. Tully Barnett, "Monstrous agents: Cyberfeminist media and activism," in *Ada: A Journal of Gender, New Media, and Technology*, 5(2014): n.p. (www consulted October 28, 2014). Available at http://adanewmedia.org/2014/07/issue5-barnett/.

27. Lingua MOO. Available at http://uts.cc.utexas.edu/~best/html/exemplars/abstracts/haynes.htm.

28. Alan Liu, *Voice of the Shuttle*. Available at http://vos.ucsb.edu/.

29. Mark Amerika, *Alt-X*. Available at http://www.altx.com/.

30. Simon Mills, *Select Internet Resources for Writers* (Nottingham, UK: NTU: trAce Online Writing Centre, 1995). Available at http://tracearchive.ntu.ac.uk/writersforthefuture/1995/.

31. Ibid. Available at http://tracearchive.ntu.ac.uk/writersforthefuture/1995/p1.htm.

32. *trAces, 2006*.

33. Ibid.

34. Virtual Futures, Warrick University, 1996. Available at http://virtualfutures.co.uk/conferences/1996/.

35. Available at http://tracearchive.ntu.ac.uk/eastm/conf.htm.

36. Now known as Nesta. Available at http://www.nesta.org.uk/.

37. Gilles Deleuze and Félix Guattari, *A Thousand Plateaus: Capitalism and Schizophrenia*, Brian Massumi trans. (Minneapolis & London: University of Minnesota Press, 2007), 12.

38. Archive of the trAce Online Writing Centre 1995–2005. Available at http://tracearchive.ntu.ac.uk/about.asp.

39. Available at http://tracearchive.ntu.ac.uk/about.asp.

40. Wolfgang Ernst, "Media Archaeography: Method and Machine versus History and Narrative of Media," in E. Huhtamo and J. Parikka, eds., *Media Archaeology: Approaches, Applications, and Implications* (Berkeley, LA, London: University of California Press, 2011), 239–255, 241.

Bibliography

Adams, Randy, ed. *trAces: A Commemoration of Ten Years of Artistic Innovation at trAce 1995–2005*. Nottingham, UK: NTU: trAce Online Writing Centre, 2006.

Archive of the trAce Online Writing Centre 1995–2005. Available at http://tracearchive.ntu.ac.uk/about.asp.

Bachmann, Ingrid. 1998. Material and the Promise of the Immaterial. In *Material Matters: The Art and Culture of Contemporary Textiles*, ed. Ingrid Bachmann and Ruth Scheuing, 23–34. Toronto: YYZ Books.

Barnett, Tully. 2014. "Monstrous agents: Cyberfeminist media and activism," in *Ada: A Journal of Gender, New Media, and Technology* 5(2014) (www consulted. *October* 28. Available at http://adanewmedia.org/2014/07/issue5-barnett/.

Deleuze, Gilles, and Félix Guattari. 2007. *A Thousand Plateaus: Capitalism and Schizophrenia, Brian Massumi trans*. Minneapolis, London: University of Minnesota Press.

Ernst, Wolfgang. 2011. Media Archaeography: Method and Machine versus History and Narrative of Media. In *Media Archaeology: Approaches, Applications, and Implications*, ed. E. Huhtamo and J. Parikka, 239–255. Berkeley, LA, London: University of California Press.

Haraway, Donna. 1997. *Modest_Witness@Second_Millennium.FemaleMan©_Meets_OncoMouseTM: Feminism and Technoscience*. New York, London: Routledge.

Mills, Simon. *Select Internet Resources for Writers*. Nottigham, UK: NTU: trAce Online Writing Centre, 1995. Available at http://tracearchive.ntu.ac.uk/writersforthefuture/1995/.

Plant, Sadie. 1996. The Future Looms: Weaving Women and Cybernetics. In *Clicking In: Hot Links to a Digital Culture*, ed. Lynn Hershman Leeson, 123–135. Seattle: Bay Press.

Sundén, Jenny. 2003. *Material Virtualitites: Approaching Online Textual Embodiment*. New York: Peter Lang.

Thomas, Sue. 2014. "In Memory of Randy Adams 1951–2014," posted May 10, 2014 (www consulted). *October* 27. Available at http://suethomasnet.wordpress.com/2014/05/10/in-memory-of-randy-adams-1951-2014/.

26   Pseudo Space: Experiments with Avatarism and Telematic Performance in Social Media

Antoinette LaFarge

There is a communitarian dimension to all media used to make art: the crowd feeling of a theater audience, the ebb and flow of a painter's opening, the collaborative spirit of a filmmaking team; and this is apart from the social dimensions of economics, class, gender, race, and so on. But by calling only certain forms (and specifically only certain forms of *software*) "social media," we do not just underline the degree to which sociality and the social are at the heart of the work produced through such media. We also mark how the affordances of such media for real-time, distributed, pseudonymous, improvisational agency have been redefining the social itself for three decades, since the first experiments in electronic community arose in the 1980s. Within the history of social media, we can mark a trajectory wherein the frozen public mask has become the lively avatar, pretense and deceit have been reformulated as role play, multivocality has displaced monologue and monolinguality, identity is giving way to impersonation, and typing has become conversation, thereby returning writing to its roots in orality.

At this particular juncture, we are hearing a great deal about the dark side of this new bargain: 2014 and 2015 were dominated by discussions of trolling—online harassment in all its forms—with many commenters calling for technical constraints such as bans on pseudonymity. I believe that such blanket solutions are, as always, the wrong approach because they do not acknowledge the deep psychosocial desire for pseudonymous action that the net has both enabled and unleashed. In order to tease out why we are so drawn to this form of being, I want to look at the way certain aspects of social media have led to the emergence of net-specific forms of relational art. Several years before Nicolas Bourriaud gave us *Relational Aesthetics* in 1998, artists in early social media were creating works that prioritized human relationship, interaction, and experience over the production of luxury objects; they were establishing experimental creative zones almost entirely outside of the museum-gallery system, producing artworks that couldn't really be exhibited, owned, or even easily located. I used the word "interaction" above, but I have always considered that the power of social media stems from the fact that they are deeply *participatory* rather than superficially interactive.

A point often overlooked is that early social media often functioned in ways that were not always aligned with their designers' intentions. The text-based social environments known as MOOs—which were the first user-built persistent virtual communities that weren't games or bulletin boards, although they descended from both (and most directly from the game-based MUDs)—offer an excellent example of this form of emergence. Xerox PARC programmer Pavel Curtis wrote the code for LambdaMOO, the first public MOO, which opened its doors in October 1990.[1] It was modeled on Curtis's own house—first-time users arrive in a coat closet and then enter the living room—and thus was explicitly mundane compared with the sword-and-sorcery–infested MUDs that had been the rule up until then. The living room metaphor suggested that LambdaMOO would be a place to hang out and chat, something like a real-time WELL with a side of programming for those who felt like building new rooms. What actually happened is that in their heyday, between 1990 and about 1996, LambdaMOO and several of the other major MOOs[2] got largely taken over by creative activities organized by pseudonymous avatars engaged in surreal programming projects, elaborate second-life identities,[3] and live performances. Following the MOOs came a succession of similarly immersive social media, nearly all of which have been heavily used by artists, from the Palace (1995, a graphical chat room used by Adriene Jenik and Lisa Brenneis's Desktop Theater for live performance) to UpStage (2004, a cyberformance space cofounded by Helen Varley Jamieson and known for its annual cyberformance festival) to Second Life (2003, a virtual world whose first performance group, Second Front, was founded in 2006). When there is a gap between what people are encouraged or expected to do and what they *actually* do, it is the latter that demands our focused attention.

More recently, the messaging app Twitter (2006) first gained attention not so much for its now-prominent role in global activism as for its fictional personae (like @GSElevator, a supposed Goldman Sachs employee) and "Twitterature," ranging from aphorisms and haiku to serialized stories. In this sense, Twitter descends spiritually not only from the MOOs but also from chatware like Internet Relay Chat (which was used by Stuart Harris and his Hamnet Players in the mid-1990s as a bare-bones performance venue) and the interactive literature and hypertext experiments of Judy Malloy (*Uncle Roger*), Ted Nelson, Michael Joyce, Shelley Jackson, Stuart Moulthrop, and others.[4]

I belong to a transitional generation that did not grow up with computers; I adopted them as my main tool and medium as soon as the desktop computer appeared in the late 1980s. But it wasn't until I discovered social media in the early 1990s that I became completely fascinated, and this was because it was only then that I saw real potential for changing the way I could be in the world as an artist and a person. Up to then, everything I had been doing with computers was just a faster or less tedious version of things I already knew how to do without computers: writing, editing, graphic design, image development, game play.

The first social media I discovered was the UNIX platform's live chat program.[5] Called, simply, "talk," it had in skeletal form most of the key characteristics of what I would come to see as a new performance medium.[6] In "talk," you had only a nominal sense of whom you were talking to because the other person wasn't necessarily in the same physical space and showed onscreen under a login name that might very well be invented. Talk conflated speaking and typing, instantly ending the long divorce between oral literature and writing by turning the latter back into the former. As writing, it encouraged one to improvise in any voice, any tense, any mood—you didn't feel compelled to keep up the singular social persona that conversation ordinarily requires. The split-screen interface kept the two halves of the conversation strictly partitioned off from each other, with the result that you didn't feel you were talking to someone else so much as monologuing with that person in parallel.[7] It encouraged you to respond to yourself (what you had just typed) as much as to the other person, as if there were actually three people in the chat space: you, your screen self, and the other person. In that plain-text environment, I experienced the infancy of the screen self, the denaturalization of my usual subjectivity through improvisation in a separate persona that was to become, under more conducive circumstances, a full avatar.

Pseudonymity, orality, multivocality, multilinguality, writing as both improvisation and role play, the emergence of distinct personae: it turned out that all of that was implicit in a simple UNIX daemon whose inventor had undoubtedly just wanted a plain vanilla messaging app. It became much easier to work with these features in MOOs, the more robust multi-user successors to the UNIX talk daemon that offered persistent, user-built virtual communities in place of discrete or time-shifted conversations.[8] When I first found my way into a MOO in 1994, I was puzzled about why so much effort was being expended in trying to make the culture and activities more like "real life" (RL, as it quickly came to be known), when it seemed to me that we should be exploring the characteristics and affordances of virtual reality (VR). It was as if whoever invented animation had made it a rule that you could never break the laws of physics in such films because those laws hold in RL.[9]

Early in 1994—that is, about half a year before the first public release of the Netscape web browser—I began experimenting with the performance possibilities of MOO with a group of fellow MOO dwellers, some of whom I had never met in person (and in some cases still have not met). I came up with the name Plaintext Players because working with "just text" was a major feature distinguishing this form of performance from theater, radio, film, television, and performance art.[10] (Elsewhere, I have written about how odd it is that the word "just" was used this way, as a marker for an underlying assumption that some other element or medium was missing. One does not, after all, refer to novels in this way as "just text.")

The first works the Plaintext Players created when they began experimenting in MOOspace were guided improvisations based on short written scenarios. The original scenario for the first series, *Christmas* (1994–1995), which ultimately ran to 18 unique performances, was written by theater director Robert Allen. It outlined a highly unstable relationship among three central archetypal characters: the not-too-bright Big Man, his friend and guardian Little Man, and the chaos-inducing Bloody Zelda. By the time this first series was over, the Players had begun working with what remain many of the central features of avatar-space, among them the affordances to:

- switch personae easily and invisibly;
- shift fluidly between the first- and third-person voices;
- share characters through the use of matching or complementary pseudonyms;
- play multiple characters simultaneously (usually through multiple windows, a practice known in the game world as multiboxing);
- make characters of ordinarily inanimate objects;
- suddenly change the course of a narrative in progress (often through importation of ripped-from-the-headlines elements);
- live-direct or conduct a performance through the equivalent of stage whispers;
- co-opt online audiences into the performance;
- interject canned material into the live stream;
- use programming to modify the performance on the fly; and
- produce a collective text outside of the traditional writing-rewriting-editing model.

Many of these features were also exploited by Harris, Jenik and Brenneis, Jamieson, and other pioneers of what Jamieson has since termed *cyberformance*. The term *cyberformance* for live networked performance entered the discourse relatively late, around 2000, after a number of other terms had been tested out and found wanting: online performance, online theater, digital theater, net performance, telematic performance, and avatar performance. The word "theater" always fit awkwardly because these projects were initially much closer to writing-as-performance, and the shared space was virtual rather than physical. Even calling it "performance" remains something of a stretch because the multiplication of avatar identities across the Internet has made it ever more difficult—in some cases impossible—to determine which avatar activities are performances and which are not. I appreciate the term *cyberformance* for its implication that cyberspace is *inherently* a *performance space*, that in this home of our pseudonymous and distributed selves, everything that happens has performative aspects. The recent spate of calls for strict identity policing on the Internet (e.g., through portable avatars, secure avatars, or required "real name" identities) testifies to how uneasy we remain with the open-ended pseudonymity that has become woven into all of our daily lives.

The first half-dozen Plaintext Players' projects took place wholly online, with viewers attending by either logging into the same virtual space or watching video projections in RL.[11] Several of these series had literary origins, including *LittleHamlet* (based on Shakespeare), *Gutter City* and *The White Whale* (both riffing on Herman Melville's *Moby Dick*), and *Orpheus* and *Silent Orpheus*, paired adaptations of the Greek myth. All of these displayed a peculiar feature in which anthropomorphization met reification: a tendency on the part of players to animate absolutely everything in sight, at the same time reifying their own thoughts, words, and actions. The director became a persistent character (Digital.Director), as did the curtain and the scenery, the stage directions, the audience, we, history, somebody's drinking problem, somebody's liver, the void, even a stray comma. This situation of protean generation was further complicated by the way players would often change personae, pronouns, verb tense and number, and even affect in the middle of a line, sometimes to the extent of contradicting themselves several times over. Here you could discern one of the boundaries of avatarism, where it opens out from its foundation in character to become a way of engaging language as an infinite toy.

In the next phase, I began adapting the Plaintext Players' performance transcripts to re-performance in other media: one improvisation from the Christmas series became a radio play and then a graphic novel (both entitled *The Cake of the Desert*), while *The Candide Campaign* became a stage play titled *Still Lies Quiet Truth*. In *The Roman Forum* (2000), transcripts from a series of online performances featuring first-century Romans pondering the U.S. presidential election were rewritten into a play of the same title that was staged in downtown Los Angeles a few blocks from where the Democratic Convention was going on. Each night during the play's run, new material was added from online improvisations responding to and commenting on events of the prior 24 hours.[12] Like the contemporaneous *Daily Show* on television, the Players were making creative use of the 24-hour news cycle in ways that anticipated the emergence of today's 24-hour blogosphere. With *The Roman Forum*, we also began to take an expansive view of the locus of the piece, moving toward a distributed-media model that underpins current game styles such as alternate-reality games, geocaching, and letterboxing, as well as reflecting the ways people distribute aspects of themselves across multiple social media platforms today. During the run of *The Roman Forum*, for example, we put the U.S. presidency up for sale on eBay in one of the earliest spoof auctions on that site, and we saw this event as part of the overall project rather than ancillary to it. eBay was as much our performance venue as the MOO or the physical space of Side Street Projects in Los Angeles, which hosted *The Roman Forum*. In this refusal to treat the network as yet another conventionally bounded space, we were following the ethos of projects such as the Art Com Electronic Network (ACEN) and Electronic Café International a decade earlier.

In those two initial phases of Plaintext Players' projects, the virtual and physical spaces were visible to one another but not yet fully confluent: RL audiences could see the VR space, or vice versa, but there was no live interaction between the two. With *The Roman Forum 2003* (a sequel to *The Roman Forum*) and *Demotic* (2004/06), the two spaces became linked through live two-way information streams that required the online and physical-space performers to engage each other in real time—through text, audio, and video. In *The Roman Forum 2003*, the physical-space performers were required to interact with their online counterparts of the same name, creating an encounter with virtual selves (figure 26.1). In *Demotic*, by contrast, the real-space performer was outnumbered by a half-dozen online personae, all sending material her way almost faster than it was humanly possible for her to respond. In addition, all communication within the piece was multimediated through textual transformations, text-to-speech synthesis, sound-processing, and randomization functions. In *Demotic*, I would argue that an expansive understanding of "social media" would embrace not merely

Figure 26.1
Antoinette LaFarge. In *The Roman Forum Project* (2003), two characters react to the improvisation being live-generated by their online counterparts around current events, especially the then-impending invasion of Iraq. Left to right: Petronius (Kevin Keaveney), Empress Poppaea (Helen Wilson). Photo courtesy of Antoinette LaFarge and Robert Allen.

Figure 26.2
Antoinette LaFarge. Part of the "social media" of *Demotic* (2006): the performance software and a customized game controller used by sound artists Maria de los Angeles Esteves and Jeff Ridenour to play with the online performers. Photo courtesy of Antoinette LaFarge and Robert Allen.

the online MOO—the only obvious piece of social media software—but also the physical space of the venue in Baltimore, the sound artists' performance software and customized game controllers, and the several interfaces linking virtual and physical performers.[13] All of these combined to create the meta-social-medium of improvisational space (figure 26.2).

As might be clear by now, a critical feature enabling this two-way improvisational flow has been custom programming to interconnect the different kinds of players and their media: text-to-speech processing, audio post-processing, video green-screening on the fly, and so on.[14] For example, in *Demotic*, we deployed a small censorship bot in one part of the performance to force the Plaintext Players to find ways to respond creatively in a censored environment. Routing around censorship—whether through variant spellings of tabooed words or through use of proxyware—remains a vital activity on social networks. So crucial did custom programming turn out to be for the Players' work that I ended up cofounding two different MOOs in order to have absolute control over programming of user features—not least because by a couple of years into our work, some of the major MOOs had already banned a key tool of our performances—what is known as "spoofing" (pretending to be someone you are not).[15] Indeed, one of the important things we have lost in the shift from user-built spaces like MOOs to commercial platforms like FaceBook and Twitter is the power to profoundly change what happens through user-initiated programming. It's not that there have ever been all that many artist-programmers at work in user-built worlds: it's that their disenfranchisement was necessary for the ascendance of monolithic forms of social media whose status as closed systems is barely disguised by pointing to "user-created content." Like many of the other writers in this book, I felt an extraordinary excitement about the potential of the Internet during its formative years that has trickled away year by year along with the loss of under-the-hood agency. Today the culture of the net strikes me as generally staid except insofar as it is being used by social activists.

Performance transcripts show the Plaintext Players behaving badly by ordinary standards, mixing their strange narratives up with insults, silliness, random comments, intemperate confessions, and gibberish in a chaotic spew. Often it looks like nothing so much as crappy writing by idiots and jesters. Yet even at its lowest points, it never became trolling, and that was only partly because the Players were a real as well as a virtual community: what one might call a practical micro-utopia. It was also because they placed their avatar agency under the frame of art, marking the cyber off as a Huizingan protected space under the same terms as ritual, play, and theater. So the question I want to ask here is: What would happen if we all took a step back from our fears and placed more of cyberspace under the same frame? Thought of it as "FR" (fictional reality) rather than a problematic version of RL? Openly developed fictional personae rather than trying to maintain singular, usually professional, identities? Openly embraced the not-knowing that is a major component of any telematic setup, no matter how much security is put in place? What, in other words, would it take for us to be able to fully embrace the performative affordances of network pseudonymity—if not absolutely everywhere on the net, at least in a more expansive and joyful way than now seems possible? Paradoxically, it will require a good deal more privacy protection for our avatars—in effect, Tor for everyone—because a common game now is to hunt down and expose the person behind the avatar. Most of all, it will ask us to make a change in our assumptions about what we are expected to do with social media. To accept that what people gravitate toward—pseudonymity rather than identity, play rather than obligation—should be encouraged and enabled in the most productive ways possible. That could, in turn, lead to the next change in the nature of social media—one that demands something more of us than existence as an industrial-scale flock of data-producing sheep.

Notes

1. For useful outlines of the early history of MOOs, see Pavel Curtis, "MUDding: Social Phenomena in Text-Based Virtual Realities," in Peter Ludlow, ed., *High Noon on the Electronic Frontier: Conceptual Issues in Cyberspace* (Cambridge, MA: MIT Press, 1996), 347–374; and Susan Sim, "How Community and Deviance on LambdaMOO and Lord Graham's Demesne Can Prepare Us for the Future of Computer Mediated Communication in Cyberspace," *Computers and Society*, University of Toronto, December 1994.

2. Among these I include Curtis's LambdaMOO, PMCMOO (which specialized in postmodern culture and was hosted by the University of Virginia), ATHEMOO (hosted by the University of Hawaii), and the New-York-based IDMOO.

3. See Julian Dibbell, *My Tiny Life* (New York: Holt, 1999).

4. For a thorough discussion of the Hamnet Players and other early cyberpformance groups, see Brenda Danet, *Cyberpl@y: Communicating Online* (New York, London: Bloomsbury, 2001).

5. Elsewhere in this book, see David Woolley's discussion of a similar talk feature in the 1960s-era PLATO software, which has been reinvented for the web. Available at http://talko.cc/.

6. I'm writing here about a version of talk introduced in the mid-1980s; there are even earlier versions that worked slightly differently.

7. I've written more about the talk daemon elsewhere in "Imposture as Improvisation," in George Lewis and Ben Piekut, eds., *Oxford Handbook of Critical Improvisation Studies* (Oxford, UK: Oxford University Press, forthcoming).

8. For detailed discussions of the experience of being in MOOspace, see Dene Grigar and Deena Larsen's chapters in this volume.

9. My first article about the potential of MOOspace was "A World Exhilarating and Wrong: Theatrical Improvisation on the Internet," *Leonardo* 28:5 (1995). Another early article on the performance potential of the Internet was Thyrza Goodeve's "Houdini's Premonition: Virtuality and Vaudeville on the Internet," *Leonardo*, 30:5 (1997).

10. Core members of the Plaintext Players who contributed to multiple performances include Marlena Corcoran, Ursula Endlicher, Joe Ferrari, Richard Foerstl, Thessy Mehrain, Lesley Mowat, Lise Patt, Richard Smoley, Heather Wagner, and Adrianne Wortzel.

11. These were *Christmas* (1994–1995), *LittleHamlet* (1995), *Gutter City* (1995), *The Candide Campaign* (1996), *The White Whale* (1997), *Orpheus* (1997), *Silent Orpheus* (1997), and *Birth of the Christ Child* (1999).

12. In addition, video excerpts from the performances were put online as quickly as possible afterward, usually within a day.

13. In several subsequent projects created by Plaintext Players Marlena Corcoran and myself for UpStage festivals, live drawing, webcams, and deployment of 2D images joined the other media used in these live cyberformances. These were Marlena Corcoran's *Salvation* (2009) and W*ater Under the Bridge* (2008) and my *Noxiterra* (2008).

14. The programmers responsible for the majority of the Plaintext Players' custom software needs were Robert Allen and Tom Ritchford (in MOO language) and Jeff Ridenour (in Max multimedia software).

15. These MOOs were IDMOO, hosted by the School of Visual Arts, New York, and YinMOO, hosted by the University of California, Irvine.

VII Responses

A Conversation and Two Epilogues

Judy Malloy

When Kit Galloway and Sherrie Rabinowitz arrived in Telluride for *Tele-Community* in the summer of 1993, it seemed as if the whole town joined them on Main Street, as using slow scan video they connected townspeople and visiting digerati with artists, universities, and cultural centers around the world. Their Electronic Café had already presented New York City pedestrians with display windows of people waving and talking real time from Los Angeles (*Hole-In-Space*, 1980) and linked cafés in the Korean, Chicano/a, African American, museum, and beach communities of Los Angeles to exchange video images and work together on an electronic writing tablet during the 1984 Olympics.

Following the trail of the Electronic Café in the days before the World Wide Web, it was possible to envision that in the twenty-first century, people would be talking to friends, families, colleagues, and strangers with Skype, exchanging photos with Snapchat, and posting videos on YouTube.

Now, as if it were an influential archeological discovery of Roman sculpture—unearthed in Italy during the Renaissance—information about the pre-web lineage of social media enhances and clarifies contemporary temporary social media. Logging on in spirit to the platforms documented in this book, it is clear that in the days before the World Wide Web, there was unprecedented online cultural energy and a concerted striving for online community for the arts and humanities.

Core questions and answers arise, are answered differently, elicit study. How did the communities they served shape each platform? How did technology and interface play a role in the user experience of each platform? How were the affordances of social media differently implemented? How was early social media funded (or not funded)? What were the reasons for the demise, sidelining, or, in some cases, evolving use of social media platforms?

Email, for instance, was once an amazing source of information. In Palo Alto, at the 1995 Author's Guild Conference, I gave a talk on what was in my email in the course of a week or so. Now email is so ubiquitous that amid the profusion of necessary correspondence, interesting informational emails may never be opened, and much of the information that email once carried has migrated to websites, blogs, and social media platforms. On my Twitter stream, for instance, in an early July 2015 week, Mitch Kapor pointed to the 25th anniversary of the founding of The Electronic Frontier Foundation; from Malaysia, Clarissa Lee posted about teaching critical gaming; in Puerto Rico, Leonardo Flores became a full professor; in California, the Napa Museum was exhibiting Alison Knowles' "Homage to Each Red Thing" and posted a memorable image of the installation; The Berkman Center announced new fellows; poet Elizabeth Alexander looked at racial injustice as expressed in contemporary art and literature; VSA invited the Twitter community to a chat on #DisabilityStories; Furtherfield invited media artists to exchange data for commodities.

And as Geert Lovink observes in *this* book, "We'd need a next to impossible alien perspective to regain the wonder of a Facebook page."[1]

Presenting the different ideas of three individuals—Geert Lovink, Judith Donath, and Gary O. Larson—with years of experience creating and/or writing about social media and the cultural Internet, in the following section are different initial responses to the platforms revealed in this book. Appropriately, we begin with a conversation in which Geert Lovink expands on his 2012 essay in *e-flux*: "What Is the Social in Social Media?"[2] Looking at cultural social media from the point of view of creating, teaching, the content it carries, and how social media content is disclosed, he responds:

As you might know I prefer the essay as form (reread Adorno and Sontag), the manifesto, the short newspaper articles Jean Baudrillard used to write for *Liberation*, the aphorism (but not the tweet as it is too rigid), the blog entry (I deeply admire Dave Winer!), the list thread (terrible!), the wild yet unfulfilled promises of the Internet forum, the 2,334 comments on the Guardian site, and the 50,000 e-books a friend of mine recently gave me on a disc, which took 20 minutes to copy.[2]

"You could imagine…"—Judith Donath

Next, Judith Donath—who, with a focus on innovative interface and design, sets forth in her 2014 book, *The Social Machine*,[3] an elegant and continually interesting approach to the pleasurable interaction of/with online social environments—focuses in *this* book on the contingent idealism and utopian approaches of early social media, suggesting that:

Nascent technologies provide a blank canvas for imaginings. And with social media, in the early days not only were the interfaces still in development, the population was small and select. You could imagine that the cooperation and creativity you saw with your small group would be magnified as hundreds, thousands, and millions more joined—that the technology itself would be capable of moderating differences, unchaining imaginations, and transforming society.[4]

Looking to the past, looking to the future, this book concludes with Gary O. Larson's in-depth look at cultural policy and funding for the arts and humanities on the Internet: "From Archaeology to Architecture: Building a Place for Noncommercial Culture Online." Proposing a five-point plan to build a sustainable cultural sector online, Larson addresses both preservation and "a coherent online cultural sector" while emphasizing the difficulties of achieving this. "And yet in the current environment—with growing numbers of "digital natives" who have essentially grown up online—the reclamation of this cultural legacy need not take so long, " he observes:

Neither archaeology nor architecture in the digital domain is anywhere near as labor intensive as their analog counterparts, and compared to unearthing ancient civilizations or erecting monumental structures, the cultural riches of the digital era, as this volume demonstrates, are much more immediate. There's still time, in other words, for another "community memory" project, not unlike the one that Lee Felsenstein and his colleagues launched in Berkeley over four decades ago, but this time casting a much wider net, and looking toward the future of online culture as well as at its past and present.[5]

Notes

1. Judy Malloy, "Expanding on 'What Is the Social in Social Media?'—A Conversation with Geert Lovink." [*Social Media Archeology and Poetics*].

2. Ibid.

3. Judith Donath, *The Social Machine, Designs for Living Online* (Cambridge, MA: MIT Press, 2014).

4. Judith Donath, "Slow Machines and Utopian Dreams." [*Social Media Archeology and Poetics*].

5. Gary O. Larson, "From Archaeology to Architecture: Building a Place for Noncommercial Culture Online." [*Social Media Archeology and Poetics*].

27 Expanding on "What Is the Social in Social Media?": A Conversation with Geert Lovink

Judy Malloy

In 1995, Dutch media theorist and Professor of New Media Geert Lovink and media artist and theorist Pit Schultz started the <*nettime*> mailing list. It expanded into many lists and languages and in the process demonstrated that English language and American-centric platforms do not have to be the lingua franca of the Internet.

Lovink's contemporary work with the Institute of Network Cultures (INC) at the Amsterdam University of Applied Sciences and INC's research networks, such as Unlike Us, has shaped a coalition that explores network architectures, the role of collective production, aesthetic tactics, and diverse, open information exchange. At the same time, he has reached out personally—for instance, recently teaching a master class on Critical Internet Culture in Indonesia.

With three questions and three answers, this conversation focuses on social media as viewed through the lens of Lovink's 2012 essay in *e-flux*—"What Is the Social in Social Media?"[1] The questions are:

What roles you envision for artists and writers in contemporary social media?

How can we create in a medium that combines text and image and interactivity?

Yes! Here are my data, show me yours. But why are we not only doing this but also enjoying it on Saturday night? And can you take this somewhat rhetorical question into the future with your personal vision? How do you envision a social media of the future?

The conversation began via email on March 2, 2015.

On March 17, 2015, it concluded with nine data objects that emerged from a list of Geert's works and projects, which I picked out and he altered and annotated.

Thus, appropriately for contemporary social media, our short conversation unexpectedly veers into an exchange of data, and each of the concluding data objects addresses in some way Geert Lovink's vision of the past, present, and future of social media.

Three Questions and Nine Data Objects

Judy Malloy (JM): A classic documentation of information art is Peter D'Agostino and Antonio Muntadas' the *Un/Necessary Image*—with papers including Hans Hacke's "Social Grease," General Idea's "Cocktail Boutique," Muntadas' "Selling the Future," and D'Agostino's "Invading the Information age."[2] The titles of the papers suffice to set an information art stage.

In "What Is the Social in Social Media?" you, a master of contemporary information art, begin with a barrage of quotations:

Headlines, 2012: "Next time you're hiring, forget personality tests, just check out the applicant's Facebook profile instead." —"Stephanie Watanabe spent nearly four hours Thursday night unfriending about 700 of her Facebook friends—and she isn't done yet."

Although in a sense you have already answered this question with your own information art-centered writing and your emphasis on collaborative process, there are other approaches. Can you expand on what roles you envision for artists and writers in contemporary social media?

Geert Lovink (GL): Interesting electronic art questions protocols. It disturbs and reverses motives. I am from The Netherlands, so I naturally think of Jodi, Peter Luining, Rosa Menkman, Constant Dullaart, Jan Robert Leegte, and others. Their rude radicalism, aimed to destroy the templates, is waking us up from the unconscious habits that come with the dominant mode of interactive consumerism and change our perspective by looking right through the surfaces, ignoring all those smooth interfaces—but without the boring PC condemnations of corporate America. It's punk, but not anti-aesthetics, and that's precisely what media critics—those who lack even the most basic form of imagination and seemingly have not read any philosophy or criticism for ages—need.

The 1990s net.art attitude hasn't changed in the age of social media. Destructive joy remains necessary and still does the job, even though the recipe hasn't changed over the past 20 years. Artworks that deal with the monopolistic Internet should be hilarious, over the top, subversive, annoying, boring—in short, everything one expects an artwork to be. Art cuts up, spreads out, stretches its material and pushes boundaries of normalcy right into the very outer corners of the hyper sphere. What's surprising here is the fact that "social media" has become a topic like any other. You could also have asked me about art and sailing, the nursing artist, the figure of the devil in contemporary arts—and that frightens me. Art and social media? Sigh. A few decades ago, we would have thought about it in a strategic manner, discussed the tools, and zoomed in on a simplistic element in order to bring out the techno sublime. No more. We'd need a next to impossible alien perspective to regain the wonder of a Facebook page.

Where is the Warhol of our time who declares Tinder an artwork? The aesthetics for us is in the shock of the historical element, the re-recognition of telnet, the messiness of MySpace, and the truth of VRML. If artists work on social media, we expect them to treat it no differently than Walmart. You wouldn't say IKEA, right? Precisely, because there is nothing Scandinavian about Facebook. We want to see artists struggling with social clay, but in the end, it's going to be bloody hard to produce something meaningful and sustainable with that material. It's sad for us that we no longer live in the conceptual 1970s. Whatever. Art. Before we know it, the art teacher throws the clay works of his pupils into the garbage bin. Today's hype circulation is damned fast. We need to be quick to salvage the historically significant bits from the bin before we move to next, post social media stage. So let's rush and assist Olia Lialina and Brewster Kahle in salvaging the situation and store the social media experience before it disappears for good and will only exist in the pages of Dave Eggers' *The Circle*.

JM: So, in "What Is the Social in Social Media?" with the images you salvaged from the bin—"Rand Corporation think tank employees brainstorming, 1958," for instance—you epitomize effective combinations of words and images, in such a way that the words speak what is difficult to image and the images speak what is difficult to put into words.

Contingently, images and sound are extraordinarily potent on the contemporary Internet, but in <nettime>, fiberculture and onward, your work and the work of your colleagues have demonstrated that words and word-based discussion continue to be important in contemporary social networks. <nettime>, for instance—first hosted by THE THING, now hosted by Ted Byfield and Felix Stalder on kein.org—continues to attract followers. Nevertheless, the visual aspects of screen-based, screen-framed information are inescapable in contemporary social media platforms.

To a certain extent, your own essay answers the question: How can we create in a medium that combines text, image, and interactivity? But how can we teach students to create in a difficult medium that so beautifully (and relentlessly) combines text, image, design, interactivity, and collaboration?

GL: Taken my historical baggage into consideration, there is no doubt: the web is a social sculpture. With this in mind, art materializes itself in the social, not in the material remnants of the action. The Internet is no different from participatory arts that Claire Bishop describes in her epic *Artificial Hells* (Verso Books, 2012). As Beuysian artwork it exists in the living connections between the users and their data. The role of teachers then is to bring together the development of a signature of the art student, through skills, and combine this (or, should we say, collide) with the world of discourse and debate. I am not saying theory because I do not believe that the mechanical ways in which theory has been instrumentalized in contemporary arts has been such an overwhelming success. It has never been a good idea to project fashionable thinkers onto an artwork in order to give a social standing. This was also the case with new media arts. We all know what happens if one applies Deleuze's idea of rhizome to the Internet and then produces a random info visualization in order to exemplify this deep insight. That didn't work in the past with Marx and doesn't work today with Rancière, which doesn't mean that art doesn't need an intellectual surrounding. I never thought that autonomy in the arts was the way to go. There is enough of that, here in Amsterdam, and it doesn't reflect the city. Art doesn't improve when it withdraws. It needs space, yes, and funding, and facilities, but that should not result in an insular culture. This is also what went wrong with new media arts, which, in particular, the United States thought it could hide and survive inside academia, and it was precisely there that it became so self-referential and, in the end, irrelevant.

JM: I might note before rushing into the next question that, at the same time, new media artists, as do any artists, need the nourishment, intellectual stimulation, and joy of working with students that are simply not available elsewhere, which does not mean that we should not be aware that there is a wider world. Indeed, you conclude "What Is the Social in Social Media?" in this way:

The Other as opportunity, channel, or obstacle? You choose. Never has it been so easy to "auto-quantify" one's personal surroundings. We follow our blog statistics and our Twitter mentions, check out friends of friends on Facebook, or go on eBay to purchase a few hundred "friends" who will then "like" our latest uploaded pictures and start a buzz about our latest outfit. Listen to how Dave Winer sees the future of news: "Start a river, aggregating the feeds of the bloggers you most admire, and the other news sources they read. Share your sources with your readers, understanding that almost no one is purely a source or purely a reader. Mix it all up. Create a soup of ideas and taste it frequently. Connect everyone that's important to you, as fast as you can, as automatically as possible, and put the pedal to the metal and take your foot off the brake." This is how programmers these days loosely glue everything together with code. Connect persons to data objects to persons. That's the social today.

Yes! Here are my data, show me yours. But why are we not only doing this but also enjoying it on Saturday night? Can you take this somewhat rhetorical question into the future with your personal vision? How do you envision a social media of the future?

GL: Before we start, I am sorry, but I have to remark that I hate academic papers; it is a dull format for people without ideas. Let's all try to stay away from it and refuse (anonymous) peer reviewing. One could defend the thesis that new media art was destroyed by "paperism" driven by *Leonardo*, *ISEA*, and so on. As you might know, I prefer the essay as form (reread Adorno and Sontag), the manifesto, the short newspaper articles Jean Baudrillard used to write for *Liberation*, the aphorism (but not the tweet as it is too rigid), the blog entry (I deeply admire Dave Winer!), the list thread (terrible!), the wild yet unfulfilled promises of the Internet forum, the 2,334 comments on the

Guardian site, and the 50,000 e-books a friend of mine recently gave me on a disc, which took 20 minutes to copy.

JM: I'm not inclined to defend your thesis. Nevertheless, I accept your essays and blogs with pleasure—observing that your words open the Epilogues to this book with the vigor and contention that at times distinguish social media.
Of course, you have—connecting "persons to data objects to persons"—already many times answered the question: "How do you envision a social media of the future?"

So, I picked out some texts with which to conclude this all too short discussion, and in response, you altered and annotated them. Readers in search of Geert Lovink's vision of the future of social media are referred to the nine data objects listed below.

1. Adilkno, *Media Archive* (Uitgeverij Ravijn, 1992/Autonomedia, 1998)

This is a collection of so-called Unidentified Theoretical Objects, media concepts, if you like, written over a decade, documenting the opening up of the collective imagination in the age of cyberspace, not in the euphoric *Wired*-style but neither with the dark attitude of Morozov. The theory crystals combined postmodern version of negative dialectics with the gayness of speculative creativity that defined the early 1990s. Its web archive can be found here: http://thing.desk.nl/bilwet/adilkno/TheMediaArchive/.

2. Geert Lovink, *Dark Fiber* (MIT Press, 2002)

The first volume in my book series on critical Internet culture that covers the roaring 1990s until the dot.com crash. Maybe it is too early to be nostalgic about that period, and the way of writing. This was, no doubt, the golden age of net criticism.

3. Geert Lovink, *The Moderation Question: Nettime and the Boundaries of Mailinglist Culture*, in: Geert Lovink, *Dark Fiber*

That's my version of the history of the mailinglist for the culture and politics of the Internet, associated with the term "net criticism," founded by me and Pit Schultz in 1995 and still going strong, thanks to Ted and Felix and others who run lists like the Dutch nettime-nl.

4. Geert Lovink, *Uncanny Networks: Dialogues with the Virtual Intelligentsia* (Cambridge, MA: MIT Press, 2004)

Late 1987, having overcome my identity crisis, reconciling with my destiny as a media theorist, I decided to start an interview series on our free radio (part of the squatters movement). I continued this in the 1990s on national radio and then moved it to nettime, where I am still posting the occasional interview. The book is a great resource of 1990s critical cyberculture and shows how theory, contemporary arts, and new media indeed can interfere.

5. Geert Lovink, co-founder of *fiberculture* (http://www.fibreculture.org)

Thanks for reminding me of this wonderful community in Australia I initiated in 2001 in Sydney with David Teh, who's now living in Singapore and I have the privilege to meet every now and then. Even though the list died and the community is rather weak, the Fiberculture journal continued and thrives, thanks to Andrew Murphy and others!

6. Geert Lovink, founder of Institute of Network Cultures (INC) (http://networkcultures.org/)

That's the reason we moved from Brisbane to Amsterdam in 2004. INC is still going strong; 11 years later, although small, we are only two to three people. I am very proud of the staff, the interns, and Sabine Niederer, who was there in the early years and is now my boss.

7. Geert Lovink, *Netcritique Blog* (http://networkcultures.org/geert/)

That's my blog. I post about two to three pieces a month. Not a lot, but regularly. I love reading blogs, and theorizing it. As the writings of Dave Winer show, there is a lot still unexplored in the blog writing formats and how it interacts with other objects on the net.

8. Geert Lovink, *Networks Without a Cause: A Critique of Social Media* (Polity Press, 2012)

My latest book in a series, which started with *Dark Fiber*. Right now I am wrapping up part 5, also for Polity. The book form inspires me to reflect on the flow of ideas and events. It condenses data and observations into concepts

9. Geert Lovink and Miriam Rasch, eds., *Unlike Us Reader: Social Media Monopolies and Their Alternatives* (Amsterdam: Institute of Network Cultures, 2013)

This publication is the result of the Unlike Us network, which we started mid-2011. It still exists, although not very active. Building social media alternatives proved difficult to do quickly, and the scene changed in June 2013 with the Edward Snowden revelations. The critique of dominant platforms remains valid.

Notes

1. Geert Lovink, "What Is the Social in Social Media?" *e-flux* (2012). Available at http://www.e-flux.com/journal/what-is-the-social-in-social-media/.

2. Peter D'Agostino and Antonio Muntadas, eds., *The Un/Necessary Image* (New York: Tanam Press, 1982).

28 Epilogue: Slow Machines and Utopian Dreams

Judith Donath

There is a special time in the development of a major new technology, when the technology is yet crude and barely developed, but exists and is real, touchable, usable, when the promises of that technology are still shiny and bright, when the future we see it bringing is a reflection of grand ideals and prayed-for salvation, when the technology is still seen as savior. It has not been used enough to be tarnished, we haven't seen the way it will be refitted for criminal use, nor the unexpected ways it will kill off a beloved culture or foul the ocean or corrupt the young.

We saw this with cars, for example, which were extolled for providing freedom and the ability to explore; this was before automotive exhaust had blackened the air and asphalt paved over the landscape, and we sat for hours in crawling traffic jams, commuting to and from our car-enabled suburban sprawls.

We saw this with plastic. Now synonymous with fake or cheap, it was once an awe-inspiring material. In the 1930s, when plastics were new, designers, fascinated by the novel ways they could be formed, crafted beautifully shaped bowls, bracelets, and sculptures. Advocates of this wonder material envisaged a future in which cheap and abundant plastic objects "would eliminate the material scarcity that had fostered inequality and war throughout history."[1] Today we see the health risk of poisonous chemicals leaching from plastic bottles, birds mortally entangled in plastic six-pack holders, oceans (and fish, whales, and on up through the food chain) clogged with tons of plastic particles: bright, colorful, indestructible.

And we saw this with social media. The early days saw tremendous optimism about the transformations that connecting people via networked computers would bring.

It's the early days of social computing—the 1970s, the 1980s, the pre-web 1990s. The promise is bright. Connectivity will deepen empathy and understanding among currently divided people; text interaction will favor rationality and elevate discourse. It is not yet realized because the machines are too primitive, the network too slow, the users too few. But it is coming. The chapters in this book are about the experience of being there: what each participant believed in and suffered through in the hopes of making his or her vision for the medium real. We are reading these recollections in a different world, where a child's cell phone is vastly more powerful than the supercomputers of those days, when the majority of the world's population is online—and where the Utopian dreams have crash-landed into a far darker landscape of hackers, scammers, viruses, omniscient government surveillance, and mind-bending advertising. What happened? How did we get here? What can we resurrect from the ruins of defunct MUDs and Usenet archives and the memories of the camaraderie born in the struggle to make a program run, a machine to connect to the net?

Multiple Nostalgias

Several strands of nostalgia run through this book.

One is the nostalgia for being an early explorer of the new medium. "I have missed the sense of being in on the ground floor of something exciting, innovative, with the potential to alter culture as part of a group"[2] (Fred Truck in Malloy). It's nostalgia for being a pioneering artist, a homesteader of strange new territories, paving the way for the later citizens, the ordinary users and commercial clients.

Novel technology attracted experimental artists and change-seeking activists, people drawn to making new tools and media. "We were exploring new frontiers— new dimensions of writing never before possible"[3] (Deena Larsen). Because there was relatively little existing infrastructure, many early computer users were, by preference or necessity, also programmers, builders of the spaces they wished to inhabit. Although they had to deal with the challenges of primitive technologies, their reward was an extraordinary ease of invention: in the early days, one could create something novel and revolutionary just by making almost anything.

Online culture, too, was nascent, its habits and customs in the process of formation. Virtual friends with fictional identities and elusive, disruptive discussion trolls were just two of the new entities for whom mores and expectations had to be developed.[4] "We were constantly struggling to establish a balance between the right and need to express yourself and how to behave in a virtual world of mostly text"[5] (Stacy Horn).

Despite the decidedly technological nature of computers and networks, this nostalgia has a pastoral, William Morris-esque air; a longing for the days of tinkering and craftsmanship, when individuals could still access the workings of the machine.[6] It is a longing for a lost village—albeit a global, electronic one—in which independent workers happily and creatively developed their skills.

As the medium became more popular, the percentage of the population that was drawn to experimentation decreased. The mainstream audience sought ease of use, not radical forms. Site-building went from hobbyist to industrial scale, as users' expectations became more sophisticated—and as the requirement grew for robust implementations that would be safe from viruses and hackers.

The inherently ephemeral nature of today's technology also triggers nostalgia. Platforms and devices are continuously reinvented, and thus artworks and applications are inevitably temporary, unless considerable effort is put into keeping them up to date. This state of being in continuous flux affects communities as well as interfaces. New generations of machines are inhospitable to projects built for their ancestors. "[W]e need to start thinking of computer related stories and art in much the same way we classify 'happenings' or 'performance art' in that there's an acknowledged temporality"[7] (A. C. Chapman in Larsen).

Continuous flux keeps everyone in the state of being a beginner. A painter can work with the same tools—canvas, paint, brushes—throughout a lifetime. The media artist cannot, for last year's displays look quickly dated or cease to function; if she insists on staying with older tools, her work inexorably becomes retro and nostalgic. The ephemerality of platforms and devices means that the focus of anyone who works with newly invented tools and materials is thus always, at least in part, on the technology itself.

There is a paradox, too, to this ephemerality: much that we had thought of as impermanent—our casual conversations, the record of our purchases and website perusals—turns out to be, in the online world, all too permanent. The indiscreet jokes of a 15-year-old haunt him again as a 25-year-old job seeker, and all of us are accumulating immense shadow data portraits, records of our daily activity, although with little knowledge of what comprises them or how they will be used and by whom.

Another strand of nostalgia is longing for lost optimism, for the hopeful time when people believed that computing would change humanity for the better. It is nostalgia for the time when it was possible to believe that the global conversations enabled by computer networks would bring about greater understanding among disparate people: "My idea was that by engaging diverse communities in conversation, we could shape online communications infrastructure in a way that would facilitate power sharing and cultural exchange and that would resolve and heal our legacies of cultural oppression"[8] (Anna Couey).

Many of the pioneers of social media felt that they were seeing the first glimpses of a new sort of society, radically remade through computer-mediated communicating. "What made ACEN so exhilarating was that it created a new kind of space where people were able to set aside their usual selves and experiment with each other"[9] (Anna Couey in Malloy). Online conversations would flatten hierarchies, giving a voice to the hitherto powerless. New media promised to create a world where identity would be freed from the harmful stereotypes of race and gender.

Social media was born in an idealistic era, and this optimism was a factor of the times as well as the technology. Tracing the impetus for creating Community Memory to the 1960s Free Speech movement at Berkeley, Lee Felsenstein remembers, "I could see that the renaissance sought by the counterculture could be brought to immanence through a sensitive application of digital technology within a social context."[10]

Yet even in 1973, as Felsenstein was writing, the counterculture's idealism was already fraying—already, itself, the subject of nostalgia. Two years earlier, Hunter Thompson had written, "San Francisco in the middle sixties was a very special time and place to be a part of ... No explanation, no mix of words or music or memories can touch that sense of knowing that you were there and alive in that corner of time and the world."[11]

Idealism, as embodied in any particular movement or technology, is short-lived, inevitably brought down to earth by messy realities, distorted and bloated by opportunists. Yet the idealistic impetus survives and finds a new cause and fresh hope. "Personal computers and the PC industry were created by young iconoclasts who had seen the LSD revolution fizzle, the political revolution fail. Computers for the people was the latest battle in the same campaign"[12] (Howard Rheingold). The nostalgia for past optimism fuels the next revolution.

The Power of the Primitive

The world that conjures up such warm nostalgic feelings was, technologically, quite primitive. The computers were tiny—today's mobile phones are more powerful than a supercomputer of the 1970s and early 1980s. Networks were excruciatingly slow. In the mid-1980s, I made educational games for personal computers, and we labored to save a bit here, a byte there, trimming a row of pixels or shortening the text of an error message to ensure the game would fit into the computer's memory. "These were the days when you got online by dialing up using your phone line, and they were very noisy back then. It didn't matter for a voice phone call, but it was hell for data transmission"[13] (Stacy Horn).

Yet while the early days of social media were a struggle technologically, the limitations were also the source of community, creativity, and idealism.

Primitive computers and software fostered community: the users of these systems needed to help each other get things to work. Working at the forefront of a technology entails working with things that are novel, untested, and buggy. "A lot of that early fraternization was necessitated by the confusing nature of the WELL's software"[14] (Howard Rheingold).

The people working at that forefront are trying to do things that have not been done before, putting together elements that were not designed to fit together. The knowledge of how to do this is often not in any documentation—but maybe someone else has tried it and has some insight about the solution. "[Online bulletin boards] provided a way for owners of modest personal computers to enter a world, which hitherto had been accessible only to those at the top military bases or university labs. They helped novice computer users learn how to manage the intricate details of digital communication. And they created a genuine sense of community among the users"[15] (Paul Ceruzzi).

Establishing yourself as a constructive participant was key to getting help from the online community. You had to contribute and assist others—or, at least, ask a good question, one that showed you had made the effort to read what others had said already and that had not already been answered in earlier discussions.

My introduction to online social interaction came in the early 1980s, when I was taking computer science classes at college, and the Usenet newsgroups were an invaluable resource for answers to all sorts of programming problems. The advent of search has changed this. Today, when faced with a technical difficulty—or, indeed, any question at all—my first thought is to consult Google rather than seek answers from other people.

The early Internet was a sweet spot for community. The network made it possible to connect with distant people, strangers with whom you shared common interests; although you never met in person, you were able to develop a virtual relationship. It was still a sufficiently small and selective group for cooperation to be the rule. This was before the web, before Google came to "organize the world's information."[16] The communities were still self-organizing, sharing information from one individual to another; algorithms had not yet supplanted personal ties and interest communities.

The limitations of primitive technology shaped artistic practice. Most obviously, the early displays were text only. Words were central, creating what Dene Grigar called a "bookish"[17] environment. But it was a garish flickering cathode-ray book, not the serene pages of typographic flow. The Minitel, for example, "was a crude computer terminal with a 9 inch screen of 25 lines by 40 columns. An alpha mosaic coding allowed the use of letters, numbers, punctuation signs and graphic or mosaic characters, all of which could also be flickering or in video inversion"[18] (Annick Bureaud). Although the content of different pieces ranged widely, such displays set the visual aesthetic of early computer artwork.

The revolutionary aspect of the new technology was that works could be programmatic and participatory. A piece could create itself, with its rules and instructions part of the artwork. Communication and audience input could be an integral part of the work; the line between artist and audience could blur or disappear entirely. Annick Bureaud describes the digital communication artist as a "context provider" rather than the traditional "content provider."[19] J. R. Carpenter notes, "the Internet introduced a hitherto unprecedented degree of atemporality, asynchronicity, and multiplicity to long-distance communication which radically and irrevocably altered a wide range of writing practices."[20] Programmable and networked were truly new elements of artistic practice, and the paucity of other things—the lack of images, the simplicity of the text display—meant that the creative endeavors could focus on experimenting with these novel aspects of the medium.

This spirit of experimentation infused the design of practical interfaces as well as artistic pieces. David Woolley's account of the design of PLATO, early in the history of social computing, shows how the now taken-for-granted form of even such basic interfaces as texting and email evolved through quickly prototyped designs. "Often a suggestion would strike me immediately as great idea, and if it was not too difficult, it might be implemented and released within a day or two."[21] The community would try the designs, rejecting some, adopting others.

The limitations could, of course, be frustrating. With the possibility of interaction and an intelligent machine, one could imagine all sorts of amazing new experiences to create—but then the limits of the actual machine and network at hand presented a rude awakening. The machines were slow, the networks slower. The struggle to make things work, to push the limit of the technology, could become the focus, distracting from other expression in the work. Hank Bull notes, "It was such an effort to get all this gear working properly, and so exciting when it did, that the matter of content became an afterthought."[22]

The type of constraint matters. Text is highly malleable and easy to create, and thus the constraint of a text-only interface fostered creativity. With no pictures, it encouraged identity play: you came to life through your own self-description, and fanciful self-presentations were as easy to create as truthful ones. Other constraints, such as slow machines or buggy software, were less conducive to innovation.

Subsequent technologies, although more powerful, have in some ways been less fertile for experimental creativity. The web made publishing text and images easy, but it made personal interaction harder. Especially in its early days, input was awkward and anonymous, limited to web forms and clickable images; there was little support for interactive and participatory art. One could still create non-web applications, but over time browsing the web became what people did online, and experimental outside applications became scarce and suspect.

Finally, the limited capabilities of early social computing gave people the freedom to imagine the brilliant future the technology would bring, once it was truly working. "[D]espite the lack of color, movement and sound, the experience we had in and with the MOO environment imbued us with a sense of the unlimited possibilities of human expression"[23] (Dene Grigar). In the meantime, disappointing experiences could be blamed on immature tools; a meager audience on the small user population.

Nascent technologies provide a blank canvas for imaginings. With social media, in the early days, not only were the interfaces still in development, but the population was small and select. You could imagine that the cooperation and creativity you saw with your small group would be magnified as hundreds, thousands, and millions more joined—that the technology would be capable of moderating differences, unchaining imaginations, and transforming society.

Epilogue

New technologies inspire both utopia and dystopian visions; they will save the world, they will destroy civilization as we know it. It is the utopians, however, who are drawn to working with the new machines and novel interfaces, to making their wondrously imagined future into something real. Their inclination is to see the brightest side.

The View from the Present

Much has changed. In the 1980s, most people were dubious if not outright dismissive of the prediction that computers would be commonplace and we would be using them to communicate—to keep up with friends, find new communities. Today, social media is ubiquitous—so much so that popular books and articles warn that virtual interactions have taken over to the point that we are in danger of forgetting how to fraternize face to face (Turkle 2012).[24]

Some of the magical excitement of the early days has disappeared simply by becoming everyday enchantment. Yet if we step back and look afresh at the ease with which we stay in touch with hundreds or thousands of people, or at how an idea sweeps a nation, spread from one personal broadcast to another, we can see that these now mundane experiences are the reification of early imaginings.

Google and other search engines have given us unprecedented access to vast troves of information—once a Utopian dream—but they have also weakened our need for community. Before networked computers, conversations were limited to proximate groups of manageable size. Social media erased the barrier of distance and greatly increased the possible scale. But as search has grown more sophisticated, the value of personal connections has decreased, and we rely instead on algorithms to harvest information. Today, our first instinct is to look to Google for an answer; although we're getting insights, advice, and tech tips from other people, the interaction is indirect. There is less relationship building. New conversation sites, such as StackExchange, automate much of the social process. The computer mediates reputation: it tallies how well someone has contributed and gives them a rank and varying privileges; other users no longer need to invest this assessment time, no need to discuss among your fellow participants. It is more efficient, but the community aspect is significantly diminished.

Spam and viruses were unheard of in the early days. The inner workings of the computers were open to programmers, and writing communication software was fairly straightforward. Today, computers are barricaded behind virtual walls of security programs, firewalls, and virus detectors. Code that allows users to interact with a machine or other people must carefully limit what they can do—an unguarded input space will allow one of the many hackers who bombard any networked computer daily to take over the machine. "[In response to hackers we] hardened the site in innumerable ways by revising table relationships, constraining permissions, turning off error reporting, and restricting access routes for editors, with the overall result that both the quantity of editor-users and the quality of their editing experience were severely curtailed"[25] (Alan Liu).

The early optimists had a rosy view of human nature. But the slippery nature of online identity emboldened many people to unleash their darkest side. Vicious arguments and harassment are common online. "Looking back, what stands out for me is how utterly unprepared I was for the role I had to assume. At the time people were still talking about the Internet in utopian terms, as the digital equivalent of a perfect world over the rainbow, forgetting that there were wicked witches and scary flying monkeys in the wonderful world of Oz. Almost immediately I was faced with the same problems that continue to plague the Internet today: trolls, harassment, love affairs gone bad, meanness, pettiness, basically all the problems and evils of society"[26] (Stacy Horn).

The feeling today is very different. Although there are certainly optimists—Silicon Valley's many adherents to The Singularity come to mind[27] (Vance 2010)—it is a far darker time for prognostication.

Online, we have massive, dense connectivity. Society has deeply embraced online communication. But there is also the constant specter of surveillance, of being tracked by the government to save us from the looming threat of hackers, cybercriminals, and terrorists, and by corporations, eager to infiltrate our minds and desires ever more subtly and insistently, selling messages of consumer longing.

Against this background, the nostalgia for a time of Utopian optimism, for when it was believed that computer networks would foster deep interpersonal empathy and international peace, is sharp and poignant.

How we got here is a complicated story. Some changes were technological. The web made publishing traditional text and pictures easy, but it was a big step back for social interaction. Some were economic. A growing online population and the opening of the Internet to commerce made advertising a profitable business model, whereas more complex sites became expensive to build and run. For social media, one result is a preponderance of sites in which the users are not customers but rather products, to be tracked, analyzed, and marketed to. Another result is the loss of creative autonomy: "Indeed, one of the important things we have lost in the shift from user-built spaces like MOOs to commercial platforms like FaceBook and Twitter is the power to profoundly change what happens through user-initiated programming"[28] (Antoinette LaFarge). There were also external changes. 9/11 put a final end to the dot.com enthusiasm and brought about an era of hair-trigger anxiety, of vast and shadowy security—and rampant surveillance.

Weaving Ideals into Culture

In the process, we have learned quite a lot. We have learned that simply connecting people does not bring about peace. "Cyberspace does not have the power to make us anything other than what we already are. Information doesn't necessarily lead to understanding or change. It is a revealing, not a transforming, medium."[29] (Stacy Horn). We have learned that effortless creation brings about lowest common denominator entertainment.

We have also seen extraordinary developments. Early visions of how people could cooperatively assemble great stores of knowledge have been fulfilled (and surpassed) by Wikipedia. Sites such as Instagram have brought about a vibrant, novel, and social way to exhibit artistic creations. We live in a truly connected world, and we are only starting to see the radical changes this will bring.

We read about the sites and projects in this book with the perspective of decades. What was the turning point when being named ceased being something you did to maintain accountability within a community and started being something that left you exposed and vulnerable to a vast data-collecting corporate/governmental apparatus? It's useful to note what happens to the Utopian dreams that welcome a new technology. Did the ideals get side-tracked because of unrealistic beliefs about the technology itself—or is there something deep in the surrounding culture that waylaid them, redirected the inventions that were supposed to save the world into new ways to exploit humans or spoil nature? Perhaps there is a cautionary tale here for assessing other techno-optimisms. Today, people claim that direct brain stimulation will perfect human nature, artificial intelligence will bring universal prosperity, or genetic engineering will cure all disease. What could possibly go wrong?

The world of dial-up modems, floppy disks, and ASCII bulletin board systems seems long ago. But the ideals of that time, despite their naiveté, indeed because of it, are valuable. Untainted by cynicism or corrupted by practicalities, they remind us of what the social net should be; they remind us of the direction to head in, even if we will not quite get there.

The nostalgia for individual craft, for example, for the online world before corporations and consumerism took over, is a call for individual creativity and social freedom. "What I am missing is the desire for autonomy and self-determination that propelled us to do what we did 20 years ago. Where is the courage for criticism and change? Where is the will to escape or defy the control and manipulation of the data suckers, be they state agencies or commercial entities, or increasingly all of them together? Why do people voluntarily allow them access into the deepest corners of their private lives?"[30] (Wolfgang Staehle in Gerber).

There are, in fact, many people who do express this desire and whose work is very much about self-determination.[31] There are open source contributors, Tor developers, and underground digital artists. There are the people who build secure tools for activists in danger, and the individuals, such as Brewster Kahle, who decided to archive the Internet and make this history freely available, or Christopher Poole (<moot>), who decided to build a site that functioned like a fun-house mirror inversion/perversion of everything sensible and tasteful (4chan). We need the beacons that point toward Utopia, even if it is unattainable, a non-place. Nostalgia is a way of inculcating ideals into a culture, by making them into myths and origin stories. "We weave stories, creating the past as much as remembering it"[32] (Fischer 2010). We create the past that we want to live up to.

Notes

1. Jeffrey L. Meikle, *American Plastic: A Cultural History* (New Brunswick, NJ: Rutgers University Press, 1995).

2. Fred Truck, in Judy Malloy, "Art Com Electronic Network: a Conversation with Fred Truck and Anna Couey" (*Social Media Archeology and Poetics*).

3. Deena Larsen, "Electronic Literature Organization Chats on LinguaMOO" (*Social Media Archeology and Poetics*).

4. Judith Donath, "Identity and Deception in the Virtual Community," in *Communities in Cyberspace*, eds. Marc Smith and Peter Kollock (New York: Routledge, 1998), 29–59.

5. Stacy Horn, "EchoNYC" (*Social Media Archeology and Poetics*).

6. Richard Coyne, *Technoromanticism: Digital Narrative, Holism, and the Romance of the Real* (Cambridge, MA: MIT Press, 2001).

7. A.C. Chapman in Larsen, 2016.

8. Anna Couey, "Cultures in Cyberspace: Communications System Design as Social Sculpture" (*Social Media Archeology and Poetics*).

9. Anna Couey, in Judy Malloy, "Art Com Electronic Network: a Conversation with Fred Truck and Anna Couey" (*Social Media Archeology and Poetics*).

10. Lee Felsenstein, "Community Memory—The First Public Access Social Media System" (*Social Media Archeology and Poetics*).

11. Hunter Thompson, *Fear and Loathing in Las Vegas: A Savage Journey to the Heart of the American Dream* (New York: Random House, 1971).

12. Howard Rheingold, "Daily Life in Cyberspace" (*Social Media Archeology and Poetics*).

13. Horn, 2016.

14. Rheingold, 2016.

15. Paul Ceruzzi, "The Personal Computer and Social Media" (*Social Media Archeology and Poetics*).

16. Google's Mission Statement is available at http://www.google.com/about/company/.

17. Dene Grigar, "MOOS and Participatory Media" (*Social Media Archeology and Poetics*).

18. Annick Bureaud, "Art and Minitel in France in the '80s" (*Social Media Archeology and Poetics*).

19. Ibid.

20. J. R. Carpenter, "TrAce Online Writing Centre Nottingham Trent University" (*Social Media Archeology and Poetics*).

21. David Woolley, "PLATO: The Emergence of Online Community" (*Social Media Archeology and Poetics*).

22. Hank Bull, "DictatiOn: A Canadian Perspective on the History of Telematic Art (*Social Media Archeology and Poetics*).

23. Grigar, 2016.

24. Sherry Turkle, *Alone Together: Why We Expect More from Technology and Less from Each Other* (New York: Basic Books, 2012).

25. Alan Liu, "Hacking the Voice of the Shuttle: The Growth and Death of a Boundary Object" (*Social Media Archeology and Poetics*).

26. Horn, 2016.

27. Ashlee Vance, "Merely human? That's so yesterday," *New York Times*, June 12, 2010.

28. Antoinette LaFarge, "Pseudo Space: Experiments with Avatarism and Telematic Performance in Social Media" (*Social Media Archeology and Poetics*).

29. Horn, 2016.

30. Wolfgang Staehle in Susanne Gerber, "Crossing Over of Art History and Media History in the Times of the Early Internet with Special Regards to THE THING NYC" (*Social Media Archeology and Poetics*).

31. Judith Donath, *The Social Machine: Designs for Living Online* (Cambridge, MA: MIT Press, 2014).

32. Claude S. Fischer, *Made in America: A Social History of American Culture and Character* (Chicago, IL: University of Chicago Press, 2010), 5.

29 From Archaeology to Architecture: Building a Place for Noncommercial Culture Online

Gary O. Larson

In the context of social media and other forms of networked communications, the terms "archaeology" and "architecture" might not immediately come to mind—the one involving the excavation of ancient sites, the other the design of new physical structures. Yet in light of the petabytes of data that have piled up online over the past 50 years, *excavation* begins to make much more sense—except when the data we're looking for have disappeared altogether. That, sad to say, is often the case with the projects discussed in this anthology—participatory experiments like Berkeley's Community Memory, Seattle's IN.S.OMNIA, and the globe-spanning Electronic Café—each of them rooted in the technology of their times, whose virtual canvas and clay, digital stage and recital hall, were eventually rendered obsolete in the face of newer platforms and faster processors. In such instances, archeology becomes more an act of recollection—of reimagining, even—in an effort to recapture the past. This anthology is full of such acts, covering a wide expanse of creative endeavors that for the most part have long since been covered over by the silicon of time. These losses, moreover, are precisely why *architecture* becomes so important—the need, that is, to conceive and build new online structures for the preservation of old and the promotion of new artistic expression.

Not all of the projects chronicled in this volume, to be sure, were destined or even designed for a permanent place in our culture. But they all made vital contributions, and a few of them, most notably Arts Wire and THE THING, set standards that should continue to inspire us today. Arts Wire (so thoughtfully recalled by the editor of this anthology) was social media with a distinct purpose: to provide a platform for the nonprofit arts community—artists, arts administrators, and arts patrons alike—to share information and ideas online. THE THING (which continues to maintain a presence on the web, although no longer operating at the breakneck pace that it managed in the 1990s) is less easy to describe. But even a cursory glance at the detailed timeline that Susanne Gerber includes with her chapter and interview with THE THING founder Wolfgang Staehle is a sufficient reminder of the staggering array of projects that it created, curated, and stimulated throughout the 1990s. The shopworn "multimedia" tag doesn't begin to capture the vast range of activities carried out under THE THING's auspices, any more than its dial-up phone number or its subsequent URL captures the art network's essence. "[T]he much larger space which connected and encompassed everything was of course the virtual space of THE THING," Staehle told Gerber. "The network of THE THING was worldwide and very much alive. ... There was this contagious enthusiasm for what was suddenly possible, combined with a sense of play and a curiosity for what would happen next." A lot of things happened next, as it turned out, but Arts Wire and THE THING, like so many other early online arts experiments chronicled here and elsewhere, have left palpable voids in the online landscape.

In the real world, admittedly, historic preservationists aren't typically mollified by the replacement of lost architectural treasures with even the most exquisitely designed contemporary buildings. Nor is there any way to recapture, in today's commercially driven online environment, the energy and excitement that surrounded the cultural experiments covered elsewhere in this volume. As Hank Bull recalls in his chapter on early telematics art, "Small, independent groups often worked outside official structures, using consumer level equipment and home-made electronics. Certainly for the events I was involved in, most of the effort was spent just getting the connection to work. To hear a voice, read a message, or see a face on the screen, beamed in from afar, seemed like some kind of miracle." Dial-up doldrums and command-line confusion notwithstanding, there *was* something genuinely exciting about forging connections with art and artists in the "prehistoric" era before the advent of the World Wide Web, a sense of discovery and wonder that today's broadband environment cannot match. That is not to say we shouldn't try, however, both to recapture that spirit and create spaces online in which such experimentation can continue to be celebrated and preserved.

From ARPANET to AOL

This would not be the first time, after all, that we've attempted to carve out space for noncommercial expression within the larger media landscape.[1] The Federal Communication Commission's (FCC's) 1945 set-aside of 20 percent of the FM spectrum (between 88 and 92 MHz) for "noncommercial educational stations" comes immediately to mind, as does that agency's 1952 allocation of 242 television channels for similar purposes. Years later, in a likeminded fashion, the Cable Communications Policy Act of 1984 required cable operators both to accommodate and support public-, educational-, and government-access channels within their systems. Similarly, in the Cable Television Consumer Protection and Competition Act of 1992, direct-broadcast satellite systems were required to allocate 4 to 7 percent of their capacity for "noncommercial programming of an educational or informational nature." So the public interest, even in the midst of private enterprise, still matters.

In all of these instances, it should be noted, the set-asides were enacted only *after* commercial broadcasters had demonstrated their inability or unwillingness to deliver anything more than the ratings-driven fare that FCC Chairman Newton Minow aptly described as a "vast wasteland" in 1961.[2] With the Internet, the process was quite the reverse. That system began humbly enough in 1969 as a wholly noncommercial service, with the launching of the Department of Defense-funded Advanced Research Projects Agency Network (ARPANET). Initially connecting three universities in California and one in Utah, the system grew to include European connections in 1973 and reached 200 sites by 1981. ARPANET was joined that year by another government-sponsored network, the Computer Science Network (CSNET), to extend networking benefits to academic and research institutions in the United States and abroad that were not directly connected to ARPANET.[3] In the mid-1980s, the National Science Foundation's (NSF's) NSFNET assumed the role of national backbone for an "Internet" of worldwide interconnected networks, which by 1992 exceeded a million nodes. That same year, however, NSF began phasing out federal support for the Internet backbone and encouraged commercial entities to set up their own private backbones for access to the Internet.

The decision to privatize the Internet was not so much a capitulation to market forces as it was a foregone conclusion.[4] Because of the NSF's "acceptable use" policy—which, reflecting the military and academic roots of NSFNET, declared that "[u]se for commercial activities by for-profit institutions is generally not acceptable"—companies had already begun to develop alternative backbone services. By the 1990s, the Internet had expanded well beyond universities and research sites to include corporate and individual users connecting through commercial ISPs and consumer online services.[5] Once the National Center for Supercomputing Applications (NCSA) unleashed the Mosaic graphical browser on the World Wide Web in early 1993, web traffic experienced a 10,000-fold increase over the next year.[6] Thus, by the time federal support for the NSFNET backbone officially ended on April 30, 1995, the commercial destiny of the Internet, by then doubling in size every year, had already been set.[7]

Throughout the early history of the public Internet, as Paul Ceruzzi explains in his chapter on early PCs and bulletin boards, smaller commercial networks had also been established—a constellation of discrete, private systems, in effect, that wouldn't fully join the "network of networks" until the mid-1990s. The first dial-up computer bulletin board system (BBS) was launched in 1978 (when a Hayes 300-baud sold for $280—more than $1,000 today, adjusted for inflation). Both The Well and EchoNYC got their start as BBSs, and by 1994, there were more than 100,000 BBSs in operation worldwide, many of them—like IN.S.OMNIA, System X, and THE THING discussed earlier in this volume—home to artists and arts projects.[8] Two early consumer online portals, The Source and CompuServe, also began in 1978, but user fees ($7.80 per hour during the day and $6.00 an hour at night for CompuServe in 1984) delayed their widespread adoption.

America Online (AOL), meanwhile, blazed its own trail across the online skies, from Quantum Link (Q-Link), an online service for Commodore personal computers in 1985, to the emergence of America Online four years later and its spectacular growth thereafter, reaching 1 million subscribers in 1994, 5 million in 1996, 10 million in 1997, and 25 million in 2000—the same year it acquired Time Warner for $162 billion. Despite its decidedly mass-market tendencies, AOL was not entirely bereft of cultural content. Featuring 19 content "channels" that attempted to simplify the Internet for its users, AOL included a handful of keyword shortcuts to the arts (many of which offered links to websites, Usenet newsgroups, and listservs as well as to AOL's own conferences and chat rooms). In AOL's Entertainment Channel, for example, there were keywords such as Critics (where users could find reviews of various disciplines, including theater reviews from the *New York Times*), Culturefinder (resources for dance, classical music, and theater), and Playbill (news and information on Broadway shows and touring productions).[9] But while AOL, which in its heyday kept the U.S. Postal Service busy with millions of promotional floppy disks and CDs, succeeded in bringing the Internet to the masses, the dial-up service eventually fell victim to the dot.com bust of the early 2000s, as well as to increased competition from cable and telephone company broadband providers, who decimated AOL's dial-up service by offering much faster connections to the World Wide Web.[10]

As this book makes clear, artists and cultural organizations managed to insinuate themselves into many of these various early networks, academic or mainstream. In the process, they launched creative projects in both the United States (such as the Art Com Electronic Network, Arts Wire, and LinguaMOO) and internationally (e.g., the telematics projects in Canada, networks in Australia, TrAce Online Writing Centre in the UK, Minitel-based experiments in France, and outposts of THE THING in Germany, Austria, England, Switzerland, and Sweden). As Hank Bull explains in his chapter, the artists behind these projects ran the gamut of possible affiliations, from academic (e.g., Bruce Breland establishing his Digital Art Exchange at Carnegie Mellon University) to corporate (availing themselves of I. P. Sharp's international network), and from governmental (gaining access to NASA satellites on more than one occasion) to institutional ("Conference on the Artist's Use of Telecommunications," organized by Bill Bartlett and La Mamelle at the San Francisco Museum of Modern Art). The community networks chronicled here, as well as scores of others, proved especially hospitable to the arts and culture.

On a smaller scale, but collectively no less important to the online culture, artists spawned any number of USENET groups, including the seminal alt.hypertext group discussed in this volume, as well as dozens of others—from alt.architecture to alt.stagecraft in the alt (alternative) hierarchy, and from rec.arts.books to rec.video.

production in the rec (recreational) hierarchy. Equally important to those with specialized cultural interests were the mailing lists started by artists and scholars in the pre-web years, such as the Humanist Listserv examined in Julianne Nyhan's chapter, as well as countless others, from the Accordion list to the Wearable Art Clothing list. Phoebe Sengers, currently an information science professor at Cornell, who started the accordion listserv while an undergraduate at Carnegie Mellon University, captured the spirit of the times when she described the purpose of her mailing list in 1995: "The general philosophy will be that we can all learn a great deal if we each share what we know with the other members of the e-mail list."[11] Such was the gift economy, part of the DNA of the Internet, that informed many of the early art projects online. But just as in the real-world communities in which artists live and work, neither their livelihood nor that of the online projects they initiated was guaranteed. The need for a more coherent organizational scheme, and a more reliable support structure, was apparent, then as now.

The Information Superhighway

The first opportunity for a more consistent and cohesive approach to online culture came in the early 1990s during the policy debates surrounding the proposed National Information Infrastructure (NII)—more popularly known at the time as the "Information Superhighway."

"This was a term for the Internet that had already been in use, but Al Gore popularized it and made it mainstream," recalls Stacy Horn in her chapter on EchoNYC. "The Internet and Echo and places like it had already gotten some attention in the press, but now articles started appearing all over the place almost daily. It wasn't long before everyone was talking about the Information Superhighway." Even the U.S. Congress was talking about it. Having addressed the future of science and engineering research networks in the passage of the U.S. High Performance Computing Act of 1991 (which authorized NSF funding for a National Research and Education Network that led eventually to the Internet2 consortium, which currently includes some 252 universities and 82 corporations among its members), U.S. policymakers turned to a broader set of networking applications, envisioning a more democratic and participatory system.[12] Vice President Gore took the lead in this effort, painting a vivid picture of the networked future. "This Administration intends to create an environment that stimulates a private system of free-flowing information conduits," he declared at a National Press Club speech in late 1993. "It will involve a variety of affordable and innovative appliances and products giving individuals and public institutions the best possible opportunity to be both information customers and providers. Anyone who wants to form a business to deliver information will have the means of reaching customers. And any person who wants information will be able to choose among competing information providers, at reasonable prices. That's what the future will look like—say, in ten or fifteen years. But how do we get from here to there?"[13]

Among the legislative roadmaps for fulfilling Gore's vision was Rep. Thomas Boucher's (D-VA) National Information Infrastructure Act of 1993, which stipulated that "the Federal Government should ensure that the applications achieved through research and development efforts such as the High-Performance Computing Program directly benefit all Americans."[14] Although Boucher's H.R. 1757 passed in the House by a 326–61 margin on July 26, 1993, it died a quiet death in the Senate Subcommittee on Education, Arts, and Humanities late that year. Not surprisingly, the arts were not included among the benefits generally enumerated for the proposed new telecommunications infrastructure, which invariably stressed the practical over the cultural: "helping speed the delivery of new services, such as distance learning, remote medical sensing, and distribution of health information," in the words of Sen. John Danforth's (R-MO) Telecommunications Infrastructure Act of 1993. Even at its most visionary, S. 1086 sounded a decidedly civic tone: "access to switched, digital telecommunications service for all segments of the population promotes the core First Amendment goal of diverse information sources by enabling individuals and organizations alike to publish and otherwise make information available in electronic form."[15]

But the door to more creative uses of the NII had not been entirely closed. The Clinton–Gore administration's 1993 NII "Agenda for Action" report, for example, asked its readers to "imagine the dramatic changes in your life if ... [t]he vast resources of art, literature, and science were available everywhere, not just in large institutions or big-city libraries and museums."[16] Later in 1993, FCC Commissioner Ervin S. Duggan (the veteran government official who would become the president and CEO of the Public Broadcasting Service the following year) thrust the old notion of noncommercial bandwidth set-asides squarely into the NII debate. "I want to defend the idea of public support for public culture," Duggan declared, a fairly bold gesture at a time when the so-called "culture wars" had put such government funding very much in doubt.[17] Duggan, to be sure, wasn't calling for the establishment of the National Endowment for the Electronic Arts or a Corporation for Public Telecommunications. Rather, he was thinking in terms of "publicly supported institutions that educate and enlighten and call every generation to high-minded ideals. Public libraries, schools, museums and universities are such vital institutions of public culture. So is public television." Still, a digitized public broadcasting system, with far more capacity than the 350 public television stations and 600 public radio stations in operation at the time, might have been a potent force in American culture, moving far beyond the laurels on which pub-casters have long rested—from *American Playhouse* to *Great Performances* to *Live from Lincoln Center*—to become a platform that accommodated local artists and regional organizations as well.

As Duggan observed, "We are entering ... an age of enormous bandwidth: an age of the information superhighway. Public television needs to ensure a place for itself on that superhighway and to preserve what might be called 'public bandwidth' for its services. ... For the first time in recent memory, the federal government has created a set of overarching goals for telecommunications, and has committed itself to support a nationwide infrastructure designed to deliver information and service to all Americans."[18] He went on to call for "reserving public bandwidth—what I will call 'lanes for public culture on the information superhighway.' ... At its core, the new information superhighway will buzz with commercial traffic—and this is fine, as far as it goes. My challenge to you is this: fight to ensure that poetry, art, science, cultural knowledge and the essential tools of education are allowed to travel that superhighway as well."[19]

Putting the matter more bluntly, the Consumer Federation's Director of Research Mark Cooper testified that the Information Superhighway would need "a very generous bike lane" for noncommercial uses, contrasting Vice President Gore's public-interest vision of the NII with the home-shopping and entertainment extravaganza that the big telecom companies inevitably touted. "When business people talk about the highway," Cooper declared, "it sounds like an arcade." The reality, according to Cooper, "is that without aggressive public policy, money will buy entertainment, not the information-rich classrooms or the electronic town hall. Providing socially useful applications requires direct public-policy intervention." Corporate philanthropy alone, Cooper insisted, would not be sufficient.[20]

Unfortunately, public-sector philanthropy wasn't up to the task of civilizing the NII, either. The beleaguered National Endowment for the Arts and its sibling for the humanities were more concerned with their own continued existence than with the launching of new online projects, and they were absent in the policy debates over the future of networked communications that occupied the rest of Washington in the 1990s. This was a challenging period for public culture, after all, framed by the Mapplethorpe-Serrano controversy that embroiled the NEA in 1989–1990, and the Senate's 99–1 vote in 1995 for a resolution condemning the NEH-funded National History Standards (and the subsequent House hearings on plans to abolish the agencies). From a high of $176 million in 1992, the NEA's annual budget had been reduced to $99 million by 1996.

While the initial report on the NII from the Clinton administration did not include the arts (although it did have chapters on the information needs of schools and libraries), a subsequent report, *The Information Infrastructure: Reaching Society's Goals*, set the stage for a much broader discussion of the Internet's cultural implications, with a chapter on "Arts, Humanities, and Culture on the NII" that raised a number of important points.[21] True, many of those points were couched in terms most likely to appeal to Congress, concerning the GNP (in which the "core copyright industries" contributed

"$206 billion in values added to the U.S. economy" in 1992), attendance figures ("museums, in particular, attract more than half a billion visits each year"), and the overall economic impact of the arts ("the direct expenditures of nonprofit arts organizations are estimated at $36.8 billion").[22] But the chapter was not without its visionary aspects as well:

> Users of a National Information Infrastructure—especially artists and scholars—will have the opportunity to be both information consumers and information providers. In the interactive environment of the NII, the opportunity for information traffic to move two ways will permit the creation of unique works and the augmentation of existing works. ... The NII could permit any one person or group of persons to have access to a large body of materials or ideas, and to study them, combine them in new ways, and make their results available to anyone else.[23]

Sounding a more realistic note in assessing the *current* state of online culture, the chapter acknowledged that "this picture of cultural computing represents a cacophony of projects with little reference to each other. As yet their development is guided by no systematic, coordinated policy. They exemplify often original and even heroic individual achievement, but cannot hope to represent the fullness of our cultural heritage and creations."[24] Unfortunately, that observation proved to be more prophetic than any of the more sanguine prognostications for the future of online culture, including the following bold prediction that remains unfulfilled: "The National Information Infrastructure promises to extend the power of the human imagination to new frontiers, and American artists and scholars will be at the forefront of this exploration. ... The NII can give all Americans, of all races, ages and locations, their cultural birthright: access to the highest quality thought and art of this and prior generations."[25]

Most public interest advocates, in any case, had higher priorities and other interests with regard to the NII, such as ensuring equitable broadband access to all Americans and wiring schools and libraries. Ultimately the Clinton–Gore administration's 1993 "Agenda for Action" took three years to play out, culminating in legislation in which the once-vaunted public NII gave way to the thoroughly privatized Internet we have today. The Telecommunications Act of 1996, the first thorough overhaul of telecommunications policy in the United States in more than 60 years, turned out to be a deregulatory feast to which only the major telephone and cable companies were invited. In three days of hearings before the House Commerce Subcommittee on Telecommunications and Finance in May 1995, for example, 49 witnesses testified on the House version of the bill that eventually became the Telecommunications Act of 1996: 36 industry representatives, seven public officials, and six members of public-interest groups. None of the witnesses spoke on behalf of the arts and culture.[26]

If there was a glimmer of hope in the otherwise discouraging policy debates that surrounded the NII in the 1990s, it was in the creation of a broad-based coalition of academic, library, and cultural organizations that formulated plans for one important aspect of the future of the Internet that had been overlooked in the rush toward e-commerce and broadband entertainment: the potential, that is, for an online cultural archive. Spearheaded by the Getty Art History Information Program, the Coalition for Networked Information, and the American Council of Learned Societies, the project—Humanities and Arts on the Information Highways: A National Initiative—was designed to "confront the issues and responsibilities connected with bringing the nation's cultural heritage into the digital environment."[27] Although at its heart a preservationist movement, the project was not without its forward-looking aspects, too. "The creation of a fully interactive and exploratory environment essential for the arts and humanities to thrive would transform the NII from a link between computers to a connection between people," it explained in a 1994 report, by "enriching a sense of community through active participation in a networked environment" and "[f]ostering intellectual and artistic collaborations that will result in new resources in the arts and humanities."[28] Most important, the initiative offered a clear-eyed assessment of the impediments—"financial under-capitalization, technological under-development, and political neglect"—that then prevented (and in many ways continue to stymie) the full development of a robust cultural sector online.[29]

Although the initiative failed to have an impact on the policy discussions that led to the passage of the 1996 Telecom Act, the three sponsoring organizations moved ahead with plans of their own, including the formation of the National Initiative for a Networked Cultural Heritage (NINCH) in 1996. Led by David Green, former director of communications at both the New York Foundation for the Arts and Arts Wire, the undertaking was an ambitious one, bringing together more than a hundred arts and humanities organizations to continue the quest for a fully networked cultural archive. Unfortunately, with the impetus of the telecom bill now long past, and with the broadband future surging ahead, NINCH's mission was rather like attempting to sell libraries and opera houses in the middle of the Oklahoma Land Rush. "We'll get to that later" was the general response. More than a decade after NINCH's Washington office closed its doors in 2003, we're still waiting.[30]

Admittedly, the power of the new technology was undeniable at the beginning of the 21st century. There was plenty of virtual space to go around, and with the annual domain registration fee for a website less than the cost of a monthly cable bill, anyone could afford to participate in the Digital Land Rush. Standing A. J. Liebling's famous dictum on its head—"Freedom of the press is guaranteed only to those who own one"—that First Amendment right could now belong to anyone.[31] "In the many-to-many environment of the Net," observed Howard Rheingold in 1999, "every desktop is a printing press, a broadcasting station, and place of assembly. Mass-media will continue to exist, and so will journalism, but these institutions will no longer monopolize attention and access to the attention of others."[32] Tech writer Dan Gillmor, who made his own transition from old media to new in 2004 by leaving the *San Jose Mercury News* to help spearhead the new citizen-journalism movement, echoed Rheingold's assessment. "Personal technology is undermining the broadcast culture of the late 20th century," Gillmor wrote. "It's putting tools that were once the preserve of Big Media into the hands of the many."[33] If only it were that simple.

The World's Biggest Popularity Contest

Traditional print journalism has certainly suffered in the digital era, with newspaper circulation dropping 10.25 percent in North America and 23 percent in Europe over the past five years (with ad revenues declining even more sharply).[34] Still, although some of the names have changed, the mass media are as massive as ever, extending their reach across traditional and new platforms alike.

The new-media democracy that Rheingold and Gillmor and scores of others proclaimed at the dawn of the twenty-first century may exist, but not without a lot of effort on the part of content creators and consumers alike. As Wolfgang Staehle suggests in Susanne Gerber's chapter, "One good thing could be to decentralize the net, to let millions of independent servers bloom. Instead what's happening is quite the opposite." That sobering truth became readily apparent in 2004, when Matthew Hindman and Kenneth Neil Cukier, both fellows at the National Center for Digital Government at Harvard's Kennedy School, took a closer look at the traffic patterns on the web, and in particular at Google's unrivaled influence on web users' search for content. "Imagine if one company controlled the card catalog of every library in the world," the researchers mused. "The influence it would have over what people see, read and discuss would be enormous. Now consider online search engines."[35] At the time, Google and Yahoo controlled 95 percent of the search market in the United States. Today, Microsoft's Bing has surpassed Yahoo, but Google's dominance—handling roughly two-thirds of all searches made in the United States (and more than 90 percent in Europe)—remains unrivaled. Therein lies the problem, according to Hindman and Cukier:

While search engines are indispensable for finding information online, the technology on which they are based serves to narrow the field of sites that people see. In this respect, Google's technology reinforces a worrisome feature of the Web: the trend toward consolidation that affects everything from politics to news to commerce.

Google's use of links to find content essentially turns the Web into the world's biggest popularity contest. ... Behind Google's complex ranking system is a simple idea: each link to a page should be considered a vote, and the pages with the most votes should be ranked first. This elegant approach uses the distributed intelligence of Web users to determine which content is most relevant.

But what is good for Google is not necessarily good for the rest of the Web. The company's technology is so strong that its competitors have adopted a similar approach to organizing online information, which means they now return similar search results. Thus popular sites become ever more popular, while obscure sites recede ever further into the ether."[36]

As the authors pointed out elsewhere, "Users may be able to choose from millions of sites, but most go to only a few. This isn't an accident or the result of savvy branding. It's because Internet traffic follows a winner-take-all pattern that is much more ruthless than people realize. Relying on links and search engines, most people are directed to a few very successful sites; the rest remain invisible to the majority of users. The result is that there's an even greater media concentration online than in the offline world."[37]

If there's a saving grace to the rich-get-richer online world that Hindman and Cukier describe, it's that smaller, less heralded sites still play an important role in promoting and preserving smaller, less heralded topics. As "web usability" consultant Jakob Nielsen observed in his critique of Hindman and Cukier, "the Web is not a mass medium. It's not broadcast. The Web is on-demand, *driven by each customer's specialized need in each moment*. ... The question here is not whether some topical sites are bigger than others. They obviously are, given the power law for the Web and its subsets. The question is whether the same few sites would always dominate, regardless of a user's goal."[38] Nielsen's research suggests that depending on the particular topic for which a user conducts a search, the results will vary considerably.[39] The top ten returns on those searches, which do indeed attract most of the traffic, are by no means uniform across a variety of different topics. In that sense, perhaps a major part of the concentration of new-media power that Hindman and Cukier decry is more a product of popular taste than of corporate oligarchy. The top ten Google searches worldwide for 2014, for example, included a celebrity death (Robin Williams), two sporting events (the World Cup and Winter Olympics), a health crisis (Ebola), airline tragedies (Malaysia Airlines Flights 17 and 370), a viral video (the ALS Ice Bucket Challenge), a mobile game (Flappy Bird), a cross-dressing singer (Conchita Wurst), a Jihadist organization (ISIS), and an animated film (*Frozen*).[40] So the World Wide Web, in all its annual four-trillion-page-view glory, might not qualify as a mass medium according to Nielsen, but at times it manages an uncanny impersonation of American television.

At the margins, however, the web is still full of surprises, and "small sites," Nielsen insists, "have two huge advantages over big sites: there are many more of them and they are more specialized and thus more targeted. Small sites speak directly to the specific needs and interests of a committed user community, and thus have much higher value per page view. A site on growing blueberries can be a must-read service for people who farm them, and thus of immense value as a place to promote blueberry-farming equipment."[41] In their time, the projects described in this book were no less vital—must-read, must-see services for those interested in contemporary art. Today, more generally, there are thousands of small, equally essential sites devoted to the arts in all their variety, from the oldest traditions to the latest developments. But while blueberry farmers presumably know exactly where to find the resources they need online, artists and arts organizations, and their audiences worldwide, face a much more daunting task.

In a perfect World Wide Web, of course, a well-defined online cultural zone would already exist, clearly demarcated by its top-level domain—.*arts*—and as easy to find as Google or Facebook. But the web, as we know, is far from perfect. Although ".org" is the reputed designation for nonprofit organizations, in fact that generic top-level domain (TLD), like .com and .net among the seven initial TLDs (and *unlike* .edu, .gov, .int, and .mil) has always been available to all without restriction. As a result, with more than 10 million .org domains currently registered, it's not an especially exclusive club.[42] In recent years, more than 700 new TLDs have been added; however, all but 12 have fewer than 100,000 domains associated with them. An ".arts" TLD was in fact proposed by the Internet Ad Hoc Committee (IAHC) in 1996, and that designation, properly managed, might have been useful as a means of identifying cultural resources online. By the time the new domains were ultimately authorized four years later, however, .arts had become .museums and was limited to institutions approved by the Museum Domain Management Association (MuseDoma). As that organization explains, "The top-level domain .museum was created by and for the global museum community. It enables museums, museums associations and museum professionals to register .museum Web site and e-mail addresses. This, in turn, makes it easy for users to recognize genuine museum activity on the Internet."[43] It is difficult for those same users to find genuine arts activity elsewhere online.

In a perfect world, moreover, support for an online arts zone would flow from a variety of sources, both public (a reinvigorated NEA and NEH and a reconfigured Corporation for Public *Media*) and private (foundations as well as the newly minted high-tech barons who have yet to emerge, Bill and Melinda Gates notwithstanding, as the twenty-first-century disciples of Carnegie, Rockefeller, Ford, et al.). However, the real world, even at its most beneficent, remains as imperfect as the virtual world. Consider the two primary public patrons of culture. "Since 2010," the NEA proclaimed in late 2014, that agency "has encouraged and invested in work at the intersection of art, science and technology. Through funding, publications and content development, and collaborations with federal agencies and departments, the NEA has sought to catalyze and support work in this realm of innovation."[44] Yet this support has been modest, piecemeal, and sporadic. Sadly, the most notable NEA experiment in arts-and-technology funding—Open Studio: The Arts Online—was comparatively short-lived. Launched in 1996 in conjunction with the Benton Foundation (with additional support from the AT&T Foundation, Microsoft Corporation, and the Ford Foundation), Open Studio awarded grants of up to $35,000 to organizations around the country "to instruct artists and arts organizations in traditionally underrepresented communities how to post artwork online, disseminate cultural information via the Internet, and link with the arts community across the country."⁴⁵ More than 1,000 artists and arts organizations in 28 states received training through Open Studio, which, had it continued, might have evolved from its bootstrapping origins to a more robust support system for sustaining the nonprofit arts online.⁴⁶

Currently, the agency explains, "There is no designated funding category for arts, science, and technology projects, so grant awards appear in several NEA's disciplines." The one NEA discipline that seems most open to web-based projects (although the majority of its budget is devoted to traditional film, video, and public broadcasting) is Media Arts. Concerning its Art Works grants under the agency's Engagement subcategory, for example, the program singles out "[t]he development of web portals, hubs, mobile and tablet apps (developed and made available for both iOS and Android devices), or other innovative uses of technology or new models to provide audiences with access to media artists and art works."⁴⁷ The endowment's Museum and Visual Arts programs, similarly, support technology projects that provide online access to collections, exhibitions, organizational history, and other programming information.⁴⁸ But overall, the NEA in recent years has not been a hotbed of new-media activity.

The National Endowment for the Humanities (NEH), in contrast, because of its long association with libraries and museums, has been much more active in support of both digital scholarship and digital preservation projects. It's difficult to imagine, for example, the NEA making the kind of sweeping statement about the impact of technology that the NEH's Office of Digital Humanities recently made: "In a short period of time, digital technology has changed our world. The ways we read, write, learn, communicate, and play have fundamentally changed due to the advent of networked digital technologies."⁴⁹ Through its Division of Preservation and Access and its Office of Digital Humanities, the NEH has supported a number of projects to develop best practices, tools, and training workshops in "… a national effort to preserve and make accessible cultural resources for research, education and the general public good."⁵⁰ Two NEH grants to Cornell University's Digital Preservation Management training project, for example, supported an online tutorial covering the basics of digital preservation along with a series of one-week workshops for advanced training in the preservation of digital materials in cultural institutions. The NEH also supported a collaborative project involving Indiana University, Bloomington's and Harvard University's "Sound Directions: Digital Preservation and Access for Global Audio Heritage" project for the development of best-practice guidelines in the digital reformatting of analog sound recordings and the investigation and testing of a distributed preservation framework.⁵¹

Such efforts as these, moreover, are all part of the larger National Digital Information Infrastructure and Preservation Program (NDIIPP), a collaborative archival effort under the direction of the Library of Congress, which in 2000 received congressional funding to archive and provide access to digital resources.⁵² More recently, in 2010, the library launched a National Digital Stewardship Alliance (NDSA) to extend the work of NDIIPP to more institutions, which now number more than 300 partners.[53] Curiously, only a handful of arts organizations and institutions belong to that alliance (*not* including the NEA), as history seems to be repeating itself: just as artists and arts organizations were mere hitchhikers on the Clinton–Gore Information Superhighway, so are they destined to become mere spectators in the design and construction of the archival rest stops, scenic overlooks, and historical markers along that thoroughfare.

For all of their good intentions, the arts and humanities endowments represent only a small, $292 million piece of the larger $2.5 billion arts-and-culture funding puzzle in the United States (and whatever leadership role the two federal agencies may have played over the first 25 years of their existence quickly eroded in the face of conservative attacks in the 1990s and the low-profile, bunker mentality they adopted thereafter). The private sector, in any case, provides roughly 88 percent of cultural funding in the United States. Unfortunately, at least from the perspective of the arts online, the nation's foundations and corporate funders have not been particularly generous to new-media projects, either. In 2012, for example, some 1,500 funders made 43,000 arts and culture grants to 12,500 recipients in the United States, for a total of $2.5 billion. Of that amount, a mere 145 grants, totaling $19.3 million—less than 1 percent of all arts and culture support—were for projects that involve (in the parlance of the Foundation Center's grants database) "web-based media" or "electronic communications/Internet."[54] These figures, needless to say, do not auger well for the development of a robust, clearly demarcated, and financially supported online arts sector anytime soon. Which brings us back to the question that Vice President Gore first posed in1993.

"How Do We Get from Here to There?"

Some will respond, and not without reason, that we *are* there, that the arts are already flourishing online, with more web pages than any one person could hope to visit in a lifetime. With Google just a click away, users can find anything and everything on the web—including the arts and culture, in an instant. Interested in the music of Ornette Coleman? Google returns some 535,000 results in less than half a second. The choreography of Martha Graham? "About 30,000,000 results," Google assures us. Internet art? No less than 1.53 *billion* results. An embarrassment of riches, it would seem.

What's embarrassing, actually, is that with the Internet having reached middle age and the web now in its mid-20s, we're still left with a grab bag of cultural resources online. An enormous grab bag, admittedly, but a haphazard affair nonetheless. A Google search *may* give us what we're looking for (especially if all we need is a Wikipedia entry or a YouTube clip). But if we're looking for a definitive cultural resource or an in-depth analysis of an artist or arts discipline, we're still better off in the real world, visiting a museum or a library. In that real world, at least, a fairly orderly, if financially undernourished, cultural sector has been delineated. The arts institutions in our midst are easy to find, as they wear their 501(c)(3) tax-exempt status on their marble sleeves (and in their year-round fundraising campaigns). Artists, of course, have long made their presence known, albeit more often than not in neighborhoods primed for the next wave of gentrification. The old media are even clearer in this regard: just tune to the left side of the FM dial for jazz, classical, folk, and NPR programming; find the PBS affiliate on local TV; and check the "Style" section of the local newspaper for what passes for arts criticism these days. Much slimmer pickings than the web's overflowing bounty, to be sure, but generally reliable. In contrast, although the web offers an abundance of content, it's not always easy to locate trusted, authentic, and professional resources amid all of the commercial clatter.

So, the question remains: How do we map the nonprofit cultural sector onto an online landscape that is still evolving—and becoming more commercial every day? The arts aren't at risk; they've existed for eons and will continue to do so. But the art that depends on the support structures that have developed around them over time, from church and court patrons in the distant past to state and private patronage today, threatens to recede farther into the margins of the World Wide Web, precisely at a time when that environment attracts increasing amounts of our time and attention. Our online access to these art forms is what is at risk in the Digital Age. The "there" we're trying to reach, however, is not simply some civilized corner of the web or even a restricted top-level domain conferred on those organizations and artists deemed to be "qualified" in some fashion. Rather the "there" is nothing more—or less—than the acknowledgment that (1) the noncommercial arts can't really thrive in the ad-driven, box-office environment that the Internet has become; and (2) a concerted, broad-based effort is needed to preserve and promote their presence online.

If the lessons from the past—the "archaeology"—are any indication of what is needed as we move forward, five things must happen to build a sustainable cultural sector online. First, and perhaps most important, a truly collaborative effort will have to be mounted, an online environmental movement, in effect, designed to identify, cultivate, and promote the digital art resources that already exist, and to weave them together in a fashion that allows them to receive the serious attention and study they deserve. Artists and arts institutions, like most individuals and organizations today, have embraced the web and its latest social media applications. But a coherent online cultural sector needs much more than these isolated efforts. It needs an extensive, reliable directory of cultural resources, an archive of past achievements (online and off), as well as a laboratory for new works. This would entail a large-scale, ongoing operation, certainly, guided by curatorial expertise but driven by individual and institutional collaboration. Rhizome's ArtBase (1999 to the present) is a model in this regard, a participatory database of web resources, but only a small first step. All of the disciplines, from dance and design to theater and the visual arts, should have comprehensive directories of online resources. The traditional arts service organizations (e.g., Dance/USA, Theater Communications Group, College Art Association, and the various music and museum associations) will have a role to play, along with national and regional arts advocacy organizations, in marshaling this effort. From a leadership standpoint, what's needed now is the kind of focused effort that NINCH once mounted, but with a greater sense of urgency. It's clear that our laissez-faire approach to the Internet, in which a page of hyperlinks is too often mistaken for true collaboration, has failed to achieve anything resembling a *collective* arts presence on the web.

Such a broadly collaborative effort will need assistance, material and otherwise, from both the public and private sectors—items two and three on our five-point wish list. The federal agencies are already in place, certainly, from the arts and humanities endowments and the Corporation for Public Broadcasting to the Smithsonian Institution and the Library of Congress. But with their budgets already stretched to the limits tending primarily to twentieth-century tasks, new revenue streams will be needed to extend their reach fully into the Digital Age. Securing new funds from a squeeze-cut-and-trim Congress is always a long shot, but the legislative models, at least, already exist. While neither the proposed Spectrum Commons and Digital Dividends Act of 2003 nor the Digital Opportunity Investment Trust Act two years later bore any fruit, their basic premise—taxing the high-profit Peters of the telecommunications industry to pay the low-budget Pauls of education and culture—still makes a lot of sense. Invoking the hallowed ground of the Morrill Land-Grant College Act of 1862 (in which the proceeds from the sale of public lands in the West were used to fund state colleges and universities), the two bills seized on the twenty-first-century equivalent of public lands—the electromagnetic spectrum—portions of which were soon to be auctioned off to corporate bidders for various wireless applications. The goal, according to the plan's originators, Newton Minnow and former PBS President Lawrence Grossman, was "to create a trust fund dedicated to innovation, experimentation, and research in utilizing new telecommunications technologies across the widest possible range of public purposes. ... It would enable schools, community colleges, universities, libraries, museums, civic organizations, and cultural, arts, and humanities centers to take advantage of the new information technologies to reach outside their walls and into homes, schools, and the workplace."[55] The plan failed, but so did the arts and humanities legislation—for more than a decade—before the two endowments were finally established in 1965. There is no reason to believe that a digital culture equivalent will progress any more swiftly, if it moves at all, that is, given the current political climate. Which makes the third item—private patronage—all the more important.

Unfortunately, for all of its largesse—and private philanthropy in the United States generated some $358 billion in contributions in 2014—those funds are highly contested.[56] Especially with the continued shrinking of the public sphere, the competition for private support among arts groups, who invariably play fifth fiddle to the Big Four of education, human services, health, and public affairs/social benefit (more than $153 billion combined in 2014), is especially keen. Nor has any one funder, from either the corporate or independent foundation arenas, emerged as a champion of online culture. But here, too, the model for such trendsetting support exists. In 1965, at a time when the brand-new NEA had just $2.5 million to spend on all of the arts, the Ford Foundation awarded $85 million (or roughly $640 million in current dollars) to 61 orchestras nationwide. The foundation's action not only revived a sector that was in severe financial straits, but it paved the way for the NEA to support other arts institutions once its budget reached eight-figure levels in 1971 and thereafter.

Even with the collective will and the public and private wherewithal in place, our virtual promised land won't be attainable without a crucial fourth ingredient to guide our journey: expertise. It turns out that after nearly a half century of online communications—from the earliest academic and military applications to the latest social (and often anti-social) media—we still don't fully understand the impact of the Internet on our lives. We have plenty of usage statistics, certainly, some of them truly astounding: that every minute, for example, Facebook members share 2,460,000 pieces of content, Twitter users send 277,000 tweets, and YouTube fans upload 72 hours of video.[57] But we're not sure what all of those data really mean. More to the point, although artists and arts organizations knew enough to fill this anthology full of fascinating projects, they didn't know enough—or, more clearly, *couldn't* know enough—to find a way to preserve those projects. This kind of expertise takes time, research, and considerable effort on the part of professional scholars. As Julianne Nyhan reminds us in her detailed analysis of the first year of the Humanist Listserv (1987), the growth and maturation of a new academic discipline like Humanities Computing takes years, even decades, to achieve institutional and professional status. But just as the interdisciplinary field of American Studies emerged in the post-WWII era in response to the new role of the United States as an economic and cultural power, perhaps the still-nascent field of Internet (or web) Studies can follow a similar path, accelerated by the swift rise of the Internet to a position of dominance in our lives. The founding of the Association of Internet Researchers in 1999 is a step in that direction, and David Gauntlett's Web. Studies publications and David Silver's work in Cyberculture Studies have already shown promise in pointing to a future in which our understanding of online culture is much more reflective of the enormous amount of time we spend consuming that culture.

Still, collaborative leadership, federal funding, private support, and curatorial expertise—all four elements, if they are to contribute to a sustainable online arts culture—will ultimately depend on the fifth and final piece of the puzzle: grassroots participation. Ironically, such participation has long been the reputed hallmark of the Internet revolution—"the people formerly known as the audience," in Jay Rosen's catchphrase for citizen-journalism—although the fruits of that revolution, beyond the sheer volume of social media, are not always readily apparent.[58] What we really need right now, rather than more viral videos, vox-populist crowd sourcing, and other vestiges of the cut-and-paste culture, is a rejuvenated, motivated "audience formerly known as passive spectators." These are the people who are passionate about what they see and do online in connection with the arts, but who don't really have anywhere to share their interests—and their online discoveries—beyond mass-market franchises such as Pinterest and Reddit. But channeled into a collaborative, funded, and curated movement of organizations and individuals committed to building an online arts sector, the wisdom of *this* crowd could be a powerful force indeed. Freed of the tyrannous majoritarian instincts of mainstream web-aggregation sites (from Google News and Epinions to Digg and StumbleUpon), a culturally focused recommendation system could spotlight the best websites, blogs, and other resources on the Internet. A small number of these resources, including THE THING's and trAce Online Writing Center's notable archival efforts, have been covered in this volume. However, many other model projects—from the specific new-media focus of Rhizome to the vast public-domain expanses of the Internet Archive—warrant much wider recognition.

Admittedly, all of this is a lot to ask—five difficult pieces for the Digital Age: collaboration not simply within art forms but across disciplinary boundaries; federal leadership and funding at a time when public culture is in retreat; innovation from private funders whose philanthropic practices often seem etched in stone; the development and maturation of a new academic discipline; and building a movement of online seekers who are concerned, not with the news and entertainment of today, but with cultural treasures past, present, and future. If there's one sure thing in what is otherwise a list of long shots, it's the Academy, which without question will create a new discipline devoted to the study of the Internet, in all of its varied facets. The real question, then, is how these scholars will ultimately judge our efforts to build a place for nonprofit culture online, which thus far have fallen far short of the mark.

Such things as these always take time, of course. More than eight decades elapsed between the release of *The Great Train Robbery* in 1903, after all, and the passage of the National Film Preservation Act of 1988, long after the great majority of silent films were irretrievably lost, and film preservation is still a work in progress. Unfortunately, as this volume makes clear, tape backups, floppy disks, and even URLs have all proved as unstable as cellulose nitrate film stock and even more vital to the overall health of our culture.

Yet in the current environment—with growing numbers of "digital natives" who have essentially grown up online—the reclamation of this cultural legacy need not take so long. Neither archeology nor architecture in the digital domain is anywhere near as labor intensive as their analog counterparts. Compared with unearthing ancient civilizations or erecting monumental structures, the cultural riches of the digital era, as this volume demonstrates, are much more immediate. There's still time, in other words, for another "community memory" project, not unlike the one that Lee Felsenstein and his colleagues launched in Berkeley more than four decades ago, but this time casting a much wider net and looking toward the future of online culture as well as at its past and present.

Notes

1. Although the emphasis here is on *noncommercial* culture, the discussion is by no means limited to 501(c)(3) nonprofit projects and organizations. Rather, the concern is for those artists and arts organizations of all stripes that are most likely to be overshadowed by popular culture and the multinational conglomerates whose elaborate cross-promotional efforts support their entertainment products. At the same time, online art should not be viewed as a substitute for, or in competition with, offline art. Indeed, the two worlds of "virtual" and "real-world" culture are inextricably linked, with the former often serving as an adjunct to and a support structure for the latter (especially as the traditional mainstream media have largely abandoned arts coverage).

2. "When television is bad, nothing is worse," declared Minow that year at the annual meeting of the National Association of Broadcasters in Washington, as he offered the following challenge to the television industry: "I invite you to sit down in front of your television set when your station goes on the air ... and keep your eyes glued to that set until the station signs off. I can assure you that you will observe a vast wasteland." Newton N. Minow and Craig L. LaMay, *Abandoned in the Wasteland: Children, Television, and the First Amendment* (New York: Hill and Wang, 1995), 188.

3. One of the progenitors of the global Internet, ARPANET was the first to use network email in 1972, and in 1983, it became the first network to implement the TCP/IP protocols that would become the transmission standard of the Internet. (For security reasons, Military Network [MILNET] was split off from ARPANET in 1983.) CSNET, which eventually hosted 180 sites around the world, was responsible for the first Internet gateways between the United States and many countries in Europe and Asia. National Science Foundation, "The Internet: Changing the Way We Communicate." Available at http://www.nsf.gov/about/history/nsf0050/pdf/internet.pdf.

4. Peter H. Lewis, "U.S. Begins Privatizing Internet's Operations," *New York Times*, October 24, 1994. Available at http://www.nytimes.com/1994/10/24/business/us-begins-privatizing-internet-s-operations.html; National Science Foundation, "Internet Moves Toward Privatization," June 24, 1997. Available at http://www.nsf.gov/news/news_summ.jsp?cntn_id=102819.

5. Robert H. Zakon, "Hobbes' Internet Timeline." Available at http://www.zakon.org/robert/internet/timeline/.

6. National Center for Supercomputing Applications, "Enabling Discovery: NCSA Mosaic." Available at http://www.ncsa.illinois.edu/enabling/mosaic/.

7. Even as it phased out its direct support of the Internet, the NSF established a Very High Speed Backbone Service (vBNS) to take the place of NSFNET in high-speed research and academic

connectivity. National Science Foundation, "Networking for Tomorrow." Available at http://www.nsf.gov/news/special_reports/cyber/futurenetworks.jsp.

8. Bernard Aboba, *The Online User's Encyclopedia: Bulletin Boards and Beyond* (Reading, MA: Addison-Wesley, 1994).

9. Other AOL channels included similar arts-related keywords: Computing Channel (with keywords for both Mac and PC music/sound and computer graphics forums); Influence Channel (literary reviews, author events, and message boards, as well as a collection of classical music resources); Interests Channel (books, *Crafts* magazine, photography forum, and writers' resources and chat rooms); and the Research & Learn Channel (various resources for architecture, art, dance, fashion, film, museums, music, opera, and theater, as well as an online edition of the *Dictionary of Cultural Literacy*). Curt Degenhart and Jen Muehlbauer, *AOL in a Nutshell: A Desktop Guide to America Online* (Sebastopol, CA: O'Reilly Media, 1998), 240–270.

10. For the decline and fall of AOL as an online service provider, see Saul Hansell, "Eleven Years of Ambition and Failure at AOL," *New York Times*, July 24, 2009. Available at http://bits.blogs.nytimes.com/2009/07/24/eleven-years-of-ambition-and-failure-at-aol/.

11. "Squeezebox," Publicly Accessible Mailing Lists, Part 20/22. Available at https://groups.google.com/forum/#!topic/news.lists/vluXru5Nr5w/.

12. Popularly known as the "Gore Bill," after its sponsor, Sen. Albert Gore (D-TN), the U.S. High Performance Computing Act was designed "to help ensure the continued leadership of the United States in high-performance computing" by (among other means) promoting "greater collaboration among government, Federal laboratories, industry, high-performance computing centers, and universities." It produced the Networking and Information Technology Research and Development (NITRD) Program, which was subsequently supported by the Next Generation Internet Research Act of 1998 and the America COMPETES Act of 2007.

13. "Remarks by Vice President Al Gore at the National Press Club," December 21, 1993. Available at http://www.ibiblio.org/nii/goremarks.html.

14. "H.R. 1757—National Information Infrastructure Act of 1993," Congress.Gov. Available at https://www.congress.gov/bill/103rd-congress/house-bill/1757/text/.

15. "S.1086—Telecommunications Infrastructure Act of 1993." Available at https://www.congress.gov/bill/103rd-congress/senate-bill/1086/text/.

16. Information Infrastructure Task Force, "The National Information Infrastructure: Agenda for Action," September 15, 1993. Available at http://www.ibiblio.org/nii/NII-Agenda-for-Action.html. The Clinton–Gore administration also assembled an Advisory Council on the National Information Infrastructure, whose 27 members, however, were bereft of any representatives from arts or cultural organizations. Department of Commerce, National Telecommunications and Information Administration, "Advisory Council on the National Information Infrastructure, Meeting; Notice," *Federal Register*, April 8, 1994. Available at http://www.gpo.gov/fdsys/pkg/FR-1994-04-08/html/94-8580.htm.

17. Federal Communications Commission, "Lanes for Public Culture on the Information Superhighway," Remarks of Commissioner Ervin S. Duggan, Federal Communications Commission, before the Southern Educational Communications Association, Atlanta, Georgia, September 27, 1993.

18. Ibid.

19. Ibid.

20. "APTS Presents Highway Bill, Subcommittees Want Revisions," *Current*, February 14, 2014. Available at http://current.org/files/archive-site/in/in403.html.

21. National Institute of Standards and Technology, "Putting the Information Infrastructure to Work: A Report of the Information Infrastructure Task Force Committee on Applications and Technology," May 1994. Available at http://files.eric.ed.gov/fulltext/ED385275.pdf.

22. National Institute of Standards and Technology, "The Information Infrastructure: Reaching Society's Goals," September 1994, 119–120. Available at http://babel.hathitrust.org/cgi/pt?id=mdp.39015033251508;view=1up;seq=3/.

23. Ibid., 121–122.

24. Ibid., 125.

25. Ibid., 142.

26. U.S. Congress, House of Representatives, 104th Congress, 1st Session, "Communications Act of 1995," Report 104–204, pt. 1, 55–56. Available at ftp://ftp.fcc.gov/pub/Bureaus/OSEC/library/legislative_histories/1755.pdf.

27. Getty Art History Information Program, "Humanities and Arts on the Information Highways," September 1994, 2. Available at https://cni.org/wp-content/uploads/2013/06/Humanities%C2%A0and%C2%A0Arts.pdf.

28. Ibid., 3–4.

29. Ibid., 7.

30. NINCH's website, with a number of useful resources, is still available online at http://www.ninch.org/.

31. A. J. Liebling, "Do You Belong in Journalism?", *The New Yorker*, May 14, 1960, 105.

32. Howard Rheingold, "Community Development in the Cybersociety of the Future," BBC Online, June 1999. Available at http://www.partnerships.org.uk/bol/howard.htm.

33. Dan Gillmor, "Democratizing the Media, and More," eJournal, January 11, 2004. Available at http://lists.gnu.org/archive/html/dmca-activists/2004-01/msg00003.html.

34. World Association of Newspaper and News Publishers, "World Press Trends: Print and Digital Together Increasing Newspaper Audiences." Available at http://www.wan-ifra.org/

press-releases/2014/06/09/world-press-trends-print-and-digital-together-increasing-newspaper-audienc.

35. Matthew Hindman and Kenneth Neil Cukier, "More Is Not Necessarily Better," *New York Times*, August 23, 2004. Available at http://www.nytimes.com/2004/08/23/opinion/23hindman.html.

36. Hindman and Cukier, "More Is Not Necessarily Better."

37. Matthew Hindman and Kenneth Neil Cukier, "More News, Less Diversity," *New York Times*, June 4, 2003. Available at http://www.nytimes.com/2003/06/02/opinion/02HIND.html.

38. Jakob Nielsen, "Diversity Is Power for Specialized Sites," Nielsen Norman Group, June 16, 2003. Available at http://www.nngroup.com/articles/diversity-is-power-for-specialized-sites/.

39. Jakob Nielsen, "Wide Spectrum of Sites Found for Sample Questions," Nielsen Norman Group, June 16, 2003. Available at http://www.nngroup.com/articles/wide-spectrum-of-sites-found-sample-questions/.

40. The list of top U.S. searches in 2014 differed slightly from the global list, with Ferguson (Missouri) taking the place of Conchita West and the Ukraine replacing the Sochi Olympics. Greg Kumparak, "These Were the Top 10 Most Popular Searches on Google In 2014," TechCrunch, December 16, 2014. Available at http://techcrunch.com/2014/12/16/these-were-the-10-top-trending-searches-on-google-in-2014/.

41. Nielsen, "Diversity Is Power for Specialized Sites."

42. By far the largest TLD, .com passed the 115 million mark in 2014 (with .net ranking second at just over 15 million domains). The four restricted TLDs among the original seven are tiny in comparison: .int (15,000 domains), .edu (13,500), .gov (fewer than 2,000), and .mil (a small number used by the U.S. armed forces). RegistrarStats, Inc., "TLD Domain Counts." Available at http://www.registrarstats.com/TLDDomainCounts.aspx.

43. MuseDoma was created by the International Council of Museums (ICOM) and the J. Paul Getty Trust. "About.Museum." Available at http://about.museum/.

44. National Endowment for the Arts, "National Endowment for the Arts Awards $29 Million for Arts Projects," December 2, 2014. Available at http://arts.gov/news/2014/national-endowment-arts-awards-29-million-arts-projects/.

45. "Open Studio Adds Ten New Regional Internet Training Sites," *Arts Wire Current*, April 6, 1999. Available at http://www.artscope.net/NEWS/news040699-7.shtml.

46. Benton Foundation, "Open Studio: The Arts Online—Lessons Learned," 1999. Available at https://www.benton.org/archive/openstudio/toolkit/lessonslearned.html/; Grantmakers in the Arts, "Proceedings from the Pre-Conference: The Digital Revolution and the Arts," October 15, 2000. Available at https://www.giarts.org/sites/default/files/Digital-Revolution-and-the-Arts.pdf.

47. National Endowment for the Arts, "Grants: Art Works Guidelines: Media Arts." Available at http://arts.gov/grants-organizations/art-works/media-arts/.

48. National Endowment for the Arts, "Grants: Art Works Guidelines: Museums." Available at http://arts.gov/grants-organizations/art-works/museums; National Endowment for the Arts, "Grants: Art Works Guidelines: Visual Arts." Available at http://arts.gov/grants-organizations/art-works/visual-arts/.

49. National Endowment for the Humanities, "About the Office of Digital Humanities." Available at http://www.neh.gov/divisions/odh/about/.

50. Barrie Howard, "NEH Grants Relating to Digital Preservation," Library of Congress Digital Preservation Blog, August 25, 2011. Available at http://blogs.loc.gov/digitalpreservation/2011/08/neh-grants-relating-to-digital-preservation/.

51. Ibid. National Endowment for the Humanities, "Division of Preservation and Access." Available at http://www.neh.gov/divisions/preservation/.

52. Library of Congress, "Digital Preservation." Available at http://www.digitalpreservation.gov; Library of Congress, "Web Archiving." Available at http://www.loc.gov/webarchiving/.

53. Library of Congress, "National Digital Stewardship Alliance." Available at http://www.digitalpreservation.gov/ndsa/.

54. Foundation Center, "Foundation Directory Online." Available at http://foundationcenter.org/findfunders/fundingsources/fdo.html (subscription required).

55. Lawrence K. Grossman and Newton N. Minow, *A Digital Gift to the Nation* (New York: The Century Foundation Press, 2001), 4.

56. These figures reflect total U.S. charitable contributions (including donations from individuals and bequests, representing 80 percent of the total), of which more than 32 percent went to religion in 2014. Giving USA, "Giving USA 2015 Highlights." Available at http://givingusa.org/product/giving-usa-2015-report-highlights/.

57. Max Knoblauch, "Internet Users Send 204 Million Emails Per Minute," Mashable, April 23, 2014. Available at http://mashable.com/2014/04/23/data-online-every-minute/?utm_content=buffer7d0eb&utm_medium=social&utm_source=twitter.com&utm_campaign=buffer/.

58. Jay Rosen, "The People Formerly Known as the Audience," PressThink, June 27, 2006. Available at http://archive.pressthink.org/2006/06/27/ppl_frmr.html.

About the Authors

Madeline Gonzalez Allen has been exploring for several decades how technology could enhance our ability to share knowledge and facilitate collaboration within communities, organizations, and corporations, in balance with sustaining a practice of conscious presence, mindfulness, *Beingness*. She helped create internationally acclaimed "community networks," and she is the author of numerous published articles, concept papers that led to future products/services, and proposals selected for funding by the Apple Library of Tomorrow, the National Telecommunications Infrastructure Administration (NTIA), and the Corporation for Public Broadcasting (CPB). She has worked/served as a consultant for IBM, AT&T Bell Labs, Apple Advanced Concepts Group, the University of Phoenix, the University of Colorado/Boulder Community Network, Association for Community Networking, the W. K. Kellogg Foundation, the Colorado Advanced Technology Institute, the Colorado Historical Society, Fort Lewis College/Southern Ute Tribe, and the Ute Mountain Ute Tribe. She was a speaker at early community networking conferences, including the *1998 Casey Foundation Technology Conference* in Atlanta, GA; the 1997 *Hypernetwork Society Conference* in Oita, Japan; *the 1997 Pacific Northwest Library Association Conference* in Seattle, WA; the *1997 Community Space & Cyberspace: What's the Connection conference* in Seattle, WA; the *1996 Rural Telecommunications Conference* in Durango, CO; a "Virtual Panelist" on the *1997 Virtual CivicNet* and the *Utne Café's 1996 Panel Discussion on Community Networking*; and a panelist at the 1993 *Ideas Festival* in Telluride, CO. She has been a consultant with IBM since 1998 and continues to be an active community volunteer.

James Blustein began studying hypertext shortly before the World Wide Web was announced in the alt.hypertext newsgroup. In his PhD thesis, he acknowledged the denizens of that group for many stimulating conversations. He is now an Associate Professor of Computer Science and Information Management and director of the Hypertext Augmenting Intelligent Knowledge Use (HAIKU) research group at Dalhousie University. He co-authored the Association for Computing Machinery's (ACM's) first-ever hypertext format conference article outside of the Hypertext conference.

Hank Bull lives and works in Vancouver. He has been associated with the Western Front, an artist-run center, since 1973. The *HP Radio Show*, with Patrick Ready, was first aired in 1976 on CFRO-FM and would run weekly for eight years. With "Canada Shadows," he performed multimedia performances across Canada and Europe. Participation in artist telecommunication projects, using slowscan, electronic text, and fax, began in the late 1970s. This work took him eventually around the world to research shadow theater and artist networks. As a curator, he has organized numerous projects with artists in Africa, Asia, and Eastern Europe. As an advocate, he has contributed to the growth of artist-run culture across Canada. These practices informed the creation of a new gallery in 1999, the Vancouver International Center for Contemporary Asian Art. His current practice takes place at the intersection of painting, sculpture, music, performance, video, and sound.

Annick Bureaud is an independent art critic, event organizer, researcher, and teacher in art and technosciences. She is the director of Leonardo/OLATS, European sister organization to Leonardo/ISAST. She has written numerous articles and contributes to the French contemporary art magazine *Art Press*. She is the co-editor of the collection of essays *Connexions: art, réseaux, media* (Ensba Press, 2002) and the author *of Les Basiques: l'art « multimédia »*, an introductory book to new media art (Leonardo/OLATS, 2004). She is the co-editor of *Water Is in the Air: The Physics, Politics and Poetics of Water in the Arts*, in the Leonardo e-Book series at MIT Press (2014). She has organized many symposia, conferences, and workshops, including *Artmedia VIII: From Aesthetics of Communication to Net Art* (Paris, 2002) and *Visibility—Legibility of Space Art*, as well as *Art and Zero Gravity: The Experience of Parabolic Flight*, a project in collaboration with Leonardo/OLATS and the International Festival @rt Outsiders (Paris, 2003). In 2009, she co-curated the exhibition *(Un)Inhabitable? Art of Extreme Environments*, Festival @rt Outsiders, MEP/European House of Photography, Paris. In 2012, she was the curator of the work *Tales of a Sea Cow* by Etienne de France at the *Parco Arte Vivente* in Torino, Italy.

J. R. Carpenter is a Canadian-born, UK-based artist, performer, poet, novelist, essayist, new media writer, and researcher. She has been using the Internet as a medium for the creation and dissemination of nonlinear narratives since 1993. Her work has been presented in journals, festivals, and museums around the world. She has been awarded research and production grants in literature and in new media from Conseil des Arts de Montréal, Conseil des arts et des lettres du Quebéc, and Canada Council for the Arts. Her first novel, *Words the Dog Knows*, won the Expozine Alternative Press Award for Best English Book. Her second book, *GENERATION[S]*, a collection of code narratives, was published by Traumawien in Vienna in 2010. She served as President of the Board of Directors of Oboro New Media Lab in Montreal 2006–2010 and as faculty for In(ter)ventions: Literary Practice at the Edge residency program at The Banff Centre

2010–2014. She was awarded a PhD in performance writing, digital literature, and media archeology from University of the Arts London, UK, in 2015. She is currently a visiting fellow at the Eccles Centre for North American Studies at The British Library. She lives in Devon, England. More information about her work can be found at http://luckysoap.com/.

Paul E. Ceruzzi is a curator in the Space History Division at the Smithsonian Institution's National Air and Space Museum in Washington, DC. He received his B.A. from Yale University in 1970 and a PhD in American Studies from the University of Kansas in 1981. Prior to joining the Smithsonian, he taught history at Clemson University in South Carolina. At the museum, he has worked on several public exhibitions, most recently *Time and Navigation*, which covers the art and science of navigation from eighteenth-century seafarers to the current systems of global positioning satellites. He is currently working on a major renovation of the Boeing Milestones of Flight Hall, the entrance hall to the National Air and Space Museum, scheduled for a summer 2016 opening. He has written several books on the history of computing and aerospace technology, including *Beyond the Limits: Flight Enters the Computer Age, Internet Alley: High Technology in Tysons Corner, A History of Modern Computing*, and, most recently, *Computing, a Concise History*.

Anna Couey works at the intersection of art, communications, information, and social justice, using participatory media tools, story-collecting methods, community building, and organizational development processes to reimagine and restructure power. During the 1980s and 1990s, she helped develop art telecommunications projects such as the Art Com Electronic Network and Arts Wire, as well as produced temporary cross-cultural communications events as social sculpture. In the mid-1990s, she made an intentional shift to explore social sculpture strategies outside an art context to address ethical and aesthetic questions related to making art with communities. Anna contributed to the development of a "research justice" framework at the DataCenter, a social justice research organization that legitimizes diverse voices and forms of knowledge. Her recent work at Legal Services for Prisoners with Children involved resource development practices as a strategy for building power and strengthening the voices of incarcerated and formerly incarcerated people and their families to dismantle a social system of excessive punishment, social exclusion, and dehumanization. Anna's communication sculptures have been presented on the Internet, at the Interactive Telecommunications Program at New York University, *International Symposium on Electronic Art*, and SIGGRAPH. More information is available at http://www.well.com/~couey/.

Amanda McDonald Crowley is a cultural worker and curator who creates new-media and contemporary art events and programs that encourage cross-disciplinary practice, collaboration, and exchange. She has recently developed exhibitions and integrated programs with New Media Scotland and the Edinburgh Science Festival, the Austrian Cultural Forum (New York City), Pixelache and the Finnish Bioart Society, Gallery CalIT2 at the University of California, San Diego, and Bemis Center for Contemporary Arts. She is past executive director of Eyebeam art + technology center in New York—recognized internationally as a model for interdisciplinary collaboration and innovation in art and technology. As past executive producer of the International Symposium of Electronic Art 2004 in Helsinki, Finland, and past associate director of the Adelaide Festival 2002 in Australia, she curated and produced integrated programs of internationally significant artist residencies, symposia, and exhibitions. As former director of the Australian Network for Art and Technology, she made significant links with art and science industries, developing a range of residencies and master classes to support artists in the production of new research and new work. Her most recent research has been at the intersection of art, food, and technology—*ArtTechFood*—which she has also been working on during residencies at the Santa Fe Art Institute, with the Helsinki International Artists Program, and as a Bogliasco Fellow.

Steve Dietz is the Founder, President, and Artistic Director of Northern Lights.mn. He was the Founding Director of the 01SJ Biennial in 2006 and served as Artistic Director again in 2008 and 2010. He is the former Curator of New Media at the Walker Art Center in Minneapolis, Minnesota, where he founded the New Media Initiatives department in 1996, the online art Gallery 9, and digital art study collection. He founded one of the earliest, museum-based, independent new media programs at the Smithsonian American Art Museum in 1992. He has organized and curated numerous contemporary and new-media art exhibitions since Beyond Interface: net art and Art on the Net (1998).

Judith Donath is a Harvard Berkman Faculty Fellow and former director of the Sociable Media Group at the MIT Media Lab. She is known internationally for her writing on identity, interface design, and social communication. She created several of the earliest social applications for the web, including the original postcard service and the first interactive juried art show, and her work with the Sociable Media Group has been shown in museums and galleries worldwide. Her current research focuses on how we signal identity in both mediated and face-to-face interactions, and she is working on a book about how the economics of honesty shape our world. She received her doctoral and master's degrees in Media Arts and Sciences from MIT and her bachelor's degree in History from Yale University. She is the author of *The Social Machine: Designs for Living Online* (MIT Press, 2014).

Steven Durland is a visual artist, writer, and designer living near Saxapahaw, North Carolina. He was a co-founder, along with his wife Linda Frye Burnham, of the Community Arts Network and former editor of *High Performance* magazine. He is co-author of the book, *The Citizen Artist: 20 Years of Art in the Public Arena*. He is currently developing a series of large-scale digital images from tiny fragments of nature for indoor or outdoor display.

Lee Felsenstein is an engineer who contributed to the early personal computer industry and helped establish the first public-access social media system in 1973. Felsenstein took a B.S. in electrical engineering from the University of California at Berkeley in 1972, working as an engineer prior to graduation for four years. While at Berkeley, he participated in the Free Speech Movement, from which he developed his interest in technology to support community formation. His exploration of various forms of community-based media in the counterculture in the late 1960s led to the conclusion that networked computers would provide an important tool to facilitate community self-organization. He joined a group that had secured a mainframe computer for counterculture-support purposes, assisting in the development of the first social media system, Community Memory, which made its debut in August 1973. Dealing with problems of maintaining computer equipment in public service, Lee published in 1974 a specification for a personal computer that system that would work as an intelligent terminal and attract involvement from computer enthusiasts. He designed a modem that could be built from a kit and the first video text display adapter for the personal computers that arrived in 1975. This design was widely copied and has become the basic display architecture of personal computers. A computer he designed around this display predated the Apple 2 and for a while was the most advanced typewriter-style personal computer. In 1981, he designed the first commercially successful portable computer, the Osborne 1, becoming a founder of Osborne Computer Corporation. Lee was instrumental in the formation and definition of the legendary Homebrew Computer Club, from which many companies arose, including Apple, running its meetings over the 11-year span of the club. He has pursued a career in electronic design engineering and product development, along with writing and speaking about the social aspects of product design and engineering. He lives in Palo Alto, California, with Lena Diethelm, his wife and partner, and a cat.

Susanne Gerber is a Berlin-based artist, writer and teacher. At the Graduate School of the Berlin University of the Arts she instructs fellows in German and the rhetorics of art. A graduate of the University of Stuttgart with a master's in biology, chemistry, and educational science, she has taught at the University of Tuebingen, Humboldt University Berlin, Max-Planck-Institute Berlin, the Graduate School of the University of Arts Berlin, and John Hopkins University, Maryland. In 2014, she was nominated for a residency at the Villa Aurora LA. Her publications, conference papers, and digital projects include *virtual station*—www.virtual-station.de 2000—an open text reflecting the World Wide Web; *KUNST.STOFF.TUETEN plasticbags* (Hatje Cantz, 2002); "Is Kazimir Malevich's Black Square the First Pixel?"(German Society of Geometry, 2013); and "Facing Fukushima Art Has to Be Something Else" ("Global

Activism," ZKM Karlsruhe, 2014). Her artworks include *red square* (2000), a computer game about art, abstraction, and digitalization; *communication is evolution is communication is …* (animation, 2002); *tell, where is the home of love …* (video, 2004); *genomics—biotopics* (drawing and code, 2007); and *demonstrating alone* (2011), a three-channel video about experiencing the Arab Spring in Berlin. More information about her work can be found at www.cucusi.de and http://www.kuukuk.de/.

Ann-Barbara Graff is Vice-President (Academic & Research) of Nova Scotia College of Art and Design University. Her areas of research expertise include Victorian fiction and prose of thought, as well as digital humanities. She explores marginality in its political and textual forms.

Dene Grigar is Professor and Director of the Creative Media & Digital Culture Program at Washington State University–Vancouver. Her research focuses on the creation, curation, preservation, and criticism of Electronic Literature, specifically building multimedia environments and experiences for live performance, installations, and curated spaces; desktop computers; and mobile media devices. She has authored 14 media works, such as "Curlew" (2014), "A Villager's Tale" (2011), "24-Hour Micro E-Lit Project" (2009), "When Ghosts Will Die" (2008), and "Fallow Field: A Story in Two Parts" (2005), as well as 52 scholarly articles. She also curates exhibits of electronic literature and media art, mounting shows at the Library of Congress and for the Modern Language Association, among other venues. With Stuart Moulthrop (University of Wisconsin–Milwaukee), she is the recipient of a 2013 NEH Start Up grant for a digital preservation project for early electronic literature, entitled *Pathfinders*, which culminates in an open source, multimedia book for scholars. She is President of the Electronic Literature Organization and Associate Editor of *Leonardo Reviews*.

Stacy Horn is the author of five nonfiction books, including *Cyberville: Clicks, Culture, and the Creation of an Online Town* and, most recently, *Imperfect Harmony: Finding Happiness Singing with Others*. She is also an occasional contributor to the NPR show, *All Things Considered*. Over the years, she has produced pieces that include the 1945 story of five missing children in West Virginia, the Vatican's search for a patron saint of the Internet, and an overview of cold case investigation in the United States. She is also founder of the New York City-based social network Echo. Echo was home to many online media firsts, including the first interactive TV show, which was co-produced with the then Sci-Fi Channel. Echo is still in operation today. More information is available on her homepage at www.stacyhorn.com/.

Antoinette LaFarge is an artist and writer working with forgery, impersonation, and virtuality as her main subjects. Her work takes form as computer-mediated performance, interactive installation, digital prints, and writing. Recent new media performance and installation projects include *Far-Flung Follows Function* (2013), *Galileo in America* (2012), *Hangmen Also Die* (2010), *WISP* (World-Integrated Social Proxy, 2009–2010), and *World* *of World* (2009). She co-curated two early exhibitions on computer games and art: *SHIFT-CTRL: Computers, Games, and Art* (2000) and *ALT+CTRL: A Festival of Independent and Alternative Games* (2003), the latter supported by an NEA grant. Her work has been seen internationally in exhibitions and festivals ranging from *UpStage* to the *Venice Biennale* and *documenta*, and her writings have appeared in a wide range of journals and anthologies devoted to art and new media. She is professor of digital media in the Art Department at Claire Trevor School of the Arts, University of California–Irvine.

Deena Larsen, a Colorado native, has been a central voice in the writing and understanding of new-media literature. Her hypertext, *Marble Springs* (Eastgate Systems 1993), about the lives of women in a Colorado mining town, was published by Eastgate Systems in 1993, and a Wikidot version, *Marble Springs* 3.0, was created in 2011. Her work has also been published by the *Iowa Review Web*, *Drunken Boat*, *Cauldron and Net*, *Riding the Meridian*, *Poems That Go*, *The Blue Moon Review*, *New River*, and *The Electronic Literature Collection*. Her current work is the Rose Project, which in her words "ascribes meaning to letters, adding nuances to language." For many years, she hosted forums and workshops for the e-literature community. She currently hosts the website *Fundamentals: Rhetorical Devices for Electronic Literature*, and her archives, *The Deena Larsen Collection*, are housed at the Maryland Institute for Technology in the Humanities at the University of Maryland.

Gary O. Larson worked in a variety of capacities at the National Endowment for the Arts between 1980 and 1996, and he was a writer/editor at the Center for Media Education (1998–2000) and the Center for Digital Democracy (2001–2007). The author of *The Reluctant Patron: The U.S. Government and the Arts, 1943–1965* and *American Canvas: An Arts Legacy for Our Communities*, he has also written numerous articles on culture, technology, and the nonprofit sector. He was guest curator of "DiverseNet: Building a Scenic Route on the Information Superhighway" at DiverseWorks Artspace in Houston in 1996; has taught at the University of Minnesota, the University of Maryland, and American University; and lectured at the Walker Art Center in Minneapolis and the Smithsonian Institution. A graduate of the University of California at Berkeley, he earned his M.A. and Ph.D. in American Studies at the University of Minnesota.

Alan Liu is Professor in the English Department at the University of California, Santa Barbara. He has published books titled *Wordsworth: The Sense of History* (1989), *The Laws of Cool: Knowledge Work and the Culture of Information* (2004), and *Local Transcendence: Essays on Postmodern Historicism and the Database* (2008). Recent essays include "The Meaning of the Digital Humanities" (2013), "From Reading to Social Computing" (2013), "Where Is Cultural Criticism in the Digital Humanities?" (2012), "The State of the Digital Humanities: A Report and a Critique" (2012), and "Friending the Past: The Sense of History and Social Computing" (2011). He started the *Voice of the*

Shuttle website for humanities research in 1994. Projects he has directed include the University of California Transliteracies Project on online reading and the Research-oriented Social Environment (RoSE) software project. He is co-leader of the 4Humanities advocacy initiative.

Richard Lowenberg, founding programs director at the Telluride Institute, directed the InfoZone until late 1996. From 1996 to 2006, he directed the still vital Davis Community Network. Since 2006, he has directed the 1st-Mile Institute and its New Mexico Broadband for All Initiative in Santa Fe. In 2008, with Andrew Cohill (founder of BEV) and Design Nine, Inc., he prepared the statewide "Integrated Strategic Broadband Initiative" for NM Governor Bill Richardson and was subsequently contracted to coordinate all ARRA broadband stimulus projects in NM. He has also consulted and advised on rural community Internetworking projects across the United States and internationally since the mid-1990s.

Judy Malloy is an electronic literature pioneer and social media poet/researcher. Recent fellowships and teaching include seminars in *Social Media History and Poetics* and *Electronic Literature: Lineage, Theory, and Contemporary Practice* at Princeton University, and she is the founding editor of *content | code | process*. In the decades since 1986, when she wrote the online hypertext *Uncle Roger* on Art Com Electronic Network (ACEN) on The WELL, she has composed an internationally exhibited and published body of electronic literature, including the generative hypertext *its name was Penelope* (Eastgate, 1993), and a series of social media-based narratives created in the Computer Science Laboratory at Xerox PARC. Recent research has focused on authoring systems for generative hyperfiction and polyphonic electronic literature. She is a former Content Coordinator for Arts Wire, a program of the New York Foundation for the Arts, former Editor of *NYFA Current/Arts Wire Current*, and Editor of the MIT Press book, *Women, Art & Technology*. Her papers are archived as *The Judy Malloy Papers* at the David M. Rubenstein Rare Book & Manuscript Library at Duke University.

Julianne Nyhan is Senior Lecturer (Associate Professor) in Digital Information Studies in the Department of Information Studies, University College London, London, UK. Her research interests include the history of computing in the humanities and most aspects of digital humanities, with special emphasis on meta-markup languages and digital lexicography. She has published widely in digital humanities. Most recently, she co-edited *Digital Humanities in Practice* (Facet 2012) and the bestselling *Digital Humanities: A Reader* (Ashgate 2013). Currently, she is at work on a book on the history of Digital Humanities; it will be published by Springer in 2016. Among other things, she is a member of the Arts and Humanities Research Council (AHRC) Peer Review College, the Communications Editor of *Interdisciplinary Science Reviews*, and a member of various other editorial and advisory boards. She also leads the "Hidden

Histories: Computing and the Humanities c.1949–1980" project. Find her on twitter @ juliannenyhan/.

Howard Rheingold's 1980s book *Tools for Thought* forecast the future of personal computers as mind-amplifiers. In the 1990s, he identified, documented, and named *The Virtual Community*. In 2001, his book *Smart Mobs* was acclaimed as a forecast of the "always-on" era. His most recent book *Net Smart: How to Thrive Online*, about essential digital literacies, was published by MIT Press in 2012. He has also worked with the Institute for the Future, initiating an interdisciplinary study of cooperation; been awarded a grant from the MacArthur Foundation to create a social media classroom, curriculum, and community of practice around the use of participative media in pedagogy; and taught digital journalism, social media literacies, and social media issues at UC Berkeley and Stanford for ten years.

Randy Ross is an enrolled member of the Ponca Tribe of Nebraska and a descendent of the Otoe-Missouria Tribe of Oklahoma, with family roots on the Rosebud Indian reservation in South Dakota. His spiritual residence is in the Black Hills of South Dakota. Currently, he serves as Development Officer for the White Eagle Health Center in White Eagle, Oklahoma. He has a long, dedicated track record working with the Native American communities in the fields of arts, technology, information services, community economic development, and health-related community-based projects across Indian Country. Appointed in 1989, he has served as Non-Trustee Board Member for the National Museum of the American Indian (NMAI), Smithsonian Institution, Washington, DC, with a specific role involving the Information Technology Committee. The NMAI opened in September 2004 next door to the Nations' Capitol featuring the "4th museum," an electronic digital initiative to build public programming and access in a virtual environment. He is an accomplished speaker/presenter on issues involving Native Americans and emerging new technologies and networking services. He served on the Advisory Panel for the 1995 Office of Technology Assessment Report "Telecommunications Technology and Native Americans: Opportunities and Challenges." In 1999, he served as a co-author for a Benton Foundation report on Native American technology. He has been and continues to be an observer and user of these systems.

Rob Wittig's background combines Literature, Graphic Design, and Digital Culture. In the early 1980s, he co-founded the legendary IN.S.OMNIA electronic bulletin board with the Surrealist-style literary and art group Invisible Seattle. IN.S.OMNIA was one of the earliest online art projects of the digital age. In 1989, he received a Fulbright grant to study the writing and graphic design of electronic literature with French philosopher Jacques Derrida in Paris. His book based on that work, *Invisible Rendezvous*, was published in 1995. Alongside his creative projects, Rob worked for 15 years as a writer, designer, and creative director in major publishing and graphic design firms in Chicago. In 2008, Rob's web project "Fall of the Site of Marsha" was among the first works of electronic literature to be archived in the Library of Congress. He is currently developing high-design, collaborative fiction projects in a form called netprov, networked improv narrative. In 2011, he earned an M.A. in Digital Culture from the University of Bergen, Norway (equivalent to an American M.F.A.), He teaches in the Departments of Art and Design and Writing Studies at the University of Minnesota, Duluth.

David R. Woolley is President of Thinkofit, where he works as a consultant in the areas of online collaboration, online communities, and social media. He has been a pioneer in social media and online communities since before those terms existed. In 1973, he created PLATO Notes, the basis of the world's first online community and the direct progenitor of Lotus Notes and many other online forum and collaborative work platforms. He was also the co-creator of Talkomatic, the world's first multi-user online chat system. He has designed conferencing facilities for many applications and has published a number of articles and book chapters on the subject. As the Executive Director of the Twin Cities Free-Net, he spearheaded the building of an online community network designed to support the neighborhoods of the Minneapolis–St. Paul metro area. He was a founding board member of E-Democracy.org, an internationally renowned organization that uses Internet technology to engage citizens in local political discussions, and he served on E-Democracy's board for 16 years. He is also a veteran in the fields of interactive media and computer-based training, having designed a variety of authoring tools and supervised the development of many CBT applications.

Index

9/11, *Arts Wire Current* coverage, 344
ACEN. *See* Art Com Electronic Network
Adams, Kim, Arts Wire, 335, 346
Adams, Randy, 365, 370
Ada'web, 147
Adilkno, 329
Adlington, Robert, *Voice of the Shuttle*, 264–265, 268
Adrian, Robert, 198, 23–24
  Telephone Music, 130
  World in 24 Hours, The, 24, 130, 131
Advanced Research Projects Agency. *See* ARPANET
Adventure, 7, 10–11
African American BBSs, 22
AfroNet, 22
AIDSWIRE, 31, 338
Alexander, Elizabeth, 390
Allen, Madeline Gonzales, 36, 286, 288
Allen, Robert, 380–382
Alliance of Artists Communities, 333
Allman, Eric, 13
Altair, 52
alt.artcom, 209
alt.cyberspace, 300
alt.hypertext, 25, 34, 119–126
alt.isea, 300
Alt-X Online Network, 369
Alves, Don, 17
America Online (AOL), 25
  chat rooms, 57
  cultural content, 414
  history, 56–57
  September That Never Ended, 121
American Indian Telecommunications, 21, 293, 297, 299, 302, 334
American Music Council, 334
Americans for the Arts, 345
Amcrika, Mark, 369, 370
Amiga computers, 213
An Xiao Studio, *Morse Code Tweets*, 217
And/Or Gallery, 334
Anderson, Benedict, *Imagined Communities*, 84
Anderson, Fortner and Henry See, *The Odyssey*, 210
Andreas, Brian, 302
Andy Warhol Foundation for the Visual Arts, 333
Anglade, Jean-Claude, *La Vallée aux images*, 142
Anne Fallis. *See* Fines, Anne
Antenna Design, 152–154
AOL. *See* America Online
Apple II, 52
Apple Advanced Concepts Group, 294
Apple Library of Tomorrow, 282, 293, 294
Applied Rural Telecommunications Investment Guide, 36
Architecture for Temporary Autonomous Sarai, 157–158
Archive of trAce Online Writing Centre, 371–372
Archiving, electronic literature, 359

Arizona State University, Deep Creek Arts Program, 285
Aronson, Jonathan, 288
ARPANET, 3–4, 16–17, 59, 413
　bread truck gestation of Internet, 16–17
　collaboration, 7–8
　computer-mediated conferencing in FTP development process, 8
　games, 10–11
　"Origins of Social Media, The," 3–50
　RFCs, 8
　role of personal computers (*See* Personal computer history)
　TCP/IP, 3
ARS Electronica, *Art Com Software*, 210–212
Art–21, 340
ART ACCES Revue, 139, 142. 143–144
ARTBASE, 21
Art Beyond Boundaries, Arts Wire, 333
Art Com Electronic Network, 6, 27–28, 297, 302
　"Art Com Electronic Network: A Conversation with Fred Truck and Anna Couey," 191–218
Art Com/La Mamelle. *See* La Mamelle/Art Com
Art Com Magazine, 191, 194, 197, 204–205, 207
Art Com Television, 205
ArtEngine. *See* Fred Truck
Art Reseaux. *See* O'Rourke, Karen
Art's Birthday, 135
ARTEX, 7, 23–24, 130, 198–201, 207. *See also* I. P. Sharp
Artist's Use of Telecommunications conference, 130, 192, 414
Artist Trust, 334
ARTLINK, 21
Artnetweb, 149
ARTQUARRY. *See* WEBBASE
Arts Challenge Fund of New Jersey, *Building Arts Audiences and Communities on the Web*, 341
Arts Council of England, 370

ArtsNet (Australia), 297, 299
Arts Wire, 6, 31–32, 297, 299, 302, 411, 412
　AIDSWIRE, 31
　"Arts Wire: The Non-Profit Arts Online," 333–351
　Arts Wire Current (later *NYFA Current*), 31, 333, 343–345
　Arts Wire News, 346
　NewMusNet, 31
　SpiderSchool, 341
Artwarez, 151
Artwords & Bookworks, 16, 205
ASCOO, 143
Ascott, Roy, 18, 24, 132, 141–142, 155, 298
　Planetary Network, 198
　La Plissure du Texte, 24, 130, 192
AsianAvenue, 23
Asian Cultural Council, 334
Aspen Institute, 284
Association for Community Networking, 36, 286, 295
Association for Computers and the Humanities, Special Interest Group for Humanities Computing Resources, 229
Association for Literary and Linguistic Computing, 229–230
Association of Independent Video and Filmmakers, 334
Association of Internet Researchers, 427
AT&T, 53–54
　RJ11 jack, 54
AT&T Bell Labs, 23, 25, 292, 293
Atelier Bow Wow, 157
Atkins, Robert, 339
Auber, Olivier, *Le Générateur poïétique*, 142
Austin, Linda, 338
Australia
　Artsnet Electronic Network, 37
　System X, 34–35, 219–223
　Third International Symposium on Electronic Art, 298
Australian Network for Art and Technology, 21, 37, 369

Bach, Gottfried
   ARTEX, 130
   I. P. Sharp, 24
Bachelard, Gaston, 136
Bachmann, Ingrid, 367–368
Bad Information, 197, 204–205
Bagels, Community Memory, 95
Baldessari, John, 321
Baldwin, George, 21, 297, 299, 302, 303–304
Ballowe, Jeff, 353
Banana, Anna, 16
BANANARD network mail utility, 13
Banff Centre for the Arts, 214
Bannigan, Phillip, 297, 299, 301
Barber, John, 254
Barnett, Tully, Australian cyberfeminism, 369
Bartle, Richard, 29
Bartlett, Bill, 7, 127, 128, 192, 414
   Artist's Use of Telecommunications, 130
BASIC programming language, 52, 92
Baxter, Iain and Ingrid, 127
BBN-TENEX, 10, 11
BBSs, 20–23, 54, 413
   African American, 22
   ARTBASE, 21
   ARTLINK, 21
   Big Sky Telegraph, 22
   Blacksburg Electronic Village, 29, 279
   Boulder Community Network, 36, 293, 294
   Buffalo Free-Net, 22
   Center for American Indian Economic Development BBS, 21
   Charlotte's Web Community Network, 279
   Cherokee BBS, 21
   Chicago Computerized Bulletin Board System or CBBS, 20
   Christensen, Ward, 20
   Cleveland Freenet, 279
   Community Memory, 89–101
   Dakota BBS, 21, 297, 299, 301
   Davis Community Network, 279
   definition, 54
   distribution of *Caper in the Castro*, 21
   diversity and rural connectivity, 21–23
   Echo, 243–249
   equipment needed, 54
   Fidonet, 55
   history, 20–23, 54–55
   IN.S.OMNIA, 179–189
   legacy, 23
   LGBT, 21
   Matrix Artists Network, 21
   Native American, 21
   people with disabilities, 22
   Project Enable, 22
   Russell County BBS, 21
   System X, 21, 219–223
   Telluride InfoZone, 277–289, 203
   THING, THE, 309–332
Bear, Liza, *Send/Receive*, 19, 129
Beck, Stefan, 324
Beecroft, Vanessa, 314
   THING, THE, 328
Bell Labs. *See* AT&T Bell Labs
Bello, Jane, 335
Bellovin, Steve, 25
Benton Foundation, Open Studio: The Arts Online, 422–423
Berens, Kathi Inman, xv
Berger, Ted, Arts Wire, 334–335
Berkeley. *See* Community Memory
Berkman Center, 390
Berners-Lee, Tim, 368
   announcement of WWW on alt.hypertext,, 25, 120, 368
Bernecky, Bob, 23
Bernstein, Mark, xiv, 355
   origins of alt.hypertext, 121
Besser, Howard, 213
Beuys, Joseph, 129
Bickley, Thomas, 339
Big Sky Telegraph, 22
Binford, Tom, 9
Bioapparatus, 214
Bischoff, John, 37
Bishop, Claire, *Artificial Hells*, 395

Bitnet/NetNorth, 228
Bitzer, Don, PLATO, 103
BlackNet, 22
BlackPlanet.com, 23
Blacksburg Electronic Village, 29, 279
Blair, Dike, 319
Blomme, Rick, PLATO games, 114
Blustein, James, 34
Bly, Bill, 355
BNS Vision, 54–55
Bódy, Gábor, 133
Bolt, Beranek and Newman (BBN), 3
Bolter, Jay David, 355
Bookchin, Natalie and Alexei Shulgin, *Homework*, 148
Born-digital writing. *See* Electronic literature
Boucher, Sen Thomas, 416
Boulder Community Network, 36, 293, 294
Boulder County Civic Center, BBS, 293
Boundary Objects, *Voice of the Shuttle*, 261–271
Bowe, Marisa, 329
Boyer, Penny, Arts Wire, 343
Braden, Bob, 8
Brand, Stewart, 27, 63–66, 71, 196
Branscomb, Anne and Lewis, 32, 283
Bread Truck, 16–17
Breland, Bruce
  DAX Dakar D'Accord, 37, 131
  Digital Art Exchange (DAX), 37, 131, 414
Brenneis, Lisa, 378, 380
Brett, George. 229
Brilliant, Larry, 27, 64–65
Brodie, Richard, 10
Brooks, Sawad, 328
Brown, Doug, 15
  PLATO Talkomatic, 106, 117
BrownPride.com, 23
Brown University, 353
  Humanist, 237
Buffalo Free-Net, 22
Bull, Hank, 5, 7, 18, 34, 147, 404, 412, 414
  and Patrick Ready—*HP Radio Show*, 129
Bulletin Board Systems. *See* BBSs

Bunting, Heath, 329
Bureaud, Annick, 5, 34, 403
Burns, Frank, 302
Burton, Matthew, archiving *Voice of the Shuttle*, 268
Busa, Fr Roberto, 227
Bushan, Abhay, 8
Bushman, Jay, *The Loose-Fish Project*, 217
Byfield, Ted, <nettime>, 395
Byte, 20

Cabarga, Paul, IN.S.OMNIA, 182
Cable Communications Policy Act of 1984, 412
Cable Television Consumer Protection and Competition Act of 1992, 412
CADRE, 212–213
Cage, John, *First Meeting of the Satie Society*, 28, 198–201
Calvino, Italo, *The Castle of Crossed Destinies*, 26–27
Camber-Roth, 336
Campanell, Robert, 302–303
Canada, 196. *See also* ARTEX; I. P. Sharp
  communications cultures, 127
  Bull, Hank, "DictatiOn, A Canadian Perspective on the History of Telematic Art," 127–138
  telematics, 23–24, 192
Carnegie Mellon University Studio for Creative Inquiry, 214
Carpenter, J. R., 26, 403
Case, Steve, AOL, 56–5
Catlow, Ruth, 38
Caucus, Arts Wire, 336
Celent, Germano, 132
Cell phone technology in native communities, 276
Center for American Indian Economic Development, BBS, 21
Center for Civic Networking, 293
Center for Programs in Contemporary Writing, Listserv, 253

Index 449

Center for Twitzease Control, netprov, 187
Cerf, Vint, xiv, 3–4, 8, 11–12, 16, 24
Ceruzzi, Paul, 20, 33, 402–403, 413
Charlotte's Web Community Network, 279
Chaunce of the Dyse, 26
Chen Zhen, 135
Cherokee BBS, 21
Chicago Computerized Bulletin Board System (CBBS), 20
China, *Shanghai Biennale,* 135–136
Christensen, Ward, 20
Cinematheque, 302
Cisler, Steve, 32, 293, 294
  Apple Library of Tomorrow, 282, 283, 303
Civille, Richard, 293
Clan Analogue, 220
Clarke, Nancy, 335
Cleveland Freenet, 279
Clinton, Bill, Music Advocacy, 346
Clinton-Gore Administration, 418
  Information Superhighway, 244–245
  National Information Infrastructure, 280
Cmielewski, Leon, 221
Coate, John, The WELL, 65, 74–75, 196, 201, 303
Cohen, Douglas, Arts Wire, 31, 338, 339, 346
Colby, Kenneth, 11–12
Collaborative Art, Minitel 141–143
Collins, Dan, xv, 32
Collins, Tim, 338
  and Reiko Goto, 27
Colorado Advanced Technology Institute, 293
  Community Networking, 291
  Rural Telecommunications Program, 281, 284
Colorado Supernet, 277
Colquitt, Clair, 186
Commodore computers, 53, 64, 57, 414
Communications sculpture, 163–177, 208
  Cultures in Cyberspace, 297–306
  THING, THE, 309–332
Community Memory, 7, 279, 428–429

"Community Memory First Public-Access Social Media System, 89–101
Community networking, 36. *See also* BBSs
  Blacksburg Electronic Village, 36
  Boulder Community Network, 36
  "Community Networking, an Evolution," 291–296
  Native American, 275–276
  Telluride InfoZone, 36, 277–289
Composer to Composer, Telluride InfoZone, 32–33. 284
CompuMentor, 337
Compuserve, 55, 56, 413–414
  CB Simulator, 55–56
Computer History Museum, 17, 96
Computer Science Network (CSNET), 413
Computer-based education. *See* PLATO
Computers and the Humanities, 230
Computers, Freedom and Privacy, 207
Conference hosting
  Arts Wire, 333–351
  Echo, 245–248
Conferencing system affordances
  WELL, The, 77–85
Conferencing System Software
  Caucus, 336–339
  PicoSpan, 197–198
Control Data. *See* PLATO
Cooley, Lisa, 215, 338
Cooper, Mark, Consumer Federation, 417
Coover, Robert, 353, 366
Cosell, Bernie, 11–12
Cosic, Vuk, 329
Costa, Mario, 141
Couey, Anna, 36, 401
  Art Com Electronic Network, 34, 191–218
  Arts Wire, 31, 335, 346, 343
  and Judy Malloy, Interactive Art Conference, 338
Council of Literary Magazines and Presses, 338
Coverley, M.D., 353, 355
  M is for Nottingham?, 365
CPYNET, 12

Craig, Kate, 128, 130, 131
Crandall, Jordan, 318
Critical Code Studies Working Group, 6
Crocker, Dave, 13–14
Crocker, Steve, 4, 8
Cross-Cultural Television, 134
Crowdsourcing, IN.S.OMNIA, 186–188
Crowley, Amanda McDonald, 34, 369
Crowther, Will, 7, 10–11
Crumlish, Christian, 357
CSNET, 24
Cubero, Angel Borrego
  "mutant bridges," 153
Cukier, Kenneth Neil, 420–421
Cultural Commons, 411–443
Cultural identity, Echo, 246
Culture, online noncommercial, 411–443
Culture Wars, 417
  Arts Wire Current, 345
Cultures in Cyberspace, 215, 221, 297–306
Culver, Andrew, 200
Curtis, Pavel, 29, 363–364, 378
Cyber performance. *See* Telematic Performance
CyberMountain 1999, 353
Cybernetic Serendipity, 280

Dadda, Luigi, 286
D'Agostino, Peter and Antonio Muntadas, *Un/Necessary Image*, 394
Dakota BBS, 21, 297, 299, 301
Dance
  satellite transmission, 165
  telematic, 4
  World Wide Simultaneous Dance, 4
Danforth, Sen. John, 416
Daniels, Dieter, *Net Pioneers*, 311, 315–316
Dannenberg, Penelope, Arts Wire, 334–335
da Rimini, Francesca, 221, 369
DARPA. *See* ARPANET
Dartmouth, early computing, 52–53
Das Casino, 208
Dashiell, David Cannon, 201, 205
Davis, Douglas, 129–130, 169

Davis, Joe, Dana Moser, and Charles Kelley, *Nicaraguan Interactions*, 38
Davis, Matthew Ross, NewMusNet, 339
Davis Community Network, 279
Day Without Art, 345
De Landa, Manuel, 331
Deep Creek School, xv, 27, 32, 285, 347
  Collins, Dan and Laurie Lundquist, xv
  Cooper, Gene, 32–33
  Dutton, Brett, 32–33
Delappe, Joseph, 217
  "Salt Satyagraha Online, The," 38
Denjean, Marc, Déambulatoire/combinatoire, 144
Denton, Cynthia, 21
Derrida, Jacques, 184
Derrida Room, IN.S.OMNIA, 183
Deutsch, L. Peter, 93
Develay, Frédéric, *ART ACCES Revue*, 143
Diao, David, 319
Dibbell, Julian, "A Rape in Cyberspace," 257
DiCarlo, Giancarlo, 286
Dietz, Steve, 5
  Exhibitions curated, 147–160
Digital Art Exchange (DAX). *See* Breland, Bruce
Digital Divide, 21–22, 279, 298, 345
  "Community Networking—The Native American Telecommunications Continuum," 275–276
  "Cultures in Cyberspace," 297–306
Digital Dividends Act of 2003, 426
Digital Equipment Corporation, 195
Digital Humanities, 26, 227–242, 423
  history as documented in Humanist, 227–242
  issues as documented in Humanist, 227–242
  Voice of the Shuttle, 261–271
Digital Opportunity Investment Trust, 426
Ding Yi, 135
Dinsmore, Claire, 355
Dissertation defense, LambdaMOO, 251–260
DiverseWorks, 334

Do While Studio, 4
DOCTOR. *See* ELIZA
Dominguez, Ricardo, "Digital Eros," 327
Don, Abbe, 199, 201, 202
  We Make Memories, 213
Donahue, Phil, *The Peoples Summit*, 170
Donath, Judith, 5, 40
  Social Machine, The, 390
Douglass, Jeremy, *Voice of the Shuttle*
    programming, 264–266, 268
Dowden, Derek, 213
Dudasek, Karel, Ponton Media, 135
Duggan, Ervin S., FCC, 416–417
Duke University, 25
Durland, Stephen, 16, 34, 335, 342
Dyson, Freeman, 148

Earnest, Les
  finger, 9
  Yumyum, 9–10
EAT: Entertainment, Art, Technology, 151
Echo, 4, 243–249
EchoNYC. *See* Echo
Eco-Tech Corsica, 286
Edgar, Robert, 199
  Memory Theatre One, 201, 202, 210, 213
Edupage, 26
Ehrenfried, Gisela, THE THING Timeline,
    316–332
Eisenhart, Mary, 73
Electra Museum of Modern Art of the City of
    Paris, 140
Electric Bank, The. See Truck, Fred
Electronic bulletin board system. *See* BBS
Electronic Café International, 7, 19, 27, 130,
    163–177, 285, 389. *See also* Galloway, Kit
    and Sherrie Rabinowitz
  Los Angeles Olympics of 1984, 130
  Russia, 170
Electronic Frontier Foundation, 390
Electronic literature. *See also* Interactive
    Fiction
  archiving, 359

Art Com Software: Digital Concepts and
    Expressions, 209–212
  "alt.hypertext: An Early Social Medium,"
    119–126
Case for the Burial of Ancestors, The, 210
First Meeting of the Satie Society, 28, 198–201
  Grigar Dene, xv, 35, 38, 403, 404
  Guyer, Carolyn, 29
IN.S.OMNIA, 179–189
LA Flood Project, 217
La Plissure du Texte, 24, 130, 192
mez, 355, 364
Minitel-based, 143
Pathfinders Project, 256
Thirty Minutes in the Late Afternoon, 207
Uncle Roger, 28, 195, 197, 202, 378
Electronic Literature Organization, 366
Electronic Literature Organization Chats, 5,
    29, 353–362
Electronic music, System X, 219–223
Electronic publishing, 194
ELIZA, 11–12
Ellard, Tom, 221
Ellis, Jim, 25
Email, 12–14
  affordances, 13
  at BBN (*see* Ray Tomlinson)
  history, 12–14
  IPSA ARTBOX, 24
  IPSA ARTEX, 130
  Mail Transport Protocol, 13
  MSG "answer" command, 13
  at MIT, 12
  native communities, 276
  PLATO Personal Notes, 107–109
  RD mail system, 13
  sendmail, 13
  Simple Mail Transfer Protocol (SMTP), 13
  SNDMSG, 12
  standards, 14
  TENEX, 12
-empyre- mailing list, 38
Encore, 259

Endlicher, Ursula, 330
Ernst, Wolfgang, 372
Eternal Network, 128. *See also* Mail art
Eyebeam Atelier, 147

Facebook, 256, 258–259, 390, 394
Falk, Bennett, 73
Fallis, Anne, 21
Farber, Dave, 24
Farrington, Carl, 99
Farm, The
  WELL, The, 65–67
FAT Magazine
  THING, THE, 325
Federal Communication Commission, 412
Felsenstein, Lee, 5, 34, 283, 400–402, 428–429
  design of personal computers, 96–97
Ferrier, Ian und Fortner Anderson, *The Heart of the Machine*, 210
fibreculture, 38
FidoNet, 22, 55, 219, 275
Figallo, Cliff, 65–67, 196
File sharing, System X, 219–223
File Transfer Protocol (FTP), 8
Filliou, Robert
  Canadian telematics, 128
  "Eternal Network," 15
FineArt Forum, 26
Fines, Ann, 297, 299, 301
First Meeting of the Satie Society. *See* John Cage
Fischer, Claude, 408
Flaming, 257, 329
Flores, Leonardo, xv, 390
Flusser, Vilem, 141
Focke, Anne, xv
  Arts Wire, 31, 299, 334–335, 336
Ford Foundation, 427
Forest, Fred
  La Bourse de L'imaginaire, 131–132
  L'Espace Communicant, 140
  Les Immatériaux, 140
  Minitel 140–141
Forresta, Don, 132

FORTRAN, 10
Fossland, Gyda, 180
Fourth National Black Writers Conference, Arts Wire 32, 340
frAme: the trAce Journal of Culture and Technology, 363
Frank Silvera Writers' Workshop Foundation, 334
Frank, Nancy, Art Com Electronic Network, 192, 194–196
Free Speech Movement, Community Memory, 93
French Telecom, 139
Frespech, Nicolas, 143
Friesenwall 116, 314
Fujiyoshi, Aki, 318
Fuller, Buckminster, 64
Funding online cultural content, 411–443
Furtherfield, 38, 390
Fusco, Coco, 314

Galántai, György, Artpool 130
Gale, Bob, 21
Galin, Jeff, 254
Galloway, Kit, 288
Galloway, Kit and Sherrie Rabinowitz, 18–19, 33, 163–177, 298, 389. *See also* Electronic Café International
  "Defining the Image as Place," 163–177
  Hole in Space, 130, 166, 176, 389
  Satellite Arts Project, 7, 18–19, 129, 166, 167, 177, 176
Games
  ARPANET, 10–11
  PLATO, 114, 116
Ganahl, Rainer, 319
Gans, David, The WELL, 73–74, 199
Garrett, Marc, 38
Gates, Bill and Paul Allen, 52
Gates, Jeff, 152, 302, 343
Gauntlett, David, online culture, 427
Gee, Jason, System X, 219–223
Gender and Identity in New Media, 345

Gerber, Susanne, 31, 36, 411–412, 420
Gibson, William, *Neuromancer*, 262
Gidney. Eric, 130
Gilchrist, Katie, 254
Gillmor, Dan, 419–420
GLBT culture online, 25
Glen Eagles Foundation, 335
Global Community Networking Congress Barcelona, 286–287
Global Village Network, 286
Goldberg, Michael, 130
Gonzalez, Madeline. *See* Allen, Madeline Gonzalez
Good Morning Mr. Orwell, 171
Goodreads, 38
Google, 405, 420–421
   Google Glass®, 216
   Hangouts, 38, 259
Gore, Al
   Internet High Performance Computing Act, 244
   National Information Infrastructure, 415, 416
Goto, Reiko, 27
Grace, Sharon, *Send/Receive*, 7, 18–19, 129, 192
Grace, Wit & Charm, 187–188
Graff, Ann-Barbara, 34
   "alt.hypertext: An Early Social Medium," 119–126
Graham, Ali, 366
Grateful Dead, 199
Green, David, 26
   Arts Wire, 335, 343
   National Initiative for a Networked Cultural Heritage, 419, 426
   New York Foundation for the Arts, 320
Greenberg, Kenny, 337
Greenblatt, Richard, 95
Grigar Dene, xv, 35, 403, 404
   24-Hr. Micro-Elit Project, 38
   early webpage, 256
Grossberger Morales, Lucia, 215

Grossman, Lawrence, Spectrum Commons, 426
Grothus, Tom, IN.S.OMNIA, 182
Group Notes. *See* PLATO
Grover, Jan Zita, 302, 303
Grubinger, Eva, 314
Gruchy, Tim, 298
Grundmann, Heidi, 18
   Kunstradio, 131
   WIENCOUVER, 131
Guyer, Carolyn, *HI-Pitched Voices*, 29
Gysin, Brion, 127

Hacker Culture, 67, 71
Hacking, *Voice of the Shuttle*, 261–271
Hahne, Freddy, Art Com Electronic Network, 199
Hakenbeck, Regine, 316
Hall, Donna, Art Com Electronic Network, 196
Hall, Jennifer, 4
Haraway. Donna, 368
Hardt, Pamela, Community Memory, 92–93
Harness, Jim, 288
Harris, Craig, 303
Harris, Mary Dee, 230
Harris, Stuart, 380
   Hamnet Players, 378
Harris, Sue, 297, 299, 301
Harris, Sue and Phillip Bannigan
   Artsnet Electronic Network, 37, 299, 300, 304
Hawkins, David, The WELL, 61–62, 72, 196
Hayes Modem, 20, 54
Hayles, Katherine, 355
Haynes, Cynthia, 254, 370
Haynes, Cynthia and Jan Rune Holmevik
   High Wired, 265
   LinguaMOO, 252, 254
Hazen, Blyth, 4
Hazen, Chris, 288
Heinz School of Arts Management, The, 340, 343
Hendricks, Dewayne, InfoZone. 282

Hicks, Greg, 8
Higgins, Dick, 128
Higgins, Ed, 16
High Performance, 335
High Performance Computing Act of 1991, 244, 415
Hindman, Matthew and Kenneth Neil Cukier, National Center for Digital Government, 420–421
H-Net, 26
Hobbs, John Maxwell, 342, 343
   Web Phases, 343
Hoffberg, Judith, 16, 205
Hoffman, Soto, 165
Hole in Space. *See* Galloway, Kit and Sherrie Rabinowitz
Hollerbach, John, 10
Holloway, Jack, 9
Holmevik, Jan Rune, 254, 370
Holzer, Jenny, 220
Homebrew Club, 14
Hope, Akua Lezli, 342
Horn, Stacy, 4, 35, 400, 402, 406, 415
Horton, Mark. *See* Mary Ann Horton
Horton, Mary Ann, 10
Hoskin, Teri, 366
Hotwire. *See* Arts Wire Current
Hovagimyan, Gerard, THE THING, 329
HP Radio Show. *See* Hank Bull
Huber, Felix and Philip Pocock, *Arctic Circle*, 322
Hudson, Heather, 288
Hughes, Dave, 22, 283, 288
Hugo, Joan, 16
Humanist, 7, 26, 253, 415, 427
   "In Search of Identities in the Digital Humanities: The Early History of Humanist," 227–242
Humanities and Arts on the Information Highways: A National Initiative, 419
HUMAN-NETS, 10
Hurst, Jeanelle, Australian Network for Art and Technology, 213

HyperCard, 222
Hypertext, 143
Hypertext Forums, alt.hypertext, 119–126

IBM, 295
index variorum, St. Thomas Aquinas, 227
personal computers, 53
Identity, 72, 176, 377–385
   Compuserve, 56
   Facebook, 56
   "In Search of Identities in the Digital Humanities: The Early History of Humanist," 227–242
   IN.S.OMNIA, 182
   Mail Art, 128
   PLATO, 112
   "Pseudo Space: Experiments with Avatarism and Telematic Performance in Social Media," 377–385
Iimura, Takahiro, 342
Ilich, Fran, 158
Illich, Ivan, networks of instruction, 95
IN.S.OMNIA, (Invisible Seattle's Omnia), 179–189
Indian nations, economies, 275–276
INDIANnet, 21
Infermental, 133–134
Information Superhighway, 244–245. *See also* National Information Infrastructure
Info-sphere, 308
InfoZone. *See* Telluride Infozone
Inman, Admiral Bobby, 280
Institute of Network Cultures, 393
Inter Dada 84, 192
Inter-Access, 131
Interactive fiction
   Adventure, 7, 10–11
   Zork, 10
Interactive Art Conference, Arts Wire, 338
Interface Message Processor, 10
International Association for Community Networking, 292

Index

International Conference on Computer Communications, 11–12
International Conference on Computers and the Humanities, 229
Internet. *See also* ARPANET
  "Origins of Social Media, The," 3–50
  policy, 411–433
  relay chat, 378
  usage statistics, 427
Internet2 consortium, 415
Interplay, 7
Invisible Seattle's Omnia. *See* IN.S.OMNIA
I. P. Sharp, 23–24, 192, 198, 201, 414
Ippolito, Jon, 156
  Solomon R. Guggenheim Museum, 148
Issacson, Walter, 7

Jackson, Shelley, 378
Jamieson, Helen Varley, 380
  UpStage, 378
Japan
  NTT ICC, 147
Jelenic, Venanzio, 123–125
Jenik, Adriene, 378, 380
Jobs, Steve, 70
John S. and James L. Knight Foundation, 335
Journal of Contemporary Art, THE THING, 318
Joyce, Michael, 378
Joyce Theater, 334

Kac, Eduardo, 37, 140, 147
  Teleporting an Unknown State, 155
Kahle, Brewster, 394, 408
Kahn, Bob, 3–4, 11, 16, 32
  Princeton, 3–4
Kanter, Beth
  Arts Wire, 335, 336, 337, 340, 342, 346–347
  Arts Wire SpiderSchool, 341
Kapor, Mitch, 70, 390
Kaypro portable computer, 182
W. K. Kellogg Foundation, 295
Kelly, Kevin, 66–67

Kendall, Robert, 353, 358, 359
Kennedy, Kathleen, 288
Kesey, Ken, 64–65
Kids on the Net TrAce, 363
Kimura, Keija, 165
Kirshner, Bruce, 293
Kitchen, The, 334
Klingenstein, Ken, 288, 293, 293
Knott, Laura, 4
Knowbotic Research, 156
Knowles, Alison, 390
Kogawa, Tetsuo, 136–137
Kolb, David, 355
Kort, Barry, MicroMuse, 303
Kossatz, Max, 325, 328
Krilanovich, Mark, 8
Kruger, Barbara, 220
Kuehler, Jack, 288
Kunstradio, 131

Lacanian Ink, THE THING, 318
LAinundacion, *LA Flood Project*, 217
La Mamelle/Art Com, 19, 191–218, 414. *See also* Art Com Electronic Network
La Plaza Community Network, 279
La Plissure du Texte. See Roy Ascott
LaFarge, Antoinette, 5, 37, 406
  Cake of the Desert, The, 381
  Christmas, 380
  Demotic, 382
  Gutter City, 381
  LittleHamlet, 380–381
  Orpheus and *Silent Orpheus*, 381
  Performance in Social Media," 377–385
  Still Lies Quiet Truth, 381
  Roman Forum, The, 381–382
  White Whale, The, 381
LambdaMOO, 27, 363–364, 369, 378. *See also* Curtis, Pavel
Landweber, Lawrence, 24
Lanier, Jaron, 214
LaPorta, Tina, *Re:mote_corp@REALitie*, 156
Larsen, Deena, 5, 29, 36, 400, 401

Larson, Gary O., 30, 40, 335
  cultural policy, 391
Lasky, Barry, Arts Wire, 335, 340, 346
Latta, Craig, 37
Lauzzana, Ray, 26
Lebkowsky, Jon
  bOING bOING, 303
  Electronic Frontier Foundation-Austin, 303
Lee, Clarissa, 390
Lehman, Dale, 285, 288
Leonardo Electronic Almanac, 26
Leonardo Electronic News, 26, 302
Leopold's Records, Community Memory, 89–92
LeRoy, Louis, 335
Levy, Dan, 201
Ley, Jennifer, 355
Lialina, Olia, 394
Library of Congress National Digital Information Infrastructure and Preservation Program, 423–424
Lifton, John, 279–280
LinguaMOO, 5, 364
  Electronic Literature Organization Chats, 29, 353–362
  Grigar, Dene, MOO-situated defense of her thesis, 29, 251–260
Lipkin, Efrem, Community Memory, 92–94, 96
Listservs. *See also* Humanist
  history, 26
  L-Soft, 26
Literary and Linguistic Computing, 230
LITNET, Arts Wire, 338
Little, Billy, 136
Liu, Alan, 5, 35, 369, 406
  Ultrabasic Guide to the Internet for Humanities Users at UCSB, 262
Local networks, continuing need for, 100
Loeffler, Carl, 17, 19, 34, 191–218, 297, 298.
  See also Art Com Electronic Network; La Mamelle/Art Com
  Send/Receive, 7, 18–19, 129, 192

Logosphere, 136
Loney, Tina, 73
Los Angeles Olympic Arts Festival, Electronic Café, 167, 389
Lotus Notes, 116–117
Lovelace, Ada, 367
Love-letter generator, 8–9
Lovinck, Geert, 38, 322, 328, 329, 393–398
  Dark Fiber, 397
  fibreculture, 398
  Institute of Network Cultures, 398
  Moderation Question, The, 397
  Networks Without a Cause, 398
  Netcritique Blog, 398
  Uncanny Networks: Dialogues with the Virtual, 397
  "What Is the Social in Social Media?," 393–398
Lovink, Geert and Miriam Rasch, *Unlike Us Reader*, 397
Lovins, Amory, 288
Lowenberg Richard, 32–33, 36, 292
Lukasik, Stephen, 13
Luesebrink, Marjorie. *See* M. D. Coverley
Lundquist, Laurie, xv
Lusitania Magazine, THE THING, 318
Lutman, Sarah, Arts Wire, 302, 335
Lynch, Robert, 345

MacDonald, Copthorne, 127
Macie, Christopher
Mahler, David, 31, 339
Mahoney, Michael, 228
Mail Art, 15–16
  fictional identities, 128
  widening circles of communication, 16
Malina, Roger, xiv
Malloy, Judy, 14, 32, 34, 36, 297, 282, 293, 298, 299–300, 301, 302, 303
  Art Com Electronic Network, 28
  Thirty Minutes in the Late Afternoon, 207
  Uncle Roger, 28, 195, 197, 202, 378
  YOU!, 208

Malloy, Judy and Cathy Marshall, *Forward Anywhere*, 342–343
Malloy, Sean, xv
Manchester University Computer, 8–9
Mancillas, Aida, 215
Mancillas, Aida and Lynn Susholtz, PROJECTARTNET, 339
Mandel, Tom, 199
  WELL, The, 62, 72, 74, 199
Mann, Jeff, 21
Marble Springs. *See* Larsen, Deena
Margolis, Neal, 201
Marino, Mark C., xv, 6, 187
Mark, Helmut, *Telephone Music*, 130
Marketou, Jenny, 156
Marras, Amerigo, 286
Marshall, Cathy, 14
Marshall, Frank, 288
Martin, Dan J., Arts Wire, 335, 343
Mast, Kim, PLATO Personal Notes, 107–108
Matheny, John, PLATO Notes Sequencer, 111
Mathews, Harry, 184
Matrix Artists Network, 21
Matrix: Women Networking, 215
Matuck, Artur, 37, 131
Matuzak, Joe, xv
  Arts Wire, 31, 297, 299, 301, 302, 335–336, 340, 341, 347
May, Jim, 299
Maxwell, Christine, xiv
Mayr, Andrea, 314
McBryan, Oliver, 293
McBurnett, Neal, 293
McCann, Michael, 338
McCarty, Willard, 26
  Humanist, 227–242
McClure, Matthew, 196
  WELL, The, 63, 65–67, 74
McCoy, Jennifer and Kevin, *Airworld*, 153
McDaid, John, 355
McGee, Art, 22
  AfroNet, 303

McInnes, Alice, "Agency of the InfoZone: Exploring the Effects of a Community Network," 287
McKee, Bill, 288
McKenzie, Alex, 8
McLuhan, Marshall, 127, 141, 173–174
McPhee, Scot, System X, 34–35, 219–223
Megabyte University, 253
Melcher, Ralph, 303
Membership Model. *See* Arts Wire
Memmott, Talan, 355
Memory Theatre One. *See* Robert Edgar
Menke, Kasee, 9
Meta Net, The, 336
Metier Magazine, 197
Metropolitan Life Foundation, 335
mez, 355, 364
Milhon, Jude, Community Memory, 92, 94, 96
Mills, Simon, 362, 369–370
MinaMora, Gilbert, *Exquisite Corpse*, 208, 210
Minc, Alain, 17, 139
Minicomputers, input/output ports, 53
Minitel
  affordances, 139–140
  "Art and Minitel in France in the '80s," 139–146
  art projects, 7, 139–146
  history, 139–140
  participatory Art, 141–143
  specifications, 139–140
Minnow, Newton
  FCC, 413
  Spectrum Commons, 426
MIT, 5
MIT *Compatible Time-Sharing System*, 12
MIT MAIL, 12
Mitchell, Bonnie, 143
MIT-ITS system, 10
Mitsueda, Mitsuko, 165
MOBILE IMAGE. *See* Galloway, Kit and Sherrie Rabinowitz
modem software, XMODEM, 20,

Modems, 20, 53, 133
 Community Memory, 91, 96
 Hayes, 54
 IN.S.OMNIA, 180–181
Modern Language Association, MLA Commons, 39
Molacek, Rudi, 318
Monk, Meredith, 342
Montfort, Nick, 355, 358
Moore, Gordon, 51, 52
"Moore's Law," 51, 149
MOOS (Muds Object Oriented), 29, 378, 379, 383. *See also* LinguaMOO; LambdaMoo
 affordances, 251–253
 dissertation defense, 251–260
 Nouspace MOO, 258
Mori, Mariko, 314
Morino Institute, 291
Morris, Mark, 345, 346
Morris, Noel, 12
Morse Telegraph, 51
Mosaic browser, 413
 announcement on alt.hypertext, 120
Moser, Dana, 38
Mouchette Suicide Kit, 143
Moulthrop, Stuart, xv, 355, 378
Mountain Area Information Network, 279
MsgGroup, 9
MUDs (Multi-user Dungeons), 29
Multiplayer Games PLATO, 114–116
Muntadas, Antoni, 134, 394
Murphy, Robbin, 152
 Project Tumbleweed, 149–150
Murthy, Prema, 31
 Bindi, 327, 332
Museum Computer Network, 26
Museum Domain Management Association, 422
Museum_L, 26
Music, System X, 219–223

Nahrada, Franz, 286
Naisbitt, John, Telluride InfoZone, 32, 283, 288

NAPLPS (North American Presentation Level Protocol Syntax), 21
Nares, James, 320
NASA. *See* National Aeronautics and Space Administration
National Aeronautics and Space Administration, satellite communications projects 19, 129, 165–168
National Association for the Advancement of Colored People, 345
National Association of Latino Arts and Cultures, 334
National Black Writers' Caucus, 348
National Center for Supercomputing Applications
 Mosaic browser, 413
National Computer Graphics Art Conference, 212–213
National Digital Information Infrastructure and Preservation Program, 423–424
National Endowment for the Arts
 Art-21, 333, 340, 348
 NEA INFO, 338
National Endowment for the Arts and Benton Foundation, Open Studio, 285, 417, 422–423
National Endowment for the Humanities, 417, 423
 Computer Uses in Research, 230
National Film Preservation Act of 1988, 428
National Information Infrastructure, 280, 415, 416, 417–418
National Information Infrastructure Act of 1993, 416
National Initiative for a Networked Cultural Heritage (NINCH), 26, 419
NINCH Announce, 26
National Performance Network, Arts Wire, 339
National Science Foundation NSFNET, 413
Native American Telecommunications, 21, 275–276
Native American Writing Program, 281
Native Americans community networking, 295

Native Arts Network Association
  Arts Wire, 31, 333, 338
Native Net, 281
Nauman, Bruce, *Double No,* 220
Nechvatal, Joseph, 319
N.E. Thing Co, 127
Neighbourhood Television, Hank Bull and Tetsuo Kogawa, 137
Neigus, Nancy, 8
Nelson, Ted, 378
Net art, 148, 150, 157, 309, 311, 329, 394
Net Pioneers. See Daniels, Dieter
NETJAM, 37
Netprovs
  Center for Twitzease Control, 187
  Grace, Wit & Charm, 187–188
  Occupy MLA, 187
  TempSpence, 187
Netscape Navigator, 58
<nettime> mailing list, 38, 148, 328, 393, 395, 397
Network performance. *See also* Netprovs
  Das Casino, 208
Neustrup, Nels, Community Memory, 92–93
New Jersey Institute of Technology, EIES conferencing system, 98
New Museum, 38, 156
  New York Correspondence School. *See* Mail art
New York Foundation for the Arts
  Arts Wire, 31, 329
  "Arts Wire: The Non-Profit Arts Online," 333–351
NEW YORK STATE, Arts Wire, 333
New York State Council on the Arts, 341
Newby, Greg, 357, 358
NewMusicBox, 345
NewMusNet, Arts Wire, 31, 333, 339
Nielsen, Jakob, 421
Nielson, Don, 17
NINCH. *See* National Initiative for a Networked Cultural Heritage
Nonprofit cultural sector, 422–433

Nonviolent Communication, 296
Nora, Simon and Alain Minc, 17
Normals, The, *Anna Couey Museum of Descriptions of Art,* 208
Northern California Osage, 304
Nottingham Trent University, trAce Online Writing Centre, 363–375
Nouspace MOO, 258
NovaNET, PLATO spinoff, 116
Nunez, Elizabeth, 340
Nunokawa, Jeff, 6
NYFA Current. *See* Arts Wire Current
Nyhan, Julianne, 5, 26, 427

Odasz, Frank, 22
Oita province, Japan, community networking, 291
Oliveros, Pauline, 31, 335, 339, 342
Onwere, Ken, 22
Open Studio: The Arts Online, 422–423
Opera America, 334
Optic Nerve, 165
Orcas Conference, Arts Wire, 334
O'Reilly WebBoard, 364
ORLAN, *ART ACCES Revue,* 143–144
O'Rourke, Karen, 132
  City Projects, 37
Oulipo, 27
Out North, 334
Ozzie, Ray, 117

Packer, Randall, 331
Paik, Nam June, 129–130, 169
  Good Morning Mr. Orwell, 171
Palace, The, 378
  early 3D chat rooms, 147
Pannucci, Cynthia, 331
ParcBIT Mallorca, 286
Parello, Bruce, PLATO, 115
PARRY, 11–12
Pascher, Stephen, THE THING, 329
Pathfinders Project, 256
Paul, Christiane, Whitney Museum of American Art, 148

PDP–10, 10
Pedagogy, Natalie Bookchin and Alexei Shulgin, *Homework*, 148
Pelton, Joseph, 280
Perec, George, 27
Performance Bank, The. See Truck, Fred
Performance Space, The (Australia), System X, 221, 222
Perkis, Tim, The Hub, 37
Personal computers, 279
   Altair, 52
   Felsenstein, Lee, 96–97
   games, 53
   history in telecommunications, 51–59
   PC revolution, 70–71
   "Personal Computer and Social Media, The," 51–59
   in relationship to development of ARPANET, 55
   role in democratizing the Internet, 59
   role in social media, 51–59
   standardization, 55
Personal freedom in electronic space, 176
Personal Notes. *See* PLATO
Peterson, Tommer
   Arts Wire, 335, 340, 341, 346
Pew Charitable Trusts, 335
Philippe, Jean-Marc, Minitel, 142
Philomela. *See Voice of the Shuttle*
Picospan conferencing software, The WELL, 27, 65, 197–198
Pierce, Julianne, 221
Plaintext Players, 29, 377–385
Planetary Network, 132, 198, 207
Plant, Sadie, 367
PLATO (Programmed Logic for Automatic Teaching Operations) educational system, 15, 51, 103–118, 404
Platzker, David, 321
Poets & Writers, 338
Point Foundation, 71
Pollack, Steven, 324
Pomeroy, Jim, 335

Pompidou Center, 140
   "Les immatériaux," 184
Poole, Christopher, 408
Portal as art, 150
Portals. See also Web Portals
Posner, Vladimir, The Peoples Summit, 170
Postel, Jonathan B., 13
Postmodern Culture, 26
Post-punk electronic music, 220
Pound, Ezra, 136
Pratt, Bill, 335
Pratt, Spencer, 187
Prodigy, 57–58
   comparison with Minitel, 57
   content monitoring, 58
Project Enable, 22
Project Shooting Star, 284
PROJECTARTNET, 339
Prolog, 230
PS122, 334
Pseudo streaming content, 147
Public access tele-computing sites, Telluride InfoZone, 277–289
Public scholarship online, 251–260

Quantum Computer Services. *See* AOL
Quarterman, John S., 297, 300, 301, 303
Queneau, Raymond, 27

Raben, Joe, 233, 237
Rabinowitz, Sherrie, 288. *See also* Galloway, Kit and Sherrie Rabinowitz
Radio Shack computers, 52
Ralph, CM, 21
Rapoport, Sonya, 201
   Digital Mudra, 210
Raqs Media Collective, project OPUS, 157
Raqs Media Collective and Atelier Bow Wow, "Architecture for Temporary Autonomous Sarai," 157
Ready, Patrick, *HP Radio Show, 129*
Real Art Ways, 345
RealAudio, 343

Index

Reed College, 64
Reggio, Godfrey, 172
Regional Network Systems. *See* Community Networking
Request for Comments (RFC), 7–8
Resource One, Community Memory, 93–94, 96, 98
Rete Civica Community Network, 279
Rettberg, Jill Walker, 122–123
Rettberg, Scott, 353, 356, 359, 366
Reuters, 130
Reynolds, Brian, *Voice of the Shuttle*, 265, 268
Rheingold, Howard, 28, 33–34, 201, 283, 293, 402, 419, 426
   NetSmart, 443
   Virtual Communities conference, 302
Rhizome, 38
   ArtBase, 425
Richards, Michael, 344
Richardson, Jeff, 293
Roberts, Larry, 3, 13
Robot 400, 127
Rockefeller Foundation, The, 335
Rosen, Jay, audience, 427–428
Rosen, Joe, 199, 201, 202, 210
   Mr Prezopinion and Vibrabones, 210
Rosenberg, Jim, xiv, 200, 201, 202, 205, 355
   Diagram Poems, 28
Ross, Randy (Ponca Tribe of Nebraska and Otoe Missouria), 21–22, 297, 301, 299, 302, 304, 335
   American Indian Telecommunications, 293
Rossetto, Lewis, 283
Rossman, Michael, Community Memory, 92, 98
Roth, Charles, Caucus software, 341
Royon Le Mée, Franck, 144
RTmark screensaver, 152
Rural Telecommunications Investment Guide, 284
Russell County BBS, 21
Russell, Mark, 339
Russia, The Peoples Summit, 170

Sadowski, Paul R., 22
San Francisco Museum of Modern Art
   Artist's Use of Telecommunications 130, 193
   telematics, 414
Sandman, Alan, 21
Satellite Arts Project. *See* Galloway, Kit and Sherrie Rabinowitz
Schmidt, Eric, 25
Schultz, Pit, 38, 393
Schuyff, Peter, 321
SDS–940 mainframe, Community Memory, 91, 93
Seattle Arts Commission, 334
Second Life, 38, *258*, 378
Selbo, Vivian, 150
Sellman, Catherine, 288
Send/Receive Project. *See* Loeffler, Carl
September That Never Ended, 120–121, 310
Seuss, Randy, 20
Seva Foundation, 65
SF-lovers, 10
Shanghai Biennale, 135–136
Shanghai Fax, 18
Sharp, Willoughby, 130
   Send/Receive, 7, 18–19, 129, 192
Shen Fa, 135
Sherman, Tom, 132
Shi Yong, 135
Shock of the View, The, 151
Shotz, Alyson, 324
SIGGRAPH 93, 214, 215
Silver, David, online culture, 427
Simon, John F., 318
Simple Mail Transfer Protocol (SMTP), 13
Sinder, Dale, PLATO Notes, 116
Slattery, Diana, 355
Sloterdijk, Peter, 310
"Slow Machines and Utopian Dreams," 399–409
Sluizer, Suzanne, 13
Smith, Anna Deveare, 305
Smith, Joan, 230

Sneve, Shirley, *A Story of Mothers and Daughters in South Dakota*, 342
Snobol, 230
Snow, Michael, *Walking Woman*, 127
Social Contracts in Cyberspace, The WELL, 77–85
Social media
  definitions, 5, 245
  information exchange, 78–85
  of the future, 215–217, 397
  origins, 1–50
  reasons for documenting, 6–7
  role of reciprocity, 80–85
  utopian, 399–409
Social media identity. *See* Identity
Sodeoka, Yoshi, 328
Software as Art, 6, 201
Sondhein, Alan, 367
Sonnier, Keith, 19, 129
Source, The, 413. *See also* AOL
  history, 56–57
Southern Ute Tribe, community networking, 294–295
Southworth, John, 130
Spectrum Commons, 426
Spencer, Henry, alt.hypertxt archives, 120
Spender, Dale, *Man Made Language*, 369
Sperberg-McQueen, Michael, 229
SpiderSchool. *See* Arts Wire SpiderSchool
Spirit Knife (New Mexico), 21
Springer, Viennese magazine, 322
Sproull, Lee and Sara Kiesler, *Connections: New Ways of Working in the Networked World*, 79
SRI. *See* Stanford Research Institute
StackExchange, 405
Staehle, Wolfgang, 309–332
  THING, THE, 5, 309–332, 407, 411–412, 420
Stalder, Felix, <nettime>, 395
Stanford Artificial Intelligence Laboratory (SAIL), 9–10
Stanford Research Institute, 17
Star, Susan Leigh, 261–262, 266–267
Starrs, Josephine, 221

Status Report, Art Com Electronic Network, 208
Stelarc, 207
Stengers, Phoebe, accordion BBS, 415
Stone, Carl, 342
Strachey, Christopher, 8–9
Strazisar, Ginny, gateway software, 16–17
Strickland, Stephanie, 355
Studio for Creative Enquiry at Carnegie Mellon University, 131
Sundén, Jenny, 365
Susholtz, Lynn, 339
Syndicus, Maria, 199
System X, 21
  "System X: Interview with Founding Sysop Scot McPhee," 219–223
Szpakowski Mark, Community Memory, 92, 94, 96

Talkomatic. *See* PLATO
Talley, Dan, Arts Wire, 335
Taylor, Lee, 288
TCP/IP, 3, 16
Technology Transfer in the Arts, Arts Wire, 333–351
Telebrations, 285
Telecommunications Act of 1996, 419
Telecommunications Infrastructure Act of 1993, 416
Telecommunications in public spaces, 163–177
Tele-Community '93, Telluride InfoZone, 32–33, 283–285, 293, 389
TELEMAT System X, 221
Telematic, definition, 17, 139
Telematic Connections: The Virtual Embrace, 154–155
Telematic performance, 377–385
Telematics, Canadian, 127–138
Telephone costs, early online, 54
Teletype terminals, 53, 94
  Community Memory, 90
Television as communications media, 166

Telluride Infozone, 32–33, 277–289, 293. *See also* Richard Lowenberg; *Tele-Community 93*
Tele/Comm/Unity, 33
Telluride Institute, 277–289, 292
Telluride Institute Ideas Festival *See Tele-Community 93*
Telluride Summer Research Center, 283–284
Tembellini, Also, 130
TempSpence, 187
Tenczar, Paul, 15
Tenhaaf, Nell, 140
Terminals, 133
  high-resolution graphics display, 103
  Minitel, 140
Term-Talk. See PLATO
Tetherless Access, 282
Texte zur Kunst, 318
Text-to-speech processing, 382
THING, THE, 7, 309–332, 411, 412. *See also* Wolfgang Staehle
  Berlin node, 318
  Dusseldorf node, 317
  Frankfurt node, 318
  Timeline, 316–332
  Vienna node
Theise, Eric, 32, 293, 297, 300, 301
Third International Symposium on Electronic Art (TISEA), 215, 221, 297
Thirty Minutes in the Late Afternoon. See Malloy, Judy
Thomas, Eric, 26
Thomas, Sue
  Correspondence, 369
  trAce Online Writing Centre, 363–375
Thompson, Jon, Community Memory, 95
Tidmus, Michael, 31, 338
Time sharing. *See also* PLATO
  Community Memory, 92–93, 98
  Dartmouth, 52
  role in development of computing, 51
Tisch School of the Arts, 199
  Interactive Telecommunications Program, 243

Toi et Moi Pour Toujours, Minitel 143
Tomlinson, Ray, network email, 12–13
Tong, Darlene, Art Com Electronic Network, 192, 194
Torimitsu, Momoyo, 330
Toronto Community Videotex. *See* Inter-Access
Tosca, Susana, 355
Tower, Jon, 320
Toywar, 314
trAce Online Writing Centre, "trAce Online Writing Centre, Nottingham Trent University, UK," 363–375
Tribe, Mark, *Rhizome,* 38
Troika Ranch, 342
Trolls, Echo, 246–247
Trubshaw, Roy, 29
Truck, Fred, 19, 20–25, 34, 191–218, 297, 302, 400 *See also* Art Com Electronic Network
  ArtEngine, 201–202, 207, 210
  Electric Bank, The, 192
  Labyrinth, The, 214
  Squared Circle, 210, 213
Truck, Lorna, 194
Truscott, Tom, 25
Tumblr, 38
Turing, Alan, 8–9
Turkle, Sherry, 6, 405
TUTOR educational software, 103
Twitter, 26, **375**
Two Bulls, Lorri Ann, 215

Un/Necessary Image, 394
Uncle Roger. See Judy Malloy
University of California, Santa Barbara, *Voice of the Shuttle,* 261–271
University of Illinois Computer-based Education Research Laboratory, 14–15
University of Texas at Dallas online thesis defense, 251–260
UNIX, 196
  "talk," 379
Unix-to-Unix Copy (UUCP), 25

Unlike Us network, 393
Usenet, 25, 297
  artists use of, 414–415
  "alt" hierarchy groups, 25, 119
  alt.hypertext, 25, 119–126
  groups archived on Google, 25
  rec.arts.fine, 25
  soc.culture.african.american, 25
  soc.culture.asian.american, 25
  soc.culture.native, 25
  soc.motss, 25
  September That Never Ended, The, 25
Ute Mountain Ute Tribe, 291

Van Vleck, Tom, 12
Vancouver Community Network, 279
Vasulka, Steina and Woody 171
Vaughn, Idette, 22
Vautier, Ben, 144
VAX, 195
Venice Biennale, 176, 198
Vergne, Philippe, 151
Vertiges, Minitel, 143
Vesna, Victoria, *n 0 time,* 154–155
Video
  split screen 171
  in telematic performance, 171
Video art, 168–169, 205–206
Videotex systems. *See* Minitel
Vimeo, 256
Virtual reality, 214–215
Vittal, John, MSG, 13
VNS Matrix, 369
Voice of the Shuttle, 5, 35, 261–271, 369
  archiving, 268
von Meister, William, The Source, 56
von Oldenburg, Helene, 318
VoS. *See Voice of the Shuttle*
VSA #DisabilityStories, 390

Wagener, Andrea, THE THING Cologne, 314
Walker Art Center Media art exhibitions, 147–160

Walker, Jill. *See* Rettberg, Jill Walker
Walker, Steve, 9
Wardrip-Fruin, Noah, 355
Warner, Gary, 298
Wasow, Omar, 23
Watts, Marcus D., 197
  PicoSpan, 27
Web Portals, *Voice of the Shuttle,* 261–271
WEBBASE, Arts Wire, 342
Websites, Arts Wire, 341–342
WebWalker, 152
Weibel, Peter, 309, 316
Weil, Benjamin, 317
  San Francisco Museum of Modern Art, 148
Weizenbaum, Joseph, 11
WELL, The, 7, 26–28, 195, 279, 293, 297, 300, 302. See also Art Com Electronic Network
  arts conferences, 28
  conferencing system, 67
  Deadhead conference, 67, 71, 73
  hacker culture, 67, 71
  influence on Echo, 243
  identity, 72
  journalist community, 71–72
  list of public conferences, 67–70
  personal computer revolution, 71
  Rheingold, Howard, 61–85
  role in working from home, 77–78
  "What it is is up to us," 67
WESTAF. The Western State Arts Foundation, 339
Western Behavioral Sciences Institute, 64
Western Front, 128, 130, 131
Western New York Disabilities Forum, 22
White, Jim, 8
White, Norman, 128, 129
  hearsay, 23–24
White, Wendel, 342
  Small Towns, Black Live, 343
Whitehead, Helen, 363, 357
  Web, Warp, and Weft, 367

Index

Whole Earth Catalog, 63–64
Whole Earth 'Lectronic Link. *See* The WELL
Whole Earth Review, 64, 67
Whole Earth Software Catalog, 64, 71
WIENCOUVER, 130
Wilks, Christine, 364
Wine lovers list, 10
Winer, Dave, 396
Winter, Mick, 197
Wireless, 282
Wisbey, Roy, 230
Wisniewski, Maciej, *netomatheque,* 152, 155
Wittig, Rob, 34
Wohlstetter, Philip
 IN.S.OMNIA, 182, 184, 186, 187
Woods, Don, 11
Woolley, David R., 15, 34, 404
 "PLATO: The Emergence of Online Community," 103–118
World in 24 Hours, The. See Robert Adrian X
World Wide Web
 impact on text-based social media, 58–59, 214–215, 341–343
 System X, 222
Wright, Fred, Community Memory, 93
Wright, Tim, *In Search Of Oldton,* 367

Xerox PARC, 10, 11, 347, 378
Xiao, An, *Morse Code Tweets @ The Brooklyn Museum,* 39
Xmodem, 20, 53

Yahoo, 420
Yamada Village, 286
Yellowlees Douglass, Jane, 353
Yonke, Marty, 13
YOU! See Judy Malloy
Youngblood, Gene, 4–6, 18, 163–177, 283
Yumyum, 9–10

Zelevansky, Paul, *Case for the Burial of Ancestors, The,* 210
Zimmermann's Café, 27

Zitt, Joseph, 339
Zoline, Pamela, 279–280
Zuni BBS, 21
Zwick, Bob, 359